Archaeological Chemistry—II

Giles F. Carter, EDITOR

Eastern Michigan University

Based on a symposium

sponsored by the Division of

the History of Chemistry at

the 174th Meeting of the

American Chemical Society

Chicago, IL, August 31–

September 1, 1977.

ADVANCES IN CHEMISTRY SERIES **171**

AMERICAN CHEMICAL SOCIETY

WASHINGTON, D. C.   1978

Library of Congress CIP Data

Symposium on Archaeological Chemistry, 6th, Chicago, 1977.
Archaeological chemistry—II.
(Advances in chemistry series; 171 ISSN 0065-2393)
"Based on a symposium sponsored by the Division of the History of Chemistry at the 174th meeting of the American Chemical Society, Chicago, II, August 31-September 1, 1977."

Includes bibliographies and index.

1. Archaeological chemistry—Congresses.
I. Carter, Giles F. II. American Chemical Society. Division of the History of Chemistry. III. Title. IV. Series.

QD1.A355 no. 171   [CC79.C5]   540'.8s   [930'.1'028]
ISBN 0-8412-0397-0   ADCSAJ   171 1–389 78.26128

Advances in Chemistry Series

Robert F. Gould, *Editor*

FOREWORD

ADVANCES IN CHEMISTRY SERIES was founded in 1949 by the American Chemical Society as an outlet for symposia and collections of data in special areas of topical interest that could not be accommodated in the Society's journals. It provides a medium for symposia that would otherwise be fragmented, their papers distributed among several journals or not published at all. Papers are reviewed critically according to ACS editorial standards and receive the careful attention and processing characteristic of ACS publications. Volumes in the ADVANCES IN CHEMISTRY SERIES maintain the integrity of the symposia on which they are based; however, verbatim reproductions of previously published papers are not accepted. Papers may include reports of research as well as reviews since symposia may embrace both types of presentation.

DEDICATION

This volume is dedicated to Professor Earle Caley, one of the first enthusiasts in archaeological chemistry and, in a real sense, the pioneer of this field of chemistry.

CONTENTS

PREFACE

Since this volume records the proceedings of the Sixth Symposium on Archaeological Chemistry, it is fair to say that archaeological chemistry is an "old" field, but unfortunately it is not yet mature. In the interval since the Fifth Symposium in 1972, advances have been made, particularly in the understanding of the origin and distribution of obsidian and pottery, principally through chemical analysis. Lead isotope studies have deduced the sources of many archaeological objects. The understanding of other types of specimens made of glass, metals, bone, pitch, and other organic materials is improving as additional analytical work unfolds. Accordingly this volume has been organized around the classification of materials—organic materials, ceramics, and metals—with five introductory chapters on perspectives and techniques.

One of the main purposes of archaeological chemistry is to deduce history from the analysis and investigation of artifacts. Other major areas of importance include authenticity studies, identification of sources, deduction of production techniques, and dating.

Archaeological chemistry involves many of the techniques of analytical chemistry admixed with patience, primarily in accumulating a large volume of data and in working with specimens frequently imperfectly preserved. As more and more data are obtained and reported by an increasing number of laboratories using various analytical techniques, the following are important problems to be addressed by archaeological chemists, archaeologists, and others interested in the field:

(1) recommended procedures for data reporting,

(2) recommended methods of specimen handling and sampling,

(3) identification of sources of standards and synthesis of new standards,

(4) computerized data storage and retrieval, and

(5) round-robin test programs to ensure inter-laboratory agreement in the analysis of archaeological objects.

To initiate progress on the above problems the group attending the Sixth Symposium on Archaeological Chemistry voted to affiliate with the Division of the History of Chemistry of the ACS and to create several task forces to begin work in the various areas listed above. Preliminary organization of several task forces has occurred, and a directory of archaeological chemists and interested archaeologists is being prepared.

Only when satisfactory progress has been made on the above problems will archaeological chemistry earn the distinction of being a mature field of chemistry. However, in spite of some present shortcomings, archaeological chemistry is a most rewarding (other than financially) field to pursue—in fact, it is just plain fun.

Finally, I suggest the formation of a U.S. government-sponsored Archaeological Data Center for the computerized and conventional storage and retrieval of all sorts of archaeological data. The field is so diverse and the data so numerous that such a structure is needed to prevent archaeology itself from becoming something of a modern tower of Babel with the attendant problem in communication.

I wish especially to thank Robert F. Gould, Editor of the ADVANCES IN CHEMISTRY SERIES, Virginia deHaven Orr, Associate Editor, and all others who have contributed to the excellence of this volume.

Chemistry Department   GILES F. CARTER
Eastern Michigan University
April 18, 1978

Perspectives and General Techniques

Chemistry and Archaeology: A Creative Bond

S. V. MESCHEL

Department of Chemistry, University of Chicago, Chicago, IL 60637

The development of archaeological chemistry has involved interaction with other scientific disciplines. Carbon-14, thermoluminescence, fission track, archaeomagnetism, obsidian hydration, and amino acid dating techniques all have contributed to the field. The geophysical problems in carbon-14 and archaeomagnetic dating illustrate this interdisciplinary character. The most often-used modern analytical techniques are summarized with respect to their use for archaeological artifacts. Nondestructiveness and alteration of equipment to suit the archaeological aim are important factors. Trace and pigment analysis are used in authenticity studies. The study of the use of the mineral huntite in ancient times is an example of how chemistry, art, and geology interrelate in archaeological chemistry.

Archaeological chemistry is in some ways an old, established field and in other ways a new endeavor just beginning to make its influence felt on science. It is old in the sense that people have been concerned with using chemistry to study the past for nearly 200 years. It is a new field in the sense that its activities have been somewhat sporadic until recently, and the scientists involved have not always been identified as archaeological chemists. Because of this apparent dichotomy as well as the greatly accelerated development of this field in the past 25 years, it is becoming essential that a more organized delineation should be attempted. While this chapter is not intended to be an exhaustive survey of the entire field of archaeological chemistry, my primary objective is to review the major areas of inquiry, to survey the roles of chemical subspecialties to show its development, and last but not least to illustrate its interaction with other scientific disciplines.

Let us start with the three most frequent questions one is confronted with: how old is the artifact, what is it made of, and is it authentic or a

0-8412-0397-0/78/33-171-**003**$05.50/1

fake? There are probably many more questions one could ask, but I feel that these are the most basic issues.

In order to tackle the question, "how old is the artifact?" one needs to know a great deal about the material as well as the topic of archaeological dating processes. The time when the carbon-14 method represented a fascinating, unique methodology has passed long ago, and today an archaeological chemist has at his fingertips nearly 12 reasonably well established dating processes to choose from. The archaeological chemist will have to judge correctly which method is best suited to the sample on hand. He must be familiar with the chemistry, physics, at times even the geophysics, as well as the statistical reliability of the pertinent processes in order to select the most appropriate methodology.

The question, "what is it made of?" is more or less the standard inquiry analytical chemists are confronted with. However, even here the archaeological chemist has to handle intriguing and extraordinary situations which rarely occur elsewhere. The sample size is usually very small, and the methodology is often restricted to nondestructive techniques.

Isotopes and Radiation Technology

Figure 1. X-Ray fluorescence analysis of a painting (90)

The museum may be unwilling to release the sample to a laboratory at all for fear of damage, and the archaeological chemist has to design an experiment which will yield precise numerical data in the museum, preferably without touching the sample. Figure 1 illustrates such a case where pigments on a painting had been analyzed by a portable x-ray fluorescence equipment. Sometimes the sample is inaccessible or out of reach. In a particularly interesting case the sample was part of the tower on a five-story pagoda, in Kyoto, Japan (*1*). Homogeneity and representativeness of the sample usually cause great concern. Trace impurity analysis has been used in a variety of situations where chemists' work helped historians, sociologists, and anthropologists to make important conclusions in their field.

The third question, "is it authentic or a fake?" is really related to the previous two; however, the aims are sufficiently different to deserve to be listed on its own. Authenticity studies are very important for a variety of human endeavors and for widely divergent reasons. For example, for the U.S. Customs Agency, Forensic Science Laboratories, museums of all types, private collections, and many other enterprises, the question of authenticity is of great practical importance. There is of course a great deal of popular fascination with fakes. From the scientific point of view an authenticity study is at times less demanding than precise dating or a complete analytical assessment. If dating methodology is used for authenticity studies, one is concerned with deciding only whether the artifact was made in the time claimed or recently manufactured. In such instances, precision and statistical reliability are less important (*2, 3*). In authenticity studies, complete analytical survey is rarely requested, rather the presence or lack of presence of a certain component is frequently the decisive factor. Of course, the archaeological chemist needs to know which component could decide authenticity.

A few well known studies illustrate the diversity of problems archaeological chemists must face. S. C. Gilfillan hypothesized (*4*) that the reason for the fall of Rome was the slow and insidious poisoning of its leadership by lead. Such a proposal is intriguing, needless to say quite speculative. In order to substantiate such a theory the expertise of archaeological chemists would be much needed. Such a project requires precise dating of bones and other items (household utensils, cosmetic products, waterpipes, medical equipment, etc.). The problem also would necessitate precise analytical determination of lead in a variety of samples, kinetic studies, x-ray studies, and so on. In such research the archaeological chemists' work would make an impact not only in the context of proving or disproving a theory but also on present-day plumbism research, pigment studies, genetics, history, and environmental pollution projects as well.

What did Napoleon I die of? This question has puzzled generations of social and medical historians. As many as six different diseases had been proposed on the basis of historical evidence, none of which fit all the symptoms. Forshufvud, Smith, and Wassen (*5*) postulated that the real cause was arsenic poisoning. They studied hair samples taken immediately after Napoleon's death by neutron activation analysis. The results clearly showed that the arsenic content in Napoleon's hair was an order of magnitude higher than that of healthy human hair (10.38 ppm compared with 0.8 ppm). The question whether the poisoning was acute or chronic has not been explicitly resolved. Neutron activation analysis had been used in archaeological chemistry to study and solve a variety of problems (6, 7). This type of study can make important impacts on medicine, history, forensic science, and even agriculture and environmental research. The latter is of current interest, for arsenic is one of the constituents in tobacco smoke which may be considered carcinogenic.

Last I would like to mention the interesting case of two bronze statues. One, "Man with a Beard," a handsome Renaissance bronze statue was owned by the Louvre in Paris, whereas the other, "Boy with a Ball," was owned by the National Gallery in Washington, D.C. (8). These two statues were remarkably similar in style, and it was suspected that at some time these may have been part of one artifact. Stylistic analysis, however, was not sufficient to make the final decision. X-ray fluorescence study of the two objects showed identical chemical composition which can happen only if they were poured from the same batch of metal. The analysis also showed that by composition the artifacts were closer to brass than to bronze. The group was reunited and is now known as "St. Christopher Carrying the Christ Child with the Globe of the World." The statue illustrates the legend in which St. Christopher carried the Christ Child across a river.

X-ray fluorescence analysis had been used for composition studies of various materials. Probably among the most important applications are research on metals and on inorganic pigments. Analyses similar to the ones I quoted are very helpful in authenticity studies and can aid the cosmetic industry, metallurgy, and so on. The demands archaeological chemistry made (nondestructiveness, small sample size, quick analysis, sensitivity) has helped significantly to develop x-ray fluorescence instrumentation.

The scientific aspects of art conservation represents a topic which is intimately related to archaeological chemistry. Most often scientists active in archaeological chemistry concern themselves with problems associated with art artifacts as well (9–15).

History

Having touched on some illustrative examples in the domain of archaeological chemistry, let us turn to the early developments of this field. Since I am no historian, I shall mention only some of the early, well known chemists who pioneered the scientific study of the past. My coverage of history thus will be far from complete for in this brief overview I intend only to pinpoint some of the intellectual and conceptual milestones archaeological chemistry has passed.

It is a matter of definition what one could call chemistry in the modern sense, and my personal bias is to start with precise, quantitative work. Following my own definition, M. H. Klaproth, a chemist who lived in the 18th century, ought to be mentioned (16). Klaproth performed the first quantitative analysis on ancient objects, among which are samples of coins, brass artifacts, and glass. Klaproth pioneered not only what we call today archaeological chemistry but the analysis of glasses in general (17, 18). Samuel Parkes (1761–1825), the author of an early chemistry textbook, published a paper titled "On the Analysis of Some Roman Coins" in which considerable emphasis was placed on the statistical reliability and precision of results (19). Other famous chemists such as Sir Humphrey Davy and J. J. Berzelius contributed to the development of archaeological chemistry by their analytical results of ancient artifacts. In 1815 Sir Humphrey Davy published a paper on ancient pigments, and his brother John Davy contributed a study on ancient metals (19). Davy also studied Roman wall paintings and the conservation problems of papyri found at Herculaneum. John Voelker, Professor of Chemistry at the Royal Agricultural College at Cirencester, represents another important turning point. Voelker was interested in the analysis of Roman glass. He was also a member of the only institution at the time where group interest seemed to have prevailed in archaeological chemistry. Voelker could be called the founding father of archaeological chemistry for he stated for the first time its aims as a distinct subspecialty (19). F. Gobel, a German chemistry professor whose name is not well known today, had been the first investigator to suggest that chemical analysis of ancient objects could help in archaeological interpretation. There is evidence for a beginning of cooperation between chemist and archaeologist by the mid 19th century (16). J. E. Wocel was the first chemist who proposed that precise composition data on ancient metal artifacts could be related to their place of origin and to their approximate age.

These were very important ideas, which certainly represent great foresight in the mid 19th century. As much as science and technology have advanced by leaps and bounds since then, dating of metal artifacts still relies more or less on Wocel's ideas (16, 20). Tracing the geographic place of origin has developed into a very important aim. Knowledge of geographic origin has helped to solve many historical, economic, and anthropological puzzles and had been of immense help in authenticity studies as well (7, 10, 11, 21). The contribution of another well known scientist, Michael Faraday, ought to be mentioned next. Faraday examined some Roman pottery and found conclusive evidence for the presence of lead in the glaze. Presence of lead could provide clues to the kinds of minerals in the glaze as well as to the type of technology used in its production. The famous French chemist M. Berthelot contributed many publications to the literature of archaeological chemistry. Berthelot was interested mainly in identifying archaeological artifacts by chemical experimentation and in analyzing their corrosion products (*16*). The work of A. Carnot represents a significant development. Carnot studied the fluorine content of ancient bones and was the first to propose that such data could be used to date human remains as well as objects associated with these. Quantitative tests to check the validity of this method have been made only within the past 25 years. Fluorine content along with uranium and nitrogen content (F, N, U) are used today for relative dating of bones (*22, 23, 24*). These analyses are referred to in Table I.

Another important milestone in the 19th century came with the systematic application of scientific principles to the conservation of art and archaeological artifacts. The State Museum of Berlin established a laboratory in 1880 where studies pertaining to conservation problems were performed. Probably the best known and still used methodology developed there was the electrolytic removal of corrosion products from ancient artifacts (*8, 9, 10, 11, 12*). The field of underwater archaeology made especially effective use of related methods originating in the first conservation laboratory (*16*). In the first quarter of the 20th century the interests of scientists in the field of archaeological chemistry continued to grow as reflected in the increased number of publications. However, no pioneering principles, ideas, or concepts were developed in this period. This seems to have been mainly a time for refinement of earlier work which improved precision and accuracy considerably. The development

Table I.   Archaeological

Approach	*Method*
Physical chemistry	carbon-14
	isotope dating other than carbon-14
	$\quad$ (K/Ar, U/Pb, ^{3}H, etc.)
	thermoluminescence
	fission track dating
	archaeomagnetism
Classical chemical changes	obsidian dating
	relative dating of bones by F, N, U content
	amino acid dating
Changes in the environment	dendrochronology
	pollen analysis
	glacial varves
	oxygen isotope ratios (ancient temperatures)

of emission spectroscopy had been a great step in the new trend of obtaining precise results with relatively small damage to the artifact.

In the past 25 years archaeological chemistry has developed with greater acceleration than ever before. New, more sophisticated equipments became available to handle special problems. Practically every type of quantitative chemistry has been applied to the problems of archaeological chemistry. We hope that this trend will continue to grow.

Dating

Let us now return to the first question of "how old is the artifact?" and examine it in greater detail. Table I lists 12 different archaeological dating processes. Depending on our definitions as to what constitutes the domain of chemistry as opposed to physics, that number may vary somewhat. The Egyptian and other well known calendars were excluded because these are considered relevant only for certain groups of people at certain parts of the world. Reference to astronomical observations as an independent dating possibility were also omitted because there are too few to meet the requirements of reliable dating methods. A comprehensive review of all the available dating methodologies is not intended. However, to illustrate the interdisciplinary aspects of archaeological chemistry, a brief description of a few selected dating techniques will follow.

Isotope Dating. Carbon-14 and the other isotope dating methods rely on our understanding of the pertinent nuclear decay processes. If the isotope in the particular artifact has a quantitatively well established decay mechanism, the sample can, in principle, be dated. This statement

Dating Processes

Approximate Time Interval (years)	Approximate Accuracy (%)
500–50,000	5–15
10^5–10^9 (except for ^3H)	5–10
300–beyond 50,000	5–15
present–10^5	5–10
present–8000 BP	5–20
present–15,000 BP	2–10
10^3–10^8	variable
present–10^5 BP	5–10
present–beyond 7,000 BP	5–10
500 AD–1900 BC	—
present–20,000 BP	2–10
present–10^5	—

is greatly oversimplified, for in reality the processes are quite complex. There is a fundamental difference between carbon-14 and the other isotope dating methods. In the carbon-14 dating technique what is measured is the time elapsed since the carbon-containing object was removed from the natural exchange reservoir, i.e., the time elapsed since the living matter's death. The time limitation is up to approximately 50,000 years. In the other isotope dating methods the natural changes in the equilibrium of the nuclear decay processes are measured since the time of the rock's formation. The measurable time can be in the order of magnitude of 10^6 years. This latter type is as much geological dating as archaeological and illustrates yet another creative interaction between fields. There is approximately one carbon-14 atom to each 10^{12} carbon-12 atoms in the natural reservoir. Uniform distribution is assumed because of relatively rapid mixing throughout the atmosphere, the biosphere, and the ocean (surface only, not necessarily depth) (25–38). The total concentration is assumed to stay approximately constant because it represents an equilibrium level between loss of carbon-14 by natural decay and replenishment from production by cosmic rays. The carbon-14 is incorporated into all living matter. When the living matter dies and ceases to replenish carbon-14, the concentration diminishes by a definite amount according to its half-life of 5730 years (average of recent values) (29) corresponding to a loss of 1% each 83 years (25–38).

Thermoluminescence. Thermoluminescence dating, fission-track dating, and archaeomagnetic dating all have one idea in common even though the principles are fundamentally different. They measure the time elapsed since the artifact was fired at relatively high temperature. Thermoluminescence dating is based on the phenomena that all earth-type materials (clay, pottery, etc.) undergo damage from radiation arising from minute amounts (1–10 ppm) of ^{238}U, ^{232}Th, and ^{40}K they contain. Electrons trapped in these materials may be excited by the radiation. When the object is heated, energy in the form of light-glow is emitted while the excited particles may return to their normal state. The amount of glow is related to the time elapsed since the last firing and, of course, to the concentration of radioactive impurities in the artifact and in its environment as well. The physical phenomena was well known as far back as the 1660s by Robert Boyle. This method is one of the most versatile at present, capable of dating within a range of 100–50,000 years with about ± 10% accuracy (39–55). The theory of the method is not particularly lucid, and advances in theory should yield a great deal of improvement. Thermoluminescence technique has also been used extensively in authenticity studies (2, 3, 21, 41, 45–48).

Fission Track Dating. Minerals and glasses which contain traces of uranium (ppm order of magnitude) undergo radiation damage caused by the spontaneous fission process of ^{238}U. Tracks of the recoiling fragments can be made visible by etching with HF, NaOH, or other suitable solvent and subsequently can be counted under a microscope. The number of tracks are related to the age as well as to the uranium content of the artifact. Heating beyond approximately 500°C anneals the tracks, and thus firing the object sets its fission track clock to zero. This method has been used extensively for geologic dating. For archaeological dating, the artifact needs to have enough ^{238}U—approximately 10 ppm in a 1000-year-old sample—to yield several tracks/cm² of sample area. Even though this poses limitations, many artifacts with uranium-rich inclusions as well as man-made objects with added uranium content have been successfully dated (28–31, 56–65).

Since my primary aim in this survey is to show interaction, numerical agreement between two dating methods is cited. The well known excavation containing humanoid remains found at the famous Olduvai Gorge by Dr. Leakey has been dated by fission track dating as well as by K/Ar isotope dating yielding the ages $2.03 \pm 0.28 \times 10^6$ years and 1.76×10^6 years, respectively (30, 31).

Archaeomagnetism. When hot, iron-containing clay cools, it acquires a weak, permanent magnetization. The magnetization has the same direction as the field within which it cooled, and the strength is proportional to the field intensity at that time. The fired object retains the memory of the earth's magnetic field vector at the time of the cooling process. If the variation of the earth's magnetic field is known at the artifact's specific locality, the age can be evaluated. It is important that the orientation of the artifact remains unchanged after the cooling process. This is a very complicated process, and much of it is not well understood (28–31, 66–70). The change of the earth's magnetic field, for instance, does not follow predictable patterns even though a quasisine functional relationship had been proposed by several researchers (28, 29, 30, 31). Calibrations hold only for specific localities, and to complicate matters even further, the same field direction may reappear after a few centuries. Moreover, there have also been two or three geomagnetic reversals of direction in the past 200,000 years creating added problems (69, 71).

The following problem which involves interaction among several scientific fields illustrates the interdisciplinary nature of archaeological chemistry. Quite apart from the fact that archaeomagnetism as a dating methodology is complex and not always the most reliable, research in this field has provided intriguing as well as useful information. The valiant efforts designed to show that there is a well defined functional relationship between the direction of the earth's magnetic field and time (and also intensity and time) provided science with a large pool of information concerning the changes in the geomagnetic field. Changes in the earth's magnetic field have been observed quantitatively since the 1660s in England (*31*). Should the functional relationship prove to be truly, explicitly cyclic, extrapolations prior to that time may be made, making the dating process more versatile. The geomagnetic field provides shielding against the onslaught of cosmic rays. If the cosmic ray intensity is low in a particular year, lower than usual carbon-14 production will take place. Therefore, in principle, high magnetic fields should correlate with the low cosmic ray intensity or low carbon-14 production. This correlation has been found quantitatively useful for it might make it possible to correct for systematic errors in the carbon-14 dating methodology (*29, 31*). Experimental data obtained in archaeomagnetism studies provides important, fundamental information about the complex behavior of the geomagnetic field. Experimental measurements also indicate that times of high sunspot activity correlate with lower than expected cosmic ray intensities (*72*). It has been proposed that at high sunspot activity, the interplanetary magnetic field carried by the solar wind is magnified, and this deflects cosmic rays from the earth's environment. According to these ideas, high sunspot activity correlates with lower-than-usual cosmic ray intensity, which in turn correlates with lower-than-average carbon-14 production. Measurements of neutron production at high altitudes show evidence for the effect (*30*). Fortunately for the carbon-14 dating methodology, large changes in the production rate of carbon-14 are not seen as severely as anticipated because of response time of the natural carbon reservoir.

Table II.   Chemical Analysis

Earth-Type Materials—Mineral Phases, Flint, Obsidian, Pottery, Clay	*Metal Artifacts—Bronze, Copper, Iron, Steel, Lead Coins (Au, Cu, Ag)*
Neutron activation analysis	optical emission spectroscopy
Atomic absorption spectroscopy X-ray diffraction	x-ray fluorescence atomic absorption spectroscopy
X-ray fluorescence	classical chemical methods (analysis of C, P, S in steel)
Optical emission spectroscopy	$^{208}Pb/^{206}Pb$ isotopic analysis
Mössbauer spectroscopy DTA	mass spectroscopy neutron activation analysis
TGA X-ray milliprobe	x-ray milliprobe classical gravimetry (coins) electron microprobe

The changes can however reach measurable magnitudes, on the order of 1–3%. There is significantly less change in carbon-14 content caused by sunspot variations if the geomagnetic field is high ($\pm$ 50 years in the interval 1000–1500 BC) as opposed to when the field is low ($\geq$ 70 years in the interval 3000–5000 BC) (30). The number of sunspots pass through maximum and minimum values in approximately 11 years. However, between cycles the heliomagnetic field changes direction, therefore the true cycle generally is defined as 22 years. For the climactic changes this number seems to be the important one. In view of the recent interest in correlating sunspot activity with climatic changes, the existing data on variations in the earth's magnetic field, sunspot activity, and changes in carbon-14 content of our atmosphere could be of interest to meteorology as well as to other scientific fields of endeavor.

There has been a renewed interest in the obsidian dating technique which was developed in the early 1960s. In this dating methodology age determination is related to the thickness of hydration layer in the obsidian artifact. The surface of a freshly carved obsidian artifact slowly absorbs moisture and builds up a hydration rim over a period of time. The rate with which the hydration rim is acquired depends greatly on the composition of the obsidian as well as on environmental factors (28, 31, 59, 60, 73–76).

Amino acid dating is one of the most recently developed methodologies. Age determination in this technique is based on relating the rate of conversion to the age of the amino acid-containing artifact (60, 77, 78).

of Archaeological Artifacts

Glass Type Artifacts—Glass, Glaze, Faiance	*Paintings—Inorganic Pigments*	*Material of Organic Origin— Amber, Resins, Vegetable Gum, Linseed Oil, Wax*
optical emission spectroscopy	x-ray diffraction	infrared spectroscopy
flame photometry	x-ray fluorescence	gas chromatography
x-ray diffraction	electron probe microanalysis	NMR spectroscopy
x-ray fluorescence	Mössbauer spectroscopy	
neutron activation analysis	neutron activation analysis	
x-ray milliprobe	x-ray milliprobe	
$^{18}O/^{16}O$ isotopic analysis	mass spectroscopy	

For discussion on further additional dating techniques the interested reader may consult references on dendrochronology (*28, 29, 31, 59, 60, 79*), tritium dating (*38, 80*), and oxygen isotope dating (*59, 81, 82*).

Chemical Analysis

Chemical analysis of practically every type is used extensively in archaeological chemistry. There are of course important differences between the typical problems which arise in analytical laboratories and those of archaeological chemistry. Ideally in studies of archaeological artifacts the analytical technique should be nondestructive, and if this is not feasible, only a very small sample is to be removed. If a sample is to be removed, a concerted effort should be made not to diminish the object's aesthetic appearance. For example, it may be necessary to drill at the bottom of a bronze vase where the damage is not visible. In the case where sample-taking is not allowed, the artifact must be accommo-

Table III.   Comparison of Analytical

	Atomic Absorption Spectroscopy	*Optical Emission Spectroscopy*	*X-Ray Fluorescence*
Damage to artifact (sample size)	slight (10–100 mg)	slight (5–100 mg)	nondestructive or slight (100 mg)
Analysis of surface (S) or interior (I)	I	I	S
Concentration range —major (M), minor (Mi), trace (T)	M, Mi, T (10 ppm—10%)	M, Mi, T (100 ppm—10%)	M, Mi, T (50 ppm—100%)
Elements or compounds analyzed (number or type)	30–40 elements	30–40 elements	elements $Z \geqq$ 22 (air) or $Z \geqq 12$ (vac. or He)
Estimated accuracy (%)	2	10	2–5
Speed of analyses	manual, automatic recording	manual, photographic recording	some automation. Quick
Relative cost—low (L), medium (M), high (H)	L—M	L	M
Applications	glasses, metal artifacts	pottery, glasses, metal artifacts	pottery, pigments, glasses, metal artifacts

dated by the instrument in its entirety (83). This may present special problems for objects with unusual shapes—long swords or large, amphora-shaped vases—which necessitate radical alteration of commercial instrumentation. The archaeological chemist must also decide which technique is suitable for surface analysis or for investigation of the object's interior. The anticipated concentration range of the pertinent component must be considered also. The technique selected depends on whether major element concentration (2–100%), minor element concentration (0.1–2%), or trace element concentration ($< 0.1\%$) is to be determined. It is of considerable importance that the methodology yield rapid results and be relatively inexpensive.

Table II summarizes the most often used techniques. The methods are grouped according to categories of archaeological artifacts for which they are considered most appropriate. Colorimetry, potentiometric titrations, UV, and visible range spectrophotometry are used whenever possible, primarily because the instruments are usually available and

Methods in Archaeological Chemistry

X-Ray Diffraction	Neutron Activation Analysis	Mössbauer Spectra	Infrared Spectroscopy	Gas Chromatography
slight (5–10 mg)	nondestructive or slight (50–100 mg)	slight (100 mg)	slight (few mg)	slight
I or S	I	I or S	I	I
M ($> 1\%$)	M, Mi, T (1 ppm—100%)	—	—	M, Mi, T
crystalline compounds or elements	40–60 elements	Fe compounds	organic molecules	organic compounds
2–5	2–5	—	—	1–2
automatic photographic recording	automatic recording and sample changing	—	rapid	rapid
M	H	M	L	L—M
pigments, pottery	pottery, glasses, metal artifacts	pigments, pottery	amber, resin, varnishes, binding media	resins, oils, waxes, varnish, binding media

relatively inexpensive. These methods were not listed under any particular type of archaeological material. These generally are considered part of preliminary studies. Specific gravity measurements, radiography, and microscopy studies generally are considered as physical rather than chemical analyses, therefore these could be summarized separately. The method of β-ray backscattering is not mentioned in Table II. This method has received a great deal of attention 15–20 years ago but has not been used as extensively lately. Its application is more limited than most of the other processes on Table II. Carbon, phosphorus, and sulfur (C, P, S) analysis is referred to in Table II. These methods need no detailed description as they are familiar techniques to professional chemists. In Table III some of the analytical methods are reorganized to illustrate how these fare with respect to the concerns of the archaeological chemist.

Archaeological chemistry is in the process of becoming more and more important in authenticity studies. Art forgery has become a lucrative business, and it is generally acknowledged that competent craftsman are able to produce excellent copies of archaeological artifacts (2, 3, 21, 84). There have been many instances when presumably ancient pottery, bronze, carved wood and obsidian artifacts, china, and other objects proved to be modern much to the distress of the art world. Scientific investigation therefore is becoming increasingly decisive in detecting forgeries (21, 41, 45–48).

One of the most widely publicized examples is that of P. G. Coremans, a Belgian chemist, who was the principal witness of the 1947 trial of Van Meegeren and whose testimony was instrumental in establishing Van Meegeren's guilt for the expert forgery of "Vermeer" paintings. The irony of the case was that Van Meegeren, who allegedly sold authentic Dutch paintings to the Nazis during the Nazi occupation of Holland, practically had to insist on his own guilt as a forger in order to avoid being convicted of the greater charge, that of being a traitor for selling national treasures. Incidentally some of the fakes were treasured by that infamous art collector, Herman Goering.

Because of the increasing concern about authenticity, pigment analysis has received considerable attention (85, 86, 87, 88, 89). Studying the history of pigment preparation one might conclude that white and green pigments could be considered good indicators of age. In older paintings lead white was conventionally used ($2PbCO_3xPb(OH)_2$), whereas zinc white (ZnO) was manufactured on a larger scale only after approximately 1832, and titanium white (TiO_2) has been used extensively only since the 1920s. Similarly, older paintings normally use malachite (also chrysocolla and verdigris) for green pigment, all of which contain copper as opposed to viridian green ($Cr_2O_3x2H_2O$) which is of more recent manufacture. Figures 1 and 2 represent a particularly interesting example. In a study by R. Frankel (*90*) two paintings had been analyzed by x-ray fluorescence spectroscopy. A small portion of flesh color was being examined in both paintings. The x-ray spectrum (Figure 3) clearly shows the relatively high concentration of lead in the 16th century Velasquez painting in contrast to the dominating zinc peak exhibited by the 19th century Gaugain painting.

Many methods are used for pigment identification. Among those x-ray diffraction, x-ray fluorescence, optical emission spectroscopy, and microscopy are frequently in demand (*8, 10, 11 21, 59, 83, 91–93*). The use of Mössbauer spectroscopy for pigment studies has received a great deal of attention in the past few years. The development of this field has been quite rapid and promising (*59, 94–103*).

Isotopes and Radiation Technology

Figure 2.   X-Ray fluorescence analysis of a painting (90)

Over and beyond the identification of the pigment, trace impurity content can also help to solve many authenticity problems. Trace impurity content can reveal the manufacturing process, the geologic origin, as well as the chronological age of the pigment-containing object (*38*). Predominantly neutron activation analysis and x-ray fluorescence spectroscopy provided many valuable clues not only in pigment research but in coin and metal artifact studies as well (*7, 10, 11, 59, 83, 90–93, 105–122*). Recently laser microprobe, nuclear magnetic resonance, mass spectroscopy, absorption spectrophotometry, infrared spectra, and thermoanalysis techniques also have been used successfully to analyze archaeological artifacts (*21, 59, 120–128*).

The following two studies concerning the mineral huntite (*129, 130*) represent interesting examples of the interface among several fields of endeavor. A shipwreck was found (*129*) at the north coast of Elba Island

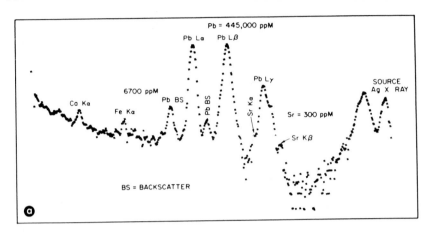

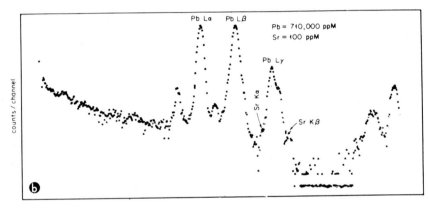

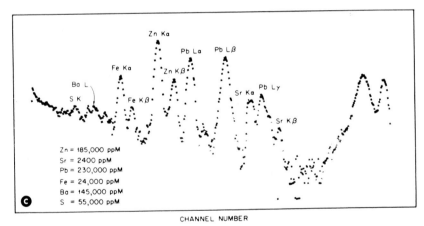

Isotopes and Radiation Technology

Figure 3.   *Comparison of x-ray fluorescence spectra of flesh-tone pigments in the paintings on Figures 1 and 2 (90)*

in the early 1970s. It was identified as a Roman ship, and was dated by the carbon-14 technique to be 2nd–3rd century AD. Among other objects in its cargo a box containing some 10 kg of white substance had been found. The chemical was suspected to be huntite, a white pigment with the composition $CaCO_3x3MgCO_3$. The mineral was studied in great detail by classical chemical analysis, TGA, DTA, x-ray diffraction, and optical microscopy as well as by electron microscopy techniques, and its structure proved to be orthorhombic. Since huntite is a relatively rare mineral, the investigators first assumed that the substance may have been formed by a reaction of seawater and some magnesium compound originally in the box. However, carbon-14 measurements established that the CO_2 component of the mineral was older than 42,000 years, therefore according to geochemical equilibrium calculations, at best only 2.6% of the huntite could be younger than 1700 years. These studies proved conclusively that the mineral huntite must have been known and traded during 2nd–3rd century AD. Its properties must have been understood and valued. In the same year another study brought new evidence concerning the use of huntite in ancient times (*130*). A thorough investigation was conducted at the Museum of Egyptian Art in Munich concerning inorganic pigments. A white pigment was found on two bowls and several sherds dated about 1600 BC and later identified as huntite. Since the chemical and physical properties of white pigments such as calcite, aragonite, dolomite, magnesite, and gypsum commonly found on ancient Egyptian artifacts are rather similar, precise x-ray diffraction technique was of the utmost importance. The x-ray diffraction patterns clearly proved the pigment to be huntite, thereby establishing the fact that this mineral was known and used by artists as early as the 16th century BC. In these studies the interaction of chemistry, art, and geology is quite apparent. The interaction of dating methodology and chemical analysis is also well illustrated.

The primary objective of this overview was to show that archaeological chemistry is an interdisciplinary endeavor—a mediator among scientific fields and between the exact sciences and the humanities. Archaeological chemists, therefore, must be flexible, versatile, and well versed in several areas of chemistry. Archaeological chemistry is a relatively new field with a bright future that should challenge the imagination of its practitioners.

Literature Cited

1. Emoto, Y., "Characteristics of Antiques and Art Objects by X-Ray Fluorescent Spectrometry," *in* "Archaeological Chemistry," M. Levey, Ed., p. 77, University of Philadelphia Press, Philadelphia, 1967.
2. Rogers, F. E., "Chemistry and Art. Thermoluminescence and Forgery," *J. Chem. Educ.* (1973) **50**, 388.

3. Fleming, S. J., "Science Detects the Forgeries," *New Sci.* (1975) **68**, 567.
4. Gilfillan, S. C., "Lead Poisoning and the Fall of Rome," *J. Occup. Med.* (1965) **7**, 53.
5. Forshufvud, S., Smith, H., Wassen, A., "Arsenic Content of Napoleon I's Hair Probably Taken Immediately After His Death," *Nature* (1961) **192**, 103.
6. Lenihan, J. M. A., "Radioactive Analysis," *Nature* (1959) **184**, 951.
7. Gordus, A., "Neutron Activation Analysis of Almost Any Old Thing," *Chemistry* (1968) **41**, 8.
8. Brill, T. B., Reilly, G. J., "Chemistry in the Museum," *Chemistry* (1972) **45**, 6.
9. Brommelle, J. P., Brommelle, N. S., "Science and Works of Art," *Nature* (1974) **250**, 767.
10. Johnson, B. B., Cairns, T., "Art Conservation: Culture Under Analysis. Part I," *Anal. Chem.* (1972) **44**(1), 24A.
11. Johnson, B. B., Cairns, T., "Art Conservation: Culture Under Analysis, Part II," *Anal. Chem.* (1972) **44**(2), 30A.
12. Zimmerman, J., "Science of Ancient Artifacts," *Chemistry* (1970) **43**(7), 28.
13. Yahia, M. A., "Chemical Examination of Archaeological Artifacts," *Sci. Chron.* (1973) **11**, 182.
14. Rainey, F., "Archaeology: New Tools for an Old Art," *Spectrum* (1976) **13**, 39.
15. Treloar, F. E., "Chemical Approaches to Archaeological Problems," *Proc. R. Aust. Chem. Inst.* (1976) **43**, 31.
16. Caley, E. R., "The Early History of Chemistry in the Service of Archaeology," *J. Chem. Educ.* (1967) **44**, 120.
17. Brill, R. H., "Ancient Glass," *Scientific American* (1963) **209**, Nov., 120.
18. Cox, G. A., Pollard, A. M., "X-Ray Fluorescence Analysis of Ancient Glass: The Importance of Sample Preparation," *Archaeometry* (1977) **19**, 45.
19. Musty, J., "Analysis in Archaeology," *Proc. Anal. Div. Chem. Soc.* (1976) **13**, 94.
20. Braidwood, R. J., Burke, J. E., Nachtrieb, N. H., "Ancient Syrian Coppers and Bronzes," *J. Chem. Educ.* (1951) **28**, 87.
21. Fleming, S. J., "Authenticity in Art. The Scientific Detection of Forgery," Crane, Russack and Co., New York, 1975.
22. Oakley, K. P., "Framework for Dating Fossil Man," Widenfeld and Nicolson, London, 1964.
23. Brothwell, D. R., "Digging Up Bones," British Museum, London, 1963.
24. Chaplin, R. E., "The Study of Animal Bones from Archaeological Sites," Seminar, New York, 1971.
25. Aitken, M. J., "Physics Applied to Archaeology, I. Dating," *Rep. Prog. Phys.* (1970) **33**, 941.
26. Libby, W. F., "Radiocarbon Dating," University of Chicago, Chicago, 1952.
27. Ralph, E. K., Michael, H. N., "Twenty-five Years of Radiocarbon Dating," *Am. Sci.* (1974) **62**, 553.
28. Michels, J. W., "Dating Methods in Archaeology," Seminar, New York, 1973.
29. Berger, R., "Scientific Methods in Medieval Archaeology," University of California, Berkeley, 1970.
30. Aitken, M. J., "Physics and Archaeology," 2nd ed., Oxford University, 1975.
31. Michael, H. N., Ralph, E. K., "Dating Techniques for the Archaeologist," MIT, Cambridge, 1971.
32. Clark, R. M., Renfrew, C., "Tree-Ring Calibration of Radiocarbon Dates and the Chronology of Ancient Egypt," *Nature* (1973) **243**, 266.

33. Harkness, D. D., Burleigh, R., "Possible Carbon-14 Enrichment in High Altitude Wood," *Archaeometry* (1974) **16**, 121.
34. Burleigh, R., "Radiocarbon Dating. Practical Considerations for the Archaeologist," *J. Archaeol. Sci.* (1974) **1**, 69.
35. Hammond, P. C., "Archaeometry and Time," *J. Field Archaeol.* (1974) **1**, 329.
36. Clark, R. M., Renfrew, C., "A Statistical Approach to the Calibration of Floating Tree-Ring Chronologies Using Radiocarbon Dates," *Archaeometry* (1972) **14**, 519.
37. Jeffreys, D., Larson, D., French, J. D., "Carbon-14 Dating with Nuclear Track Emulsions," *Am. J. Phys.* (1972) **40**, 1400.
38. Muller, R. A., "Radioisotope Dating with a Cyclotron," *Science* (1977) **196**, 489.
39. Cairns, T., "Archaeological Dating by Thermoluminescence," *Anal. Chem.* (1976) **48**(3), 266A.
40. Levy, P. W., "Physical Principles of Thermoluminescence," presented at an international seminar on "Application of Science to the Dating of Works of Art," Museum of Fine Arts, Boston, 1974.
41. Fleming, S. J., Jucker, H., Riederer, J., "Etruscan Wall Painting on Terracotta: A Study in Authenticity," *Archaeometry* (1971) **13**, 143.
42. Aitken, M. J., Alldred, J. C., "The Assessment of Error Limits in Thermoluminescence Dating," *Archaeometry* (1972) **14**, 257.
43. Aitken, M. J., Zimmerman, D. W., Fleming, S. J., "Thermoluminescent Dating of Ancient Pottery," *Nature* (1968) **219**, 442.
44. Ralph, E. K., Han, M. C., "Dating of Pottery by Thermoluminescence," *Nature* (1966) **210**, 245.
45. Fleming, S. J., "Thermoluminescent Authenticity Testing of a Pontic Amphora," *Archaeometry* (1970) **12**, 129.
46. Fagg, B. E. B., Fleming, S. J., "Thermoluminescent Dating of a Terracotta of the Nok Culture, Nigeria," *Archaeometry* (1970) **12**, 53.
47. Fleming, S. J., Sampson, E. H., "The Authenticity of Figurines, Animals and Pottery Facsimiles of Bronzes in the Hui Hsien Style," *Archaeometry* (1972) **14**, 237.
48. Fleming, S. J., Stoneham, D., "Thermoluminescent Authenticity Study and Dating of Renaissance Terracottas," *Archaeometry* (1973) **15**, 239.
49. McDougall, D. J., "Thermoluminescence of Geologic Materials," Academic, New York, 1968.
50. Wintle, A. G., Murray, A. S., "Thermoluminescence Dating: Reassessment of the Fine Grain Dose-Rate," *Archaeometry* (1977) **19**, 95.
51. Bell, W. T., "Thermoluminescence Dating: Revised Dose-Rate Data," *Archaeometry* (1977) **19**, 99.
52. Seeley, M. A., "Thermoluminescent Dating in Its Application to Archaeology. Review," *J. Archaeol. Sci.* (1975) **2**, 17.
53. Bell, W. T., "The Assessment of the Radiation Dose-Rate for Thermoluminescence Dating," *Archaeometry* (1975) **17**, 107.
54. Whittle, E. H., "Thermoluminescent Dating of Egyptian Predynastic Pottery from Hemamieh and Qurna-Tarff," *Archaeometry* (1975) **17**, 119.
55. Wagner, G., "Radiation Damage Dating of Rocks and Artifacts," *Endeavor* (1976) **35**, 3.
56. Huang, W. H., Walker, R. M., "Fossil Alpha Particle Recoil Tracks: A New Method of Age Determination," *Science* (1967) **155**, 1103.
57. Durrani, S. A., Khan, H. A., Taj, M., Renfrew, C., "Obsidian Source Identification by Fission Track Analysis," *Nature* (1971) **233**, 242.
58. Brill, R. H., "Application of Fission Track Dating to Historic and Prehistoric Glasses," *Archaeometry* (1964) **7**, 51.
59. Tite, M. S., "Methods of Physical Examination in Archaeology," Seminar, New York, 1972.

60. Fleming, S. J., "Dating in Archaeology. A Guide to Scientific Techniques," St. Martin, New York, 1977.
61. Fleischer, R. L., Price, P. B., Walker, R. M., "Fission Track Dating of a Mesolithic Knife," *Nature* (1965) **205**, 1138.
62. Fleischer, R. L., Price, P. B., Walker, R. M., "Tracks of Charged Particles in Solids," *Science* (1965) **149**, 383.
63. McDougall, D., Price, P. B., "Attempt to Date Early South African Hominids by Using Fission Tracks in Calcite," *Science* (1974) **46**, 943.
64. Fleischer, R. L., Price, P. B., Walker, R. M., "Nuclear Tracks in Solids," University of California, 1975.
65. Scott, B. C., "The Possible Application of Fission Track Counting to the Dating of Bloomery Slag and Iron," *Hist. Metall.* (1976) **10**, 87.
66. Weaver, K. F., "Magnetic Clues Help Date the Past," *Natl. Geogr.* (1967) **131**, 696.
67. Allibone, T. E., Wheeler, M., Edwards, I. E. S., Hall, E. T., Werner, A. E. A., "The Impact of the Natural Sciences on Archaeology," Oxford University, London, 1970.
68. Creer, K. M., Kopper, J. S., "Paleomagnetic Dating of Cave Paintings in Tito Bustillo Cave, Asturias, Spain," *Science* (1974) **186**, 348.
69. Huxtable, J., Aitken, M. J., "Thermoluminescent Dating of Lake Mungo Geomagnetic Polarity Excursion," *Nature* (1977) **265**, 40.
70. Longworth, G., Tite, M. S., "Mossbauer and Magnetic Susceptibility Studies of Iron Oxides in Soils from Archaeological Sites," *Archaeometry* (1977) **19**, 3.
71. Cox, A., "Geomagnetic Reversals," *Science* (1969) **163**, 237.
72. Bray, J. R., "Variation in Atmospheric Carbon-14 Activity Relative to a Sunspot Auroral Solar Index," *Science* (1967) **156**, 640.
73. Ericson, J. E., Berger, R., "Physics and Chemistry of the Hydration Process in Obsidians. II. Experiments and Measurements," *Adv. Obsidian Glass Stud.*, R. E. Taylor, Ed., p. 46, Noyes, Park Ridge, NJ 1976.
74. Lee, R. R., Leich, D. A., Tombrello, T. A., "Obsidian Hydration Profile Measurements Using a Nuclear Reaction Technique," *Nature* (1974) **250**, 44.
75. Lanford, W. A., "Glass Hydration: A Method of Dating Glass Objects," *Science* (1977) **196**, 975.
76. Meighan, C. W., Foote, L. J., Aiello, P. V., "Obsidian Dating in West Mexican Archaeology," *Science* (1968) **160**, 1069.
77. Bada, J. L., Schroeder, R. A., Carter, G. F., "New Evidence for the Antiquity of Man in North America Deduced from Aspartic Acid Racemization," *Science* (1974) **184**, 791.
78. Hare, P. E., "Amino Acid Dating—A History and an Evaluation," *MASCA Newslet.* (1974) **10**, No. 1.
79. Bauch, J., Eckstein, D., "Dendrochronological Dating of Oak Panels of Dutch Seventeenth Century Paintings," *Stud. Conserv.* (1970) **15**, 4550.
80. Melcher, C. L., Zimmerman, D. W., "Tritium–Helium Dating in the Sargasso Sea: A Measurement of Oxygen Utilization Rates," *Science* (1977) **196**, 291.
81. Shackleton, N. J., "Oxygen Isotope Analysis as a Means of Determining Season of Occupation of Prehistoric Midden Sites," *Archaeometry* (1973) **15**, 133.
82. Dansgaard, W., Johnson, S. J., Moller, J., "One Thousand Centuries of Climactic Record from Camp Century on the Greenland Ice Sheet," *Science* (1969) **166**, 377.
83. "Archaeological Chemistry," M. Levey, Ed., University of Pennsylvania, Philadelphia, 1967.

84. Rieth, A., "Archaeological Fakes," Praeger, New York, 1970.
85. Gettens, J., Feller, R. L., Chase, W. T., "Vermillion and Cinnabar," *Stud. Conserv.* (1972) **17**, 45.
86. Gettens, R. J., West Fitzhugh, E., "I. Azurite and Blue Verditer," *Stud. Conserv.* (1966) **11**, 54.
87. Plesters, J,. "Ultramarine Blue, Natural and Artificial," *Stud. Conserv.* (1966) **11**, 62.
88. Gettens, R. J., West Fitzhugh, E., Feller, R. L., "Calcium Carbonate Whites," *Stud. Conserv.* (1974) **19**, 157.
89. Gettens, R. J., West Fitzhugh, E., "Malachite and Green Verditer," *Stud. Conserv.* (1974) **19**, 2.
90. Frankel, R., "Detection of Art Forgeries by X-Ray Fluorescence Spectroscopy," *Isot. Radiat. Technol.* (1970) **8**, No. 1.
91. Brommelle, V., Smith, P., "Conservation and Restoration of Pictorial Art," Butterworths, London, 1976.
92. Cesareo, R., Frazzoli, F. V., Mancini, C., Sciuti, S., Marabelli, M., Mora, P., Rotondi, P., Urbani, G., "Non-Destructive Analysis of Chemical Elements in Paintings and Enamels," *Archaeometry* (1972) **14**, 65.
93. Hall, E. T., Schweizer, F., Toller, P. A., "X-Ray Fluorescence Analysis of Museum Objects: A New Instrument," *Archaeometry* (1973) **15**, 53.
94. Cousins, D. R., Dharmawardena, K. G., "Use of Mossbauer Spectroscopy in the Study of Ancient Pottery," *Nature* (1969) **223**, 732.
95. Keisch, B., "Mossbauer Effect Studies of Fine Arts," *Journal de Physique* (1974) **35**, C6.
96. Eissa, N. A., Sallam, H. A., "Mossbauer Effect Study of Ancient Egyptian Pottery," *Acta Phys. Acad. Sci. Hung.* (1973) **34**, 337.
97. Keisch, B., "Mossbauer Effect Studies in the Fine Arts," *Archaeometry* (1973) **15**, Part 1, 79.
98. Kostikas, A., Gangas, N. H., "Analysis of Archaeological Artifacts," *Appl. Mossbauer Spectrosc.* (1976) **1**, 241.
99. Keisch, B., "Analysis of Works of Art," *Appl. Mossbauer Spectrosc.* (1976) **1**, 263.
100. Gauges, N. H. J., Sigales, I., Moukarika, A., "Is the History of an Ancient Pottery Ware Correlated with Its Mossbauer Spectrum," *J. Phys. (Paris)* Colloq. (1976) **6**, 867.
101. Takeda, M., Mabuchi, H., Tominago, T., "A Tin—119 Mossbauer Study of Chinese Bronze Coins," *Radiochem. Radioanal. Lett.* (1977) **29**, 191.
102. Tominaga, T., Takeda, M., Mabuchi, H., Emoto, Y., "A Mossbauer Study of Ancient Japanese Artifacts," *Radiochem. Radioanal. Lett.* (1977) **28**, 221.
103. Keisch, B., "Mossbauer Effect Spectroscopy Without Sampling: Application to Art and Archaeology," *in* "Archaeological Chemistry," C. W. Beck, Ed., ADV. CHEM. SER. (1974) **138**, 186.
104. DeBruin, M., Korthoven, P. J. M., VanderSteen, A. J., Houtman, J. P. W., Duin, R. P. W., "The Use of Trace Element Concentrations in the Identification of Objects," *Archaeometry* (1976) **18**, 75.
105. Carter, G. F., "Ancient Coins as Records of the Past," *Chemistry* (1966) **39**, 14.
106. Carter, G. F., "X-Ray Fluorescence Analysis of Roman Coins," *Anal. Chem.* (1964) **36**, 1264.
107. Carter, G. F., "Reproducibility of X-Ray Fluorescence Analysis of Septimus Severus Denarii," *Archaeometry* (1977) **19**, 67.
108. Stern, W. B., Descoeudres, J. P., "X-Ray Fluorescence Analysis of Archaic Greek Pottery," *Archaeometry* (1977) **19**, 73.
109. Artzy, M., "Archaeologist Looks at X-Ray Fluorescence vs. Neutron Activation Analysis," Report 1976, LBL-5017, *ERDA Energy Res. Abst.* (1977) **2**, No. 3562.

110. Shenberg, C., Boazi, M., "Rapid Qualitative Determination of Archaeological Samples by X-Ray Fluorescence," *J. Radioanal. Chem.* (1975) **27**, 457.
111. Bateson, J. D., Hedges, R. E. M., "The Scientific Analysis of a Group of Roman-Age Enamelled Brooches," *Archaeometry* (1975) **17**(2), 177.
112. Francaviglia, V., Minardi, M. E., Palmieri, A., "Comparative Study of Various Samples of Etruscan Bucchero by X-Ray Diffraction, X-Ray Spectrometry and Thermoanalysis," *Archaeometry* (1975) **17**(2), 223.
113. Nelson, D. E., D'Auria, J. M., Bennett, R. B., "Characterization of Pacific Northwest Coast Obsidian by X-Ray Fluorescence Analysis," *Archaeometry* (1975) **17**(1), 85.
114. Allen, R. O., Luckenbach, A. H., Holland, C. G., "The Application of Instrumental Neutron Activation Analysis to a Study of Prehistoric Steatite Artifacts and Source Materials," *Archaeometry* (1975) **17**(1), 69.
115. Gilmore, G. R., "Activation Analysis and Archaeometry," *Proc. Anal. Div. Chem. Soc.* (1976) **13**, 99.
116. Harbottle, G., "Activation Analysis in Archaeology," *Radiochemistry (London)* (1976) **3**, 33.
117. Ericson, J. E., Kimberlin, J., "Obsidian Sources, Chemical Characterization and Hydration Rates in West Mexico," *Archaeometry* (1977) **19**, 157.
118. Wenen, G., Ruddy, F. H., Gustafson, C. E., Irwin, H., "Characterization of Archaeological Bone by Neutron Activation Analysis," *Archaeometry* (1977) **19**, 200.
119. "Art and Technology: A Symposium on Classical Bronzes," S. Doeringer, Ed., MIT, Cambridge, 1970.
120. "Archaeological Chemistry," C. W. Beck, Ed., ADV. CHEM. SER. (1974) **138**.
121. "Science and Archaeology," R. H. Brill, Ed., MIT, Cambridge, 1971.
122. "Application of Science in Examination of Works of Art," (Symposium), Museum of Fine Arts, Boston, 1965.
123. Hughes, M. J., Cowell, M. R., Craddock, P. T., "Atomic Absorption Techniques in Archaeology," *Archaeometry* (1976) **18**, 19.
124. Wheeler, M. E., Clark, D. W., "Elemental Characterization of Obsidian from the Koyukuk River, Alaska, by Atomic Absorption Spectrophotometry," *Archaeometry* (1977) **19**, 15.
125. Bowen, N. W., Bromund, R. H., Smith, R. H., "Atomic Absorption for the Archaeologist: An Application to Pottery from Pella of the Decapolis," *J. Field Archaeol.* (1975) **2**, 389.
126. Sismayer, B., Giebelhausen, A., Zambelli, J., Riederer, J., "Application of Infrared Spectroscopy to the Examination of Mineral Pigments of Historical Works of Art in Comparison with Recent European Deposits," *Fresenius' Z. Anal. Chem.* (1975) **277**, 193.
127. Beck, C. W., Wilbur, E., Meret, S., Kossove, D., Kermani, K., "The Infrared Spectra of Amber and the Identification of Baltic Amber," *Archaeometry* (1965) **9**, 96.
128. Tylecote, R. F., "Uses of Thermoanalysis in Archaeometallurgy," *Hist. Metall.* (1975) **9**, 26.
129. Barbieri, M., Calderoni, G., Cortesi, C., Fornaseri, M., "Huntite, a Mineral Used in Antiquity," *Archaeometry* (1974) **16**(1), 211.
130. Riederer, J., "Recently Identified Egyptian Pigments," *Archaeometry* (1974) **16**, 102.

RECEIVED September 19, 1977.

Chemical Aspects of the Conservation of Archaeological Materials

N. S. BAER

Conservation Center of the Institute of Fine Arts, New York University,
1 East 78th Street, New York, NY 10021

Long term burial of artifacts recovered in archaeological excavations often leads to friability, salt encrustation, physical damage, and severe corrosion. Field conservation is limited to such measures as are required to preserve the artifact until it may receive the attention of specialists in the museum laboratory. Typical conservation treatments for textiles, waterlogged wood, bone and ivory, cuneiform tablets, and cast and wrought marine iron are reviewed with particular emphasis on the effects such treatments may have on the subsequent technical examination of the artifact.

Archaeological artifacts seldom are recovered in a perfectly preserved state. Often throughout many years of burial the artifact has been subjected to chemical and biological attack, leaving it in a state of extreme friability. In order to recover and in some cases to save the object, conservation measures may be required in the field (*1, 2*). For example, waterlogged wood unable to bear its own weight must be supported, dried, and consolidated; pottery whose decorations may be significant to the progress of an excavation but which are obscured by mud and deposits of lime and chalk must be cleaned; some cuneiform clay tablets, sun-dried but not baked, must be fired before they may be studied.

Once the artifact has been brought to the museum laboratory, a wide range of conservation procedures are undertaken not only to assure the long term survival of the artifact but also to return it to an approximation of its original state.

All of these field and laboratory treatments will in some way change the artifact, and some treatments may have profound effects on the suitability of the artifact for technological examination. It is therefore essential for the scientist engaged in analyzing and examining archaeo-

0-8412-0397-0/78/33-171-**025**$05.00/1
© 1978 American Chemical Society

logical materials to be familiar with conservation practice. Similarly, the conservator has a professional obligation to limit his initial treatments to those essential for the survival of the artifact; to keep detailed treatment records; and, where possible, to retain some representative material in an untreated state. Under the heading, "Safeguarding Evidence," Plenderleith and Werner (3) consider the problem of record keeping.

It must be clear that laboratory treatment of antiquities and works of art carries with it a great responsibility. The excitement of discovery and the urge to reveal hidden interests must be held in check and kept subservient to the duty of maintaining full laboratory records of the work as it proceeds, illustrated, where necessary, with sketches and photographs for incorporation in a permanent archive.

Similarly, Jedrejewska (4), considering the ethics of conservation, writes about the treatment process:

During this time something is removed, something is added, and something changed in the object. In consequence some of the historical evidence inherent in the object can be damaged, changed, or lost. It may also happen that new materials introduced to the object, or used for treatment, may prove harmful immediately, or after a lapse of time.

In the following discussion a number of conservation methods commonly used to treat archaeological artifacts are examined for their effects on subsequent technical examination. The preservation of archaeological textiles, waterlogged wood, archaeological bone and ivory, cuneiform tablets, and marine iron are among the problems considered.

Archaeological Textiles

Archaeological textile artifacts often are recovered in a severely deteriorated state, embrittled and extremely friable (5, 6). However, the art historical and anthropological importance of materials such as Coptic and Precolumbian textiles requires their conservation treatment to make them available for study and display. Among the operations variously used by textile conservators are bleaching; brushing; washing with detergents; dry cleaning; use of enzymes; exposure to fungicides, insecticides, and moth-proofing agents; stain removal; steam cleaning; washing; and use of wetting agents (1).

On washing and drying, ancient textiles often become still more brittle and so are treated with lubricating materials, most often glycerine added to the final aqueous rinse (7, 8). In the most severe cases, methods of consolidation and reinforcement are used to facilitate the handling of these artifacts (6, 9, 10, 11). That these consolidative procedures are irreversible is readily acknowledged, although it is suggested that consolidation may offer the only hope for the survival of the artifact (6).

Once restored, the textile is then often subjected to the vicissitudes of the museum environment. Airborne dirt, sulfur dioxide, ozone, light, and microorganisms all affect additional change on the object (*12*). Such natural dyestuffs as madder, cochineal, and indigo can all be expected to undergo substantial fading under long term exposure to average museum lighting conditions (*13, 14*). Museum illumination may cause even further deterioration in the fibers themselves (*15, 16*).

Thus it is not surprising that archaeological textiles subject to chemical investigation present serious problems of contamination and alteration requiring extensive pretreatment for even the most routine analyses (*17*).

Waterlogged Wood

Waterlogged wood artifacts, especially prehistoric dug-out canoes, boats, and ships, have been recovered for museum display for more than a century (*18*). In recent years, the growth of underwater archaeology has accelerated the rate of discovery of such materials. The conservation process usually involves removal of excess water to reduce shrinkage and distortion followed by consolidation to strengthen the artifact (*18–24*). The major concerns of the conservator are the elimination of shrinkage and control of the surface appearance after consolidation (*24*). The most commonly used procedure involves immersing the waterlogged wood in a 10% polyethylene glycol solution in distilled water at 60°C with controlled evaporation over several months (*1, 2, 19, 23*). In a case of wooden writing tablets where the main objective was to render the ink inscriptions as legible as possible, the excess water was removed by an alcohol/ether dehydration process (*20*). Although the treatment was considered to be successful, the wood was very fragile and still might require backing or consolidation. Consolidative methods include alcohol/ether drying followed by dammar resin consolidation (*1, 24*); heating the wood at 96°C with molten potassium alum (*1, 24*); washing in water, impregnation with aqueous melamine/formaldehyde which is then polymerized in situ (*1, 24*); and impregnation with epoxy resins (*21*).

That none of the above methods is completely satisfactory as a conservation procedure is generally agreed. One interesting recent development involves the application of freeze-drying technology (*24, 25*) combined with polyethylene glycol impregnation.

In an interesting comparative study using scanning electron microscopy Oddy (*24*) demonstrated the substantial changes in cell structure which accompany the several treatments. Clearly most, if not all, of the consolidation methods described are irreversible, introducing materials which may obscure subsequent examination.

Bone and Ivory

Fossil tooth and bone material and artifacts of bone and ivory are among the most important materials recovered by anthropologists and archaeologists. A substantial literature (26, 27) exists describing the many forms of examination applied to these materials. The degree of preservation of these materials is a function of many environmental variables (28, 29). On long term burial systematic changes may occur involving replacement of hydroxyl groups in the hydroxyapatite with fluoride to form fluorapatite. In the proteinaceous component, significant change will occur in the amino acid composition (27, 28, 29, 30). Since the middle of the 19th century researchers have sought to take advantage of these changes in order to develop methods for dating tooth and fossil material (26, 27, 28, 32, 33). Most recently, considerable attention has been focused on racemization of the constituent amino acids as an indicator of age (34, 35). In some cases radiocarbon methods have been applied with success (26, 27, 30).

With the exception of radiocarbon dating, all of the methods rely on chemical changes occurring in the original bone or tooth material. However, these same changes can lead to such severe deterioration that the bone itself may disappear almost completely, leaving only a silhouette or stain in the ground (36). When an artifact rather than a simple skeletal remain is found, substantial intervention may occur. Layard, writing in 1849, provides an extreme example of the treatment some ivory specimens have received (31).

> The chamber V is remarkable for the discovery of a number of ivory ornaments, of considerable beauty and interest. These ivories, when uncovered, adhered so firmly to the soil, and were in so forward a state of decomposition, that I had the greatest difficulty in extracting them, even in fragments. I spent hours lying on the ground, separating them, with a penknife, from the rubbish by which they were surrounded. The ivory separated itself in flakes. With all the care that I could devote to the collection of fragments, many were lost, or remained unperceived, in the immense heap of rubbish under which they were buried. Since they have been in England, they have been admirably restored and cleaned, and the ornaments have regained the appearance and consistency of recent ivory, and may be handled without risk of injury.

More common are consolidation procedures (1, 2, 37) where various waxes and resins are introduced into the interstices produced by the deterioration process. The cleaning and consolidation processes may interfere seriously with any or all of the various dating methods. New organic matter will bias radiocarbon dates; animal glue consolidants will introduce extraneous proteinaceous material, thus confusing dating methods based on amino acid analysis; and nitrogen-bearing consolidants such as cellulose nitrate, soluble nylon, and animal glues will introduce errors in nitrogen measurements. Though one may use separation techniques in an attempt to isolate the original matrix materials, it is likely that these techniques will be only moderately successful.

Cuneiform Tablets

Artifacts of unbaked clay recovered in archaeological excavations present a singular series of problems. If washed, they may turn to mud; if left unfired, they may be damaged in transit. One particular class of material, cuneiform tablets, has received particularly severe treatment. Dowman describes accurately the general attitude (*38*).

> Tablets come into a somewhat different category. All that is ever wanted of them is their inscriptions and it is common practice to fire tablets so that they can be handled indefinitely. Should only clay be wanted for analysis an unscribed section can be set aside and left unfired. Any treatment to be given to tablets in the field depends on whether there is a kiln on the site or not and, if not, on how desperate the epigraphist is to read them on the spot.

Several authors have reported the general methods for firing these tablets in the field as well as in the museum laboratory (*1, 38, 39, 40*). The specific problems associated with removing soluble salts are considered in detail by Organ (*39*).

Those whose research interests involve the examination of trace element distribution in ceramic materials as an indication of provenance and those making thermoluminescence measurements will find few useful clay samples among the materials preserved by archaeologists. Thus, a potentially valuable source of˙documented clay materials is lost when unbaked clay tablets are routinely fired and cleaned without systematic sorting to preserve in their original state those specimens which do not require firing for their survival.

Cast and Wrought Marine Iron

In wrought and cast irons the examination of the metallurgical structure may provide significant insight into the method of fabrication of the artifact. Cold working in wrought iron, chill casting, and annealing in cast iron may all be observed on metallographic examination (*43, 44, 45*). Iron samples recovered from shipwreck are typical of materials usefully examined by such methods. Unfortunately, iron artifacts recovered from the sea deteriorate extremely rapidly because of the presence of large amounts of chlorides in the corrosion products (*45*). Thermal stabilization processes involving heating in air at 860°C (*41*) or heating in a hydrogen reduction furnace at 800°C and 1060°C (*42*) have been used to drive off the metallic chlorides. It has been demonstrated (43) that such heating destroys the original metallurgical structure in the iron vitiating a significant part of the evidence provided by the artifact. It was shown that hydrogen reduction at temperatures above 400°C could not be used without serious loss of microstructure. Unfortunately, at this temperature the treatment is only partially successful, allowing sufficient chlorides to remain to leave the artifact unstable. Secondary treatments will have to be developed to remove the remaining chlorides.

It is interesting to note a recent announcement of the installation in the Portsmouth City Museum, United Kingdom of a large reduction furnace to stabilize ion antiquities, cannon in particular (41). It is indicated that "experiments will be conducted to determine the feasibility of treating archaeological iron using the furnace, but not destroying the metal's microstructural characteristics."

Conclusion

Conservators have long recognized that their professional responsibilities include the information inherent in the structure as well as in the composition of the artifact under treatment. Unfortunately, the need to preserve an artifact or to prepare it for display has often required treatments which significantly changed the artifact. Cooperation between scientists engaged in the study of archaeological materials and conservators is required to develop conservation treatments which minimize structural and compositional alteration. Guidelines must be developed for the preservation of representative untreated materials where satisfactory treatments are not yet available. It is also essential that in the examination of archaeological artifacts the previous treatment history be considered.

Acknowledgment

This project was sponsored by the National Museum Act (administered by the Smithsonian Institution).

Literature Cited

1. Plenderleith, H. J., Werner, A. E. A., "The Conservation of Antiquities and Works of Art," 2nd ed., Oxford, London, 1971.
2. Dowman, E. A., "Conservation in Field Archaeology," Methuen, London, 1970.
3. Plenderleith, H. J., Werner, A. E. A., "The Conservation of Antiquities and Works of Art," 2nd ed., pp. 16–17, Oxford, London, 1971.
4. Jedrzejewska, H., "Ethics in Conservation," p. 7, Institutet för Materialkunskap, Stockholm, 1976.

5. Bery, G. M., Hersh, S. P., Tucker, P. A., Walsh, W. K., "Reinforcing Degraded Textiles. Part I: Properties of Naturally and Artificially Aged Cotton Textiles," ADV. CHEM. SER. (1977) **164**, 228–248.

6. Jedrzejewska, H., "Some New Techniques for Archaeological Textiles," *in* "Textile Conservation," J. E. Leene, Ed., Chap. 22, Butterworths, London, 1972.

7. Delacorte, M., Sayre, E. V., Indictor, N., "Lubrication of Deteriorated Wool," *Stud. Conserv.* (1971) **16**(1), 9–17.

8. *Textile Museum Workshop Notes,* The Textile Museum, Washington, D.C., papers no. 4, 8, and 24.

9. "The Conservation of Cultural Property," UNESCO, Paris, 1968.

10. Berger, G. A., "Testing Adhesives for the Consolidation of Paintings," *Stud. Conserv.* (1972) **17**, 173–194.

11. Berry, G. M., Hersh, S. P., Tucker, P. A., Walsh, W. K., "Reinforcing Degraded Textiles. Part II: Properties of Resin-Treated, Artificially Aged Cotton Textiles," ADV. CHEM. SER. (1977) **164**, 249–260.

12. Thomson, G., "Textiles in the Museum Environment," *in* "Textile Conservation," J. E. Leene, Ed., Chap. 7, Butterworths, London, 1972.

13. Padfield, T., Landi, S., "The Lightfastness of Natural Dyes," *Stud. Conserv.* (1966) **11**, 181–196.

14. Thomson, G., "A New Look at Colour Rendering, Level of Illumination, and Protection from Ultraviolet," *Stud. Conserv.* (1961) **6**, 49–70.

15. Little, A. H., "Deterioration of Textile Materials," 1964 Delft Confernce on the Conservation of Textiles, 2nd edition, pp. 67–78, International Institute for Conservation, London, 1965.

16. Padfield, T., "The Deterioration of Cellulose," *in* "Problems of Conservation in Museums," p. 119–164, Allen and Unwin, 1969.

17. Baer, N. S., Delacorte, M., Indictor, N., "Chemical Investigations on Pre-Columbian Archaeological Textile Specimens," ADV. CHEM. SER. (1977) **164**, 261–271.

18. Barker, H., "Early Work on the Conservation of Waterlogged Wood in the UK," *in* "Problems in the Conservation of Waterlogged Wood," W. A. Oddy, Ed., pp. 61–63, National Maritime Museum, Greenwich, London, 1975.

19. Stark, B. L., "Waterlogged Wood Preservation with Polyethylene Glycol," *Stud. Conserv.* (1976) **21**(3), 154–158.

20. Blackshaw, S. M., "The Conservation of the Wooden Writing-Tablets from Vindolanda Roman Fort, Northumberland," *Stud. Conserv.* (1974) **19**(4), 244–246.

21. Munnikendam, R. A., "Low Molecular Weight Epoxy Resins for the Consolidation of Decayed Wooden Objects," *Stud. Conserv.* (1972) **17**(4), 202–204.

22. Oddy, W. A., Van Geersdaele, P. C., "The Recovery of the Graveney Boat," *Stud. Conserv.* (1972) **17**, 30–38.

23. Gregson, C. N., "Progress on the Conservation of the Graveney Boat," *in* "Problems in the Conservation of Waterlogged Wood," W. A. Oddy, Ed., pp. 113–114, National Maritime Museum, Greenwich, London, 1975.

24. Oddy, W. A., "Comparison of Different Methods of Treating Waterlogged Wood as Revealed by Stereoscan Examination and Thoughts on the Conservation of Waterlogged Boats," *in* "Problems in the Conservation of Waterlogged Wood," W. A. Oddy, Ed., pp. 45–49, National Maritime Museum, Greenwich, London, 1975.

25. Rosenqvist, A. M., "Experiments on the Conservation of Waterlogged Wood and Leather by Freeze-Drying," *in* "Problems in the Conservation of Waterlogged Wood," W. A. Oddy, Ed., pp. 9–23, National Maritime Museum, Greenwich, London, 1975.

26. Baer, N. S., Majewski, L. J., "Ivory and Related Materials in Art and Archaeology: An Annotated Bibliography," *Art and Archaeol. Tech. Abstr.* (1970) **8**(2), 229–275.
27. Ibid., (1971) **8**(3), 189–228.
28. Hare, P. E., "Geochemistry of Proteins, Peptides, and Amino Acids," *in* "Organic Geochemistry," Eglinton and Murphy, Eds., Chap. 18, Springer Verlag, New York, 1969.
29. Hare, P. E., "Organic Geochemistry of Bone and Its Relation to the Survival of Bone in the Natural Environment," in press.
30. Hassan, A. A., Ortner, D. J., "Inclusion in Bone Material as a Source of Error in Radiocarbon Dating," *Archaeometry* (1977) **19**(2), 131–135.
31. Layard, A. H., "Nineva and Its Remains," H. W. F. Saggs, Ed., p. 245, Praeger, New York, 1970.
32. Baer, N. S., Indictor, N., "Chemical Investigations of Ancient Near Eastern Archaeological Ivory Artifacts," ADV. CHEM. SER. (1975) **138**, 236–245.
33. Baer, N. S., Jochsberger, T., Indictor, N., "Chemical Investigations on Ancient Near Eastern Ivory Artifacts: III. Fluorine and Nitrogen Composition," ADV. CHEM. SER. (1978) **171**, 139.
34. Bada, J. L., "Racemization of Isoleucine in Calcareous Marine Sediments: Kinetics and Mechanism," *Earth Planet. Sci. Lett.* (1972) **15**, 1–11.
35. Bada, J. L., "The Dating of Fossil Bones Using the Racemization of Isoleucine," *Earth Planet. Sci. Lett.* (1972) **15**, 223–231.
36. Keeley, H. C. M., Hudson, G. E., Evans, J., "Trace Element Contents of Human Bones in Various States of Preservation. I. The Soil Silhouette," *J. Archaeol. Sci.* (1977) **4**, 19–24.
37. Crawford, V. E., "Ivories from the Earth," *Metropolitan Museum of Art Bulletin* (1962) **21**(4), 141–148.
38. Dowman, E. A., "Conservation in Field Archaeology," p. 122–123, Methuen, London, 1970.
39. Organ, R. M., "The Conservation of Cuneiform Tablets," *British Museum Quarterly* (1961) **23**(2), 52–57.
40. Crawford, V. E., "Processing Clay Tablets in the Field," *in* "The Preservation and Reproduction of Clay Tablets and Conservation of Wall Paintings," *Colt Archaeol. Monogr. Ser.* (1966) **3**.
41. Eriksen, E., Thegel, S., "Conservation of Iron Recovered from the Sea," *Tojhusmuseets Skrifter* (1966) **8**.
42. Arrhenius, O., Barkman, L., Sjostrand, E., "Conservation of Old Rusty Iron Objects," *Swed. Corros. Inst. Bull.* (1973) **61E**.
43. North, N., Owens, M., Pearson, C., "Thermal Stability of Cast and Wrought Marine Iron," *Stud. Conserv.* (1976) **21**, 192–197.
44. North, N., Pearson, C., "Alkaline Sulphite Reduction of Marine Iron," *Proceedings, ICOM Committee for Conservation, 4th Triennial Meeting, Venice, 1975*, **75/13/3**, 1–14.
45. North, N. A., Pearson, C., "Thermal Decomposition of FeOCl and Marine Cast Iron Corrosion Products," *Stud. Conserv.* (1977) **22**, 146–157.
46. *IIC Conservation News* (March 1977) **2**.

RECEIVED December 19, 1977.

Radiocarbon Dating: An Archaeological Perspective

R. E. TAYLOR

Radiocarbon Laboratory, Department of Anthropology, Institute of Geophysics and Planetary Physics, University of California, Riverside, CA 92512

A model describing the production, distribution, and decay of the radioactive isotope of carbon (^{14}C) led to a proposal that the ^{14}C content of an organic sample could be used as an index of age. Between 1947 and 1949 Willard F. Libby conducted the critical experiments which established the basic validity of the method. Since that time all of the six basic physical assumptions of the method have been the subject of continuing scrutiny. Archaeologists who rely on ^{14}C values to construct their chronologies are especially concerned with errors that may be introduced as a result of violations of one or more of the fundamental assumptions. Error reduction strategies are best implemented by close interdisciplinary collaboration and cooperation betweeen radiocarbon specialists and those engaged in archaeological research studies.

On December 12, 1960, the Nobel Prize for Chemistry was bestowed on Willard F. Libby for, in the words of the citation, "his method to use carbon-14 for age determination in archaeology, geology, geophysics, and other branches of science." The chairman for that year of the Nobel Committee for Chemistry reported that one of the scientists who suggested Libby as a candidate for the Nobel Laureate characterized his work in the following terms: "Seldom has a single discovery in chemistry had such an impact on the thinking in so many fields of human endeavour. Seldom has a single discovery generated such wide public interest" (1).

Among most archaeologists and probably all prehistorians, it is now well accepted that the development of the ^{14}C method marked a critical watershed in the history of the discipline. One prominent senior British archaeologist has gone so far as to rank the discovery and application of

0-8412-0397-0/78/33-171-**033**$09.25/1
© 1978 American Chemical Society

the radiocarbon method with the discovery of the antiquity of the human species in the 19th century (2). A well known American archaeologist divided modern prehistoric chronological studies into pre-^{14}C and ^{14}C stages (3).

Over the past quarter century, the basis of the technique has been discussed in great detail and needs little explanation (4,5). Figure 1 sketches in diagrammatic form some of the basic elements of the radiocarbon model. The natural production of radiocarbon is a secondary effect of cosmic ray bombardment of the upper atmosphere. As $^{14}CO_2$, radiocarbon is distributed differentially into various atmospheric, biospheric, and hydrospheric reservoirs. Metabolic processes maintain the radiocarbon content of living organisms at an essentially constant level.

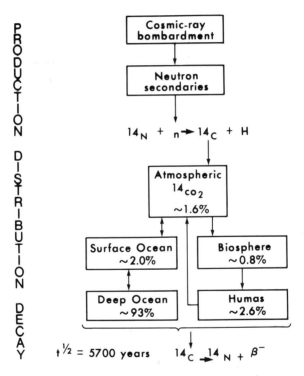

Figure 1. Diagrammatic outline of basic elements of the radiocarbon method (after Fleming, Ref. 5)

However, once such processes cease, the amount of radiocarbon decreases by decay as measured by the isotope's half-life. Thus, a radiocarbon date is based on the measurement of the residual ^{14}C contained in the sample. However, for a radiocarbon age of a sample to be equivalent to its real or calendar age, certain fundamental parameters and assumptions must hold within somewhat narrow limits. In retrospect, one reason for the rapid success of the radiocarbon method was that the original predicted model turned out to be amazingly accurate and required few major modifications. The problems that emerged as the technique was developed and applied to a large number of sample types and geochemical environments resulted from situations where there was some deviation from the assumptions around which the general radiocarbon model had been constructed originally.

Nobel prizes are given to those few who can see through the almost infinite complexity of the physical world and abstract out elements which constitute a unique way of looking at the relations of certain physical phenomena. But even Nobel prize winners operate within preexisting and on-going intellectual traditions which mold and condition the direction and focus of their studies (6). Before we review some of the current issues and problems in the application of radiocarbon dates to archaeological studies and project its future course, it may be helpful to look in retrospect at the steps that brought us this most powerful and now, at least in prehistoric studies, almost indispensable method of dating.

Radiocarbon Dating in Retrospect

Radiocarbon is one of a number of isotopes which had been produced artificially in the laboratory before being detected in natural concentrations. Martin Kamen (7) has discussed the early history of radiocarbon from the point of view of one intimately associated with its discovery and application as a tracer in biochemical studies (Figure 2). Both Kamen and Libby have pointed out that the existence of ^{14}C was suggested first in a 1934 paper by Franz Kurie as, at that time, a highly unlikely interpretation of the effect of neutron bombardment on nitrogen in a cloud chamber (8). This was the year following Libby's receipt of his doctoral degree at Berkeley and his appointment as an assistant professor in the chemistry department there. (In addition to Ref. 7, data used to review the history of development of the radiocarbon method has come from comments by Libby published in volumes of the proceedings of the various radiocarbon conferences; Refs. 9, 10, and 11).

Beginning in 1937, both Kamen and Kurie became associated with E. O. Lawrence's newly organized radiation laboratory. At that time, ^{14}C recoil tracks were used to calibrate cloud chamber experiments, but little was known about the isotope's physical characteristics. It was assumed that it was radioactive with a half-life of a few hours or, at most, a few days based on an analogy with the 0.8-sec half-life of ^{6}He (12). In view of its assumed short half-life, no determined effort to isolate it was undertaken immediately since the Lawrence laboratory between 1937 and

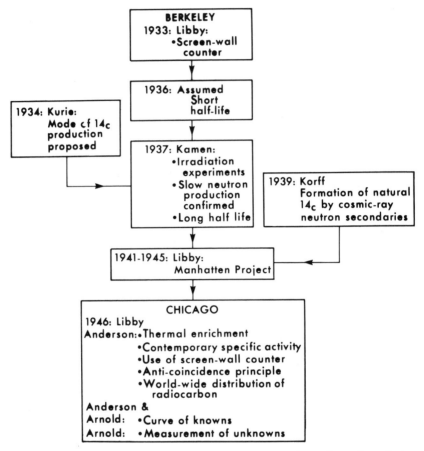

Figure 2. Representation of the development of the radiocarbon dating model. Based on information supplied in Kamen (7) and Libby (9, 10, 11, 19).

1939 was focused on producing isotopes for medical and biological research. Lawrence believed that this promised the best chance of long-term funding to support the operation and expansion of cyclotron research at Berkeley. A question of the practical value of radioisotopes in biological and medical research was raised, causing Lawrence in late 1939 to order a maximum effort to determine once and for all whether long-lived radioisotopes existed for any of the elements in the first row of the periodic table, especially hydrogen, nitrogen, oxygen, and carbon.

Kamen and his associates, notably Samuel Ruben, began the search for ^{14}C by exposing graphite to deuteron bombardment in the 37-in. Berkeley cyclotron. Following the bombardment, the graphite was burned to CO_2, precipitated out as $CaCO_2$, and used to coat the inside of a screen-wall counter of a type that Libby had used in his doctoral research

(*13, 14*). This same screen-wall counter design would be used six years later in the early work on the radiocarbon dating method. In 1940, however, this counter registered the presence of artificial ^{14}C for the first time, and the surprised researchers determined quickly that the half-life had to be orders of magnitude longer than previously assumed; the tentative value was 25,000 years.

All of these initial experiments had been based on the assumption that the reaction of deuterons with ^{13}C would be the most likely reaction in forming ^{14}C. To eliminate the possibility of a favored reaction of neutrons with ^{14}N, an experiment was devised in which a solution of saturated ammonium nitrate was irradiated with neutrons. It was expected that no detectable amount of radiocarbon would be produced. Instead, a relatively small amount of the precipitate paralyzed the screen-wall counter! Within a few month's time, despite strong theoretical arguments to the contrary, it was demonstrated experimentally that the thermal neutron mode of ^{14}C production was heavily favored and that the half-life was on the order of years or millenia (*15*).

Meanwhile, Serge Korff had sent sensitive Geiger counters aloft in balloons and had noted an increase in neutrons with altitude until, at about 16 km, the neutron flux dropped off rapidly. Korff interpreted these results to suggest that primary cosmic rays were producing secondary neutrons. He further suggested that these neutrons would disappear principally through the formation of radiocarbon (*16*).

The entrance of the U.S. into World War II redirected quickly the pursuit of nuclear studies. Many of those affiliated with Lawrence at Berkeley became attached to the Manhattan Project (*17, 18*). Libby, on leave from Berkeley, interrupted his tenure as a Guggenheim Fellow at Princeton to join the Manhattan project group at Columbia University. When not occupied with the principal matter at hand, he continued to investigate aspects of radiocarbon chemistry. One obvious issue was the question of the half-life. The initial estimates were based on values which varied by an order of magnitude or more. Attempts to obtain more precise estimates were thus very much in order. Unfortunately, the values obtained (26,000 ± 13,000 and 21,000 ± 4,000 years) were, in retrospect, seriously in error (*19*). Only with the use of mass spectrometric data to determine the isotopic composition of the samples used in the experiments would more accurate values become available.

Such investigations were begun immediately following the end of the war and, by 1950, some 10 figures had been published, ranging from 7,200 ± 500 to 4,700 ± 470 years (*19*). Initially, Libby and his collaborators in the development of the radiocarbon technique used the value 5720 ± 47 as the half-life but soon adopted the weighted average of three independently obtained measurements using isotope dilution techniques in gas counters (20). The average value obtained was 5568 ± 30, and this became identified as the Libby half-life. Many archaeologists and others would have been spared considerable confusion if the original value of 5720 had been retained since a consensus in 1962 at the Fifth Radiocarbon Dating Conference at Cambridge suggested that 5730 ± 40 probably more closely approximates the actual half-life (21). Despite this, the official Libby half-life of 5568 (5730) years continues to be used among radiocarbon laboratories around the world. The stated reason for this policy is that, since 1950, tens of thousands of radiocarbon values have been published, and any changes in the half-life constant would introduce unneeded confusion (22).

At the conclusion of the war in 1945, Libby accepted an appointment as professor of chemistry in the Department of Chemistry and Institute for Nuclear Studies (now Enrico Fermi Institute for Nuclear Studies) at the University of Chicago. The years with the Manhattan Project had provided an environment within which the knowledge gained during the pre-War years concerning the mode of production, distribution, and decay of radiocarbon within the carbon cycle could be reviewed. The fact that a long-lived carbon isotope was being produced naturally high in the atmosphere by cosmic ray-enduced neutron bombardment of nitrogen led to several obvious questions.

First of all, what happened to the ^{14}C atom following its production? This could be quickly answered. Within a few minutes, or hours at the most, it would have been oxidized to a carbon dioxide molecule. Some intermediate pathway was possible, but the high probability was that all cosmic-ray produced ^{14}C would be oxidized relatively rapidly and completely to $^{14}CO_2$. Since cosmic ray intensities varied by a factor of 3.5 between the equator and poles, the production of ^{14}C was expected to vary likewise on a geographic basis. However, since production takes place in the stratosphere, any production gradient would be destroyed rapidly by the action of the winds. The effect of this would be to mix the $^{14}CO_2$ molecule rapidly throughout the atmosphere. Since the photosynthetic process used CO_2, all terrestrial plants would absorb radiocarbon. Since all animals depend ultimately on plants for food resources, all terrestrial living organisms would be slightly radioactive with $^{14}CO_2$. However, because the largest percentage of the earth's surface is water, it was estimated that about 87% of all radiocarbon would find its way into the oceans as dissolved carbon dioxide, bicarbonate, and carbonate. This large concentration of ^{14}C in the oceans would confer a major benefit. Variations of up to 10–20% in the amount of organic carbon in terrestrial organisms would not affect significantly the total amount of radiocarbon on earth, and hopefully the thermal heat of the earth would help to mix the principal ^{14}C reservoir, the oceans. The only significant variation in the ^{14}C content in the oceans could be induced by changes in volume, temperature, or pH. These parameters were assumed to have been essentially constant at least into the Pleistocene. The rate with which atmospheric radiocarbon was mixed into the oceans was determined by assuming that it was not in excess of 500 years. The average turn-over time for any single ^{14}C atom would then not exceed 1000 years. Since this was much less than the average life of radiocarbon (ca. 8,000 years), the distribution of radiocarbon throughout all reservoirs was predicted to be relatively uniform.

Probably the most crucial question had to do with the constancy of the cosmic ray flux over time. Unfortunately in 1946, there was no experimental data to use in estimating variability in this parameter and thus how long the present production rate of radiocarbon had been maintained at approximately its present level. It was assumed that it had been essentially constant for the last 20,000–30,000 years. If one assumed also that the total radiocarbon on earth (meaning essentially the amount in carbonates in the oceans) had not changed appreciably during that time, then it followed that a balance would exist between the rate of disintegration of radiocarbon in the atmosphere and the rate with which new radiocarbon atoms were being absorbed in living systems. It was calculated that there were about 8.3 g of carbon in exchange with atmospheric carbon dioxide for each square centimeter of the earth's surface. Based on a limited amount of experimental data, it was estimated that the average number of neutrons produced by cosmic ray bombardment per sq cm of the earth's surface per sec was 2.6 and that 100% of these neutrons were converted to ^{14}C. A simple calculation ($2.6 \times 60/ 8.3$) resulted in the figure of 18.8 disintegrations per min per g as the specific activity of ^{14}C in modern carbon-containing samples (*23, 24, 25*).

In 1946, the problem was demonstrating that the most fundamental assumptions did in fact hold. Initially, this meant obtaining measurements of the natural radiocarbon concentrations in living organics to see if it occurred in the amount expected and if the worldwide distribution of radiocarbon was essentially constant. An experiment was devised whereby biological methane gas derived from the sewage disposal plant at Baltimore, MD and petroleum methane from the Sun Oil Co. refinery were each enriched by a similar factor in a thermal diffusion column. It was assumed that the petroleum methane contained no ^{14}C because of its age in excess of many tens of millions of years whereas the biological methane contained about 17–18 dpm radiocarbon per g of carbon. The experiment was conducted, and the results confirmed the calculations (*26*).

Unfortunately, the use of a thermal diffusion column to measure samples routinely was obviously impractical since each sample would cost several thousand dollars to process. A method to measure directly natural ^{14}C levels without costly procedures had to be devised. Although the use of a proportional gas counter using methane or carbon dioxide was considered, this approach was rejected in favor of the original screen-wall counter previously developed by Libby in the pre-War years at Berkeley. The sensitive counting region of the detector was defined by a series or screen of wires parallel to and surrounding the central wire about half way between the metal wall of the counting cylinder and the central wire. From this arrangement, the term screen-wall counter was derived. Samples were prepared by combustion or acidification to CO_2 and reduction by magnesium. The final product was a finely powdered lamp black. Such elementary carbon could be contaminated easily by airborne sources, and extreme precautions had to be taken to prevent such occurrences. For counting purposes, the detector was placed inside a concentric ring of Geiger tubes, and the whole assembly was placed within a shield of 8-in. steel plates surrounding the counters completely. Later, mercury also was used to shield the central detector.

The development of this detector surrounded by a ring of guard tubes was critical in the advancement of routine low-level ^{14}C work. It allowed the pulses from the detector to be compared as to their coincidence or non-coincidence with the pulses coming from one or more of the surrounding Geiger tubes. If such a coincidence occurred, the event was ignored as it resulted from an external cause. If, however, a signal from the screen-wall counter did not occur at the same time as one received from the Geiger ring, then this event was deemed to have been caused by radioactive decay in the sensitive volume of the counter.

The effect of the various strategies used to shield the main detector from external environmental radiation effects can be appreciated if one notes the various counting rates found during the development of the early detectors. Libby reports the background rate for an unshielded screen-wall counter at about 500 counts per min (cpm). Placing the detector in the 8-in. steel shield assembly reduced the rate to about 100 cpm. The use of the coincidence principle reduced the background rate to about 5 cpm (27).

Using this instrument, E. C. Anderson, a doctoral candidate working under Libby, measured the natural radiocarbon activity of a series of contemporary terrestrial biospheric samples collected from localities ranging from 60°N to 65°S geomagnetic latitude. The average specific activity was 15.3 dpm/gm with a maximum vaiation from 14.47 to 15.69 dpm/gm. The close correspondence to the estimated value (18–19 dpm/gm) and the lack of significant variation of contemporary ^{14}C with latitude provided strong support for the validity of the initial assumptions. These findings were presented in Anderson's doctoral dissertation at Chicago in 1952 (29).

The final segment of Libby's studies were coordinated by James Arnold, who had arrived at Chicago from Princeton as a postdoctoral fellow in 1948. His initial responsibility was to measure the samples that would comprise the curve of knowns. These samples were to determine whether the known age and the radiocarbon age of a given sample were similar. The state of communication between the physical science community and archaeologists and historians at that time becomes immediately apparent when Libby reported that it came as something of a surprise that historically documented samples could provide known ages back only about 5000 years at the beginning of the Egyptian historic period (Dynasty I). The first known-age sample dated was the famous tomb of King Zoser of the IVth Dynasty. According to Egyptologists, the age of the acacia wooden beam from the tomb was about 4700 years old or about 2700 B.C. The average radiocarbon age of the sample was 3979 ± 350 years B.P. (Before Present). This was considered a good agreement considering the problems associated with solid carbon counting. The results of the second sample measured, which was also from Egypt, was never published in the Chicago date lists because its count rate was essentially the same as that of modern Baltimore sewage, i.e., it was modern wood. When asked about this "known-age sample," the museum curator remarked that the museum had always wondered whether the sample was authentic since it had been purchased from a dealer of antiquities (29). Fortunately, the rest of the known-age samples which were measured were much more reliable, and the results were published by Libby, Anderson, and Arnold in 1949 (30). Rounded off to 1950, this year came to mark the zero point in the conversion of radiocarbon values from their B.P. to A.D./B.C. equivalents (31).

During its period of operation from 1947 to 1954, the Chicago laboratory provided radiocarbon values on some 375 samples. Most of these samples were selected and secured under the auspices of committees appointed by the American Anthropological Association and the Geological Society of America. The results obtained on these samples caused both delight and anguish depending on whether the values supported or conflicted with the estimates, guesses, or prejudices of those concerned with the outcome of the analyses. Most of the samples were selected to throw light on some of the then-critical problems facing geologists and historians as well as archaeologists. Some of the most interesting results of the early Chicago determinations included a significant reduction in the estimated age of the Egyptian pre-dynastic period, an indication of a somewhat surprising early age of village life in Western Asia, support for the low Mesopotamian historic chronology, the 1st Century A.D. age for the Dead Sea Scrolls, and the 8th millenium B.C. date for the termination of the Pleistocene simultaneously in Europe and North America (32).

The relatively rapid acceptance of the general validity of the radio-carbon method by most researchers is indicated by the rapid establishment of other laboratories to perform ^{14}C analyses. Many who attempted to duplicate the procedures required to obtain acceptable values using the solid carbon technique experienced moderate-to-severe difficulties. The result was the substitution of gas or liquid scintillation counting methods. Because of the greater efficiency of such detection systems, these developments permitted the maximum age range to be extended from about 25,000 years with the solid carbon system to between 40,000 and 50,000 years (70,000 with isotopic enrichment) as well as a decrease in the stated one sigma counting errors (33). When the journal *Radiocarbon* was inaugurated in 1959, there were 36 radiocarbon laboratories operating worldwide. Throughout the past two decades, the number of research and service facilities has continued to increase to over 100 in 1977.

Current Issues and Problems

The ability of the radiocarbon method to provide accurate and precise determinations of the actual or calendar age of organic samples is obviously a function of the degree to which each sample fulfills the set of basic conditions on which the validity of the method itself rests. These basic assumptions can be summarized as follows:

(1) The production of radiocarbon by cosmic rays has remained constant long enough to establish a steady-state or equilibrium in the $^{14}C/^{12}C$ ratio in the atmosphere over at least the last 50,000–70,000 years.

(2) There has been complete and rapid mixing of radiocarbon throughout the various atmospheric, hydrospheric, and biospheric reservoirs.

(3) The total amount of carbon in these reservoirs has remained essentially constant.

(4) The carbon isotope ratio has not been altered except by ^{14}C decay within sample materials since they ceased to be a part of one of the carbon reservoirs.

(5) The decay constant (or half-life) of ^{14}C is known.

(6) Natural levels of ^{14}C can be measured to appropriate levels of precision and accuracy (34).

In addition, a seventh often unstated assumption maintains that there is a direct and specific association between the sample to be analyzed and the event or phenomena to be dated. The first six assumptions are the subject of continuing study by radiocarbon specialists (35, 36, 37). It is the unique responsibility of the field archaeologist to determine if the seventh assumption holds for each sample submitted for analyses (38). We will return to this point in the conclusion. Surprisingly, however, one of the earliest and continuing problems in ^{14}C applications in archaeology had nothing to do with the physical parameters of the method itself but simply reflected the general lack of formal scientific and mathematical training among most archaeologists. The problem resulted from the manner in which ^{14}C dates are expressed. For example, a date of 5600 ± 80 radiocarbon years B.P. reflects the fact that the count rate of the sample is about 50% of the modern reference standard and that the total one sigma counting error is about 1% or ± 80 years. In many cases, what failed to be communicated was the fact that a date (i.e., in this case, the value 5600 years) had little meaning apart from the one sigma error value. The expression 5800 ± 80 is only a shorthand manner of stating that there are two chances out of three that the age equivalent of the count rate obtained on this sample will be contained within the range 5520–5680 years before the present (39). Many archaeologists followed the suggestions contained in an early excellent article by Spaulding which succinctly outlined the significance of the statistical character of radiocarbon data (40, 41).

Another equally obvious, but unfortunately still often overlooked point was that a radiocarbon date dated the time that a sample was isolated or removed from the contemporary carbon reservoir of which it had been a part, usually at the time of death. One would probably have some difficulty in determining when, for example, a structure was destroyed if one dated the wooden beams used in its construction. The date would indicate only the time that the tree was cut down. If timber happened to be in short supply and beams tended to be reused, the ^{14}C determination might not even date the construction of the building, let alone its destruction. An important by-product of early work with samples of wood was the recognition that each annual ring contained a sample of ^{14}C from only the year in which it had grown. Thus a wooden beam containing 250 rings contained wood with a 250 year range in ^{14}C values. The outermost ring would date to the time when the tree died while the innermost ring would date 250 years earlier. Obviously, in working with wood or charcoal samples from species of trees with characteristic long lifetimes, it could be important to identify the portion of the tree from which the sample had been cut to permit the estimation of the number of rings separating the sample from the outer ring.

The isotopic isolation of ^{14}C in each tree ring had crucial implications for future radiocarbon studies and continues to be very much a part of current research. Specifically, the examination of the radiocarbon contents in tree-ring-dated samples of wood provided the principal data that permits an in-depth examination of the assumption that the production of radiocarbon by cosmic rays had been maintained at a constant level. For archaeologists especially, it is important to know if there have been any anomalies in radiocarbon values which would artifically compress or expand their temporal frameworks.

Variability in Radiocarbon Production Rates. Hints of systematic discrepancies between radiocarbon values and known-age samples were quickly reported in the ^{14}C literature (42). Almost all radiocarbon determinations obtained on early Egyptian archaeological materials yielded age values consistantly too young by up to 700–800 years. However, even in the early 1960s, it was inferred that such values suggested that the Egyptian chronology was in error rather than that there was any systematic error in the ^{14}C values themselves (43). In the late 1950s, the Cambridge, Copenhagen, and Heidelberg laboratories, using the then-newly introduced CO_2 gas counting technique, obtained radiocarbon determinations on a series of tree ring samples which showed consistent variation in ^{14}C activity in the atmosphere on the order of several percent over the last 1300 years (44). These data confirmed the suggestion of Hessel de Vries that "radiocarbon years" and calendar or sidereal years should not be assumed to be equal (45). Primarily because of the geophysical implications of these deviations, several laboratories and researchers directed their attention to the magnitude and extent of what came to be called the DeVries effect or, now more commonly, secular variation.

The data base which has contributed most directly to the intensive study of secular variation was the development of the dendrochronological-scale for the bristlecone pines (*P. aristata*) of the White Mountains of eastern California by Wesley Fergusson of the Tree-ring Laboratory of the University of Arizona (46, 47). Published and unpublished bristlecone pine dendrochronological data now exist back to almost 6300 B.C., and an additional 1500-year segment may be added soon (48). An independently developed bristlecone pine tree ring chronology from a different locality in the southern portion of the White Monutains developed by LaMarche and Harlan supports the accuracy of the Fergusson chronology at least as far back as ca. 3500 B.C. (49). High-precision radiocarbon determinations of the bristlecone pine series as well as tree ring data taken from sequoia (*S. gigantea*) samples were undertaken by the Arizona (50, 51), La Jolla (52), and Pennsylvania (53, 54) laboratories, and published radiocarbon determinations on dendrochronological samples now provide data back to ca. 5500 B.C.

Significant interpretative problems in the use of radiocarbon values are apparent immediately when the ^{14}C-dated tree ring data are examined (Figures 3 and 4). If we ignore for the moment the short-term perturbations, the radiocarbon and dendrochronological values approximately agree ($\pm$ 1–3%) for the last 2000 years. However, as one moves back in time from about the beginning of the Christian era, there is an increasing deviation in the radiocarbon values so that by ca. 2000 B.C., ^{14}C values are registering ca. 200–300 years too young, and between 4000 and 5000 B.C. the deviation may be as much as 800 years.

The question of the nature of the major secular radiocarbon anomalies before ca. 5500 B.C. cannot, at present, be documented by the bristlecone pine data. It has been inferred, however, that the long-term characteristics of the secular variations may be estimated by reference to geomagnetic intensity data, since variations in the intensity of the earth's geomagnetic dipole field seem to have been identified as the principal cause of the long-term component of the effect. Previous studies had determined that a decrease in the intensity of the earth's dipole field was followed by an increase of the cosmic ray flux in the vicinity of the earth and therefore by an increase in the production rate of ^{14}C (55). Geomagnetic data strongly supported a correlation between the intensity of the earth's magnetic dipole moment and deviations observed in the bristlecone pine radiocarbon values (56, 57). The evidence seems to support the view that the long-term component of variability observed in the bristlecone ^{14}C data could best be characterized as describing a wave function, and thus, the temporal intervals of maximum and minimum deviation in atmospheric radiocarbon activity could be extrapolated back potentially to the present limit of the method at ca. 50,000–70,000 years. Any implementation of these extrapolations to provide corrections to the radiocarbon values on the order of the bristlecone pine data, however, would be premature and must await the development of a clearer consensus of the specific nature and causes of the variability in ^{14}C activity over time, especially in view of the possibility of large excursion(s) of the geomagnetic field during the late Pleistocene (58, 59). In addition to the variation in the earth's field intensity, other factors being considered include fluctuations in atmospheric mixing rates, variations in solar cosmic ray intensity, and variations in the solar wind (60).

Two calibration curves using summaries of the dendrochronology/ radiocarbon data are presented here. The first is based on data generated by the La Jolla radiocarbon facility (52) (Figure 3), and a second plot was assembled by University of Pennsylvania's Museum Applied Science Center for Archaeology, MASCA (53, 54) (Figure 4). In both the data plots, the long-term component of the anomalies can be identified easily. In both cases, the solid horizontal line represents the ideal situation in which radiocarbon and calendar years would be equivalent, i.e., any data point lying on that line would indicate that at that time radiocarbon and tree ring (calendar year) values would coincide. Any point lying below the line indicates that the radiocarbon values are too young with respect to tree ring values. Values above the line indicate that the radiocarbon values there are too old. In both sets of data, the long-term component

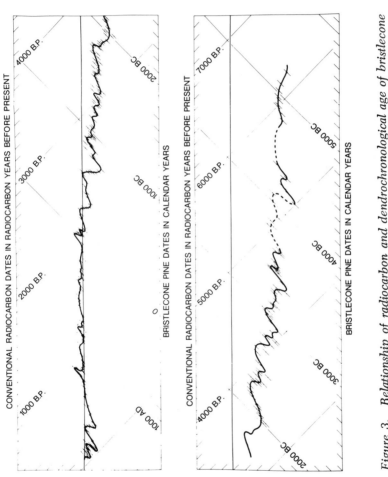

Figure 3.   Relationship of radiocarbon and dendrochronological age of bristlecone pines after Suess (52). Scale of plot has been inverted from original citation to permit comparison with Figure 4. 14C half-life used, 5568 years; points of origin of plot, A.D. 1950.

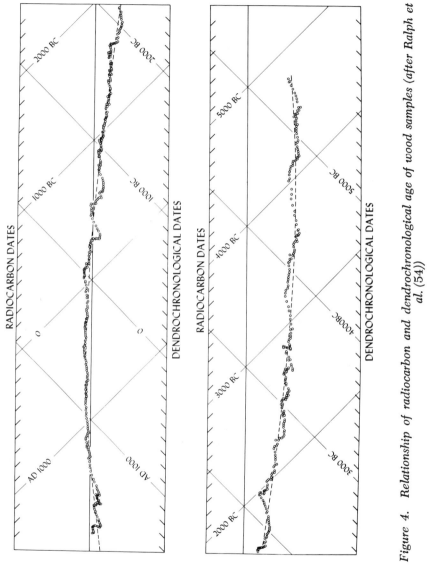

Figure 4. Relationship of radiocarbon and dendrochronological age of wood samples (after Ralph et al. (54))

exhibits the characteristics of a wave function with a period of about 8500–9000 years and with a maximum deviation of ca. 8–10% centering about 4500–5000 B.C. calendar years. After smoothing the short-term oscillations, both sets of data identify crossover points or nodes where radiocarbon years and sidereal years coincide at ca. A.D. 700 ± 100 and 500 ± 100 B.C., although the later intersection coincides with a short-term excurses.

It is in the documentation of the nature of the short-term perturbations in radiocarbon values that the two sets of data do not agree completely. Although there is a consensus that short-term episodes do exist, their magnitude is in dispute. There is also a consensus that the short-term variations generate an additional set of problems for the archaeologist in the use and interpretation of radiocarbon values (61). In the Suess (52) and MASCA (53, 54) plots, the interval of time from A.D. 1800 to the middle of the 4th millenium exhibits 12 major temporal episodes where radiocarbon values have multiple age equivalents. In these intervals, a given radiocarbon value may reflect two or more points in real time. Figure 5 illustrates this problem on an expanded scale.

The period in Figure 5 represents an interval on the radiocarbon scale between 1950 B.C. (3900 B.P.) and 2350 B.C. radiocarbon years based on the La Jolla data. For example, if the radiocarbon value of 2050 B.C. was equal to the value given by the dendrochronological calibration, then the data points would lie along the horizontal reference line in Figures 3 and 4, i.e., 2050 B.C. on the radiocarbon scale would equal 2050 B.C. on the bristlecone pine scale. Considering only a single calibrated equivalent point, a radiocarbon value of 2050 B.C. can be calibrated to a value of 2525 B.C. calendar or sidereal years. However, in looking again at the expanded scale in Figure 5, a radiocarbon value of 2150 (±100) years B.C. has been obtained on tree ring dated wood samples which range in true age from ca. 2500 to 2950 B.C. This indicates that if a sample yields an age of 4100 radiocarbon years, then the true age of the sample material cannot be isolated to better than a time interval of ca. 450 years. Thus, the tree ring/radiocarbon values may be used not only to identify the degree of deviation of the radiocarbon values from dendrochronologically determined values but also to identify the degree of maximum precision which is possible for a given temporal interval. The uncertain characteristics and magnitude of the short-term perturbations in radiocarbon values may make their use for chronological reconstruction somewhat problematical. However, Berger has demonstrated the utility and necessity for short-term type corrections for accurate radiocarbon data analysis when applied to 12th–14th century European medieval archaeological materials (62).

The question of whether variations in ¹⁴C content in wood samples taken from western North America can be used to document worldwide secular variations in radiocarbon values seemingly has been answered by studies of the ¹⁴C concentrations in tree rings from Patagonia, Argentina, Canada, and Europe. The maximum contemporaneous deviation noted was between the southern and northern hemisphere, but it did not exceed

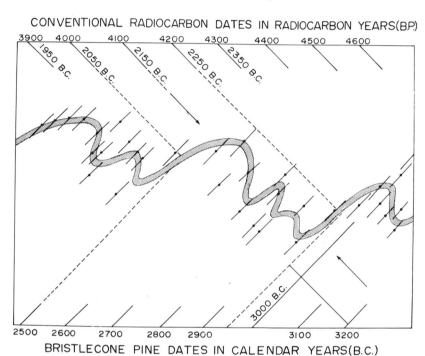

CONVENTIONAL RADIOCARBON DATES IN RADIOCARBON YEARS(B.P.)

BRISTLECONE PINE DATES IN CALENDAR YEARS(B.C.)

Figure 5.   Relationship of radiocarbon and dendrochronological age of bristlecone pine samples for 3900–4300 radiocarbon years B. P. (after Suess (52)). Scale of plot expanded from Figure 3. B. C. values associated with radiocarbon B. P. scale by subtraction of 1950 years. Value of 2150 (±110) B. C. radiocarbon years (→) cannot be calibrated to better than a range of 2500– 2950 B. C. calendar years B. C. (– – –) based on bristlecone-pine tree ring data.

0.5% (63). Thus the original assumption of Libby that there was a complete worldwide mixing of radiocarbon has been essentially vindicated, at least in wood samples, except for a small latitudinal dependence. For sample types other than wood, however, studies over the past decade have indicated that this assumption does not always hold, and significant problems can result.

Variability in Radiocarbon Distribution. One of the early major promises of the radiocarbon method was its potential to provide direct comparable age determinations on a worldwide basis for all organic samples. For this potential to be realized, radiocarbon had to be mixed rapidly and completely throughout all of the carbon-containing reservoirs. If such conditions prevailed, the contemporary radiocarbon content of all organic samples would be essentially identical. It was quickly determined, however, that in a number of situations such an assumption could not be made. The classic illustration of the breakdown of this assumption was the determination that living organic samples from a fresh water lake with a limestone bed exhibited apparent radiocarbon ages of approximately 2000 years (64). Another example was that trees growing near volcanic fumerole emissions will have ^{14}C concentrations suggesting apparent ages on the order of 1000 years (65).

One effect of the recognition that a sample's general and specific geochemical environment can affect initial ^{14}C concentrations has been to cast doubt on the reliability of particular types of samples. The use of shell in radiocarbon work has been perhaps the most seriously affected since a tradition arose that its use should be discouraged except as a last resort where no other samples were available (66, 67). Terrestrial shells from fresh water sources certainly merited this negative evaluation since it was shown that they typically take up carbonate which is not in equilibrium with atmospheric radiocarbon (68, 69). In the case of marine shell, however, it acquired a negative reputation principally as a result of early experiences with shell taken from several archaeological sites in Peru. Charcoal and marine shell samples assumed to be contemporaneous were found to have radiocarbon ages which differed by as much as 1000 years. On the basis of the data then available it was logical to assume that the marine shell was yielding the grossly erroneous values (70, 71).

However, because marine shell occurs so commonly in coastal archaeological deposits, several investigators were not content to accept the established tradition without additional evidence. Several avenues of inquiry were pursued to determine more precisely the nature of the problems surrounding marine shell. One approach continued to measure terrestrial organic/marine shell sample pairs in cases where it was assumed that they were deposited at about the same time. As more and more paired values became available, it became clear that except for the Peruvian suite of values, the shell values differed from standard terrestrial values by at most a few percent in practically all cases (72). Others approached the marine shell problems by looking at contemporary marine shells to determine if any anomalous effects could be identified. Significant variations in radiocarbon values for living specimens were noted quickly. For example, in a study of contemporary radiocarbon concentrations in marine shells along the western continental margins of North and South America (Figure 6), the ^{14}C activity ranged from a maximum

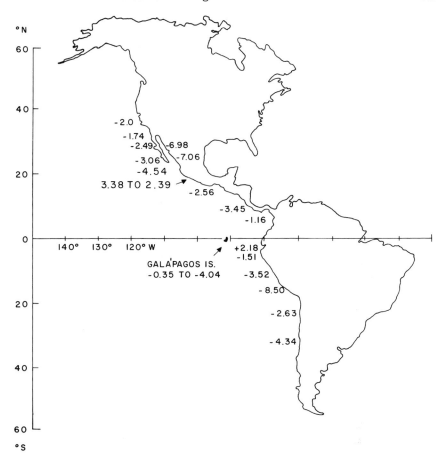

Figure 6.   Radiocarbon content of modern pre-bomb marine shell from the west coast of North and South America. Data from Refs. 73 and 74. Values have been expressed with respect to 0.95 NBS oxalic acid standard and corrected in light of their $\delta^{13}C$ values.

of −8.5% along the Peruvian coast to a +2.18% value for the closely adjacent Ecuadorian coast (73, 74). In this case, a value of −8.5% equates to an apparent age of about 700 years for modern shells growing along the Peruvian coast and begins to explain the initial problematical values from the area. The probable reason for this effect is the fact that ocean water which has been depleted of ${}^{14}C$ by long residence times in the deeper part of the ocean is periodically upwelled and mixed with surface ocean water.

An examination of Figure 6 also illustrates other problems with marine shell samples. Four samples from the environment of the Galapagos Islands ranged from a little more than −4.0% to a little less than +0.5%. This equates to an apparent variation of ca. 350 years between living organisms in close proximity to each other. Such fluctuation in a relatively small area emphasizes the fact that broad generalizations about the magnitude of upwelling effects for any major oceanic region must be scrutinized carefully. An additional example of this problem is highlighted in samples obtained from the Gulf of California (Sea of Cortez) where deviations as high as −7.06% were measured. This would equate to an apparent age of about 600 years. By contrast, five values obtained from closely adjacent localities along the west Mexican coast did not vary by more than about 1%.

This type of data raises the issue of the constancy of the oceanic circulation patterns over the region in question. In this case the question is to what degree the values cited in Figure 6 could be used to estimate oceanic upwelling effects along the continental margins of the Western Hemisphere as one moves back into the early Holocene and terminal Pleistocene. Table I lists a series of paired terrestrial organics/marine

Table I. Paired Terrestrial Organics/Marine Shell Samples from the West Coast of North America

Lab. No.	Provenience	Material	^{14}C Age[a]
GSC-24	Vancouver Is., B.C.	wood	12,200 ± 160
GSC-38		shell	12,360 ± 140
L-391D	Vancouver Is., B.C.	wood	12,150 ± 250
L-291E		shell	12,350 ± 250
UCR-509	Newport Bay, CA	charcoal	8,445 ± 280
UCR-510		shell	8,045 ± 270
LJ-923	Baja Calif.	charcoal	5,480 ± 200
LJ-924		shell	6,140 ± 250
W-27	Baja Calif.	charcoal	2,500 ± 200
W-26		shell	2,540 ± 200
LJ-922	Baja Calif.	charcoal	2,200 ± 200
LJ-925		shell	2,600 ± 200
LJ-645	Baja Calif.	charcoal	1,660 ± 200
LJ-611		shell	1,800 ± 300
UCLA-187	Morett Site, Colima, Mexico	charcoal	1,500 ± 80
UCLA-1035		shell	1,480 ± 80
UCLA-1034		shell	1,630 ± 80
LJ-85	Baja Calif.	charcoal	960 ± 150
LJ-84		shell	1,060 ± 150
W-155	La Jolla, CA	charcoal	600 ± 200
W-154		shell	580 ± 200

[a] Radiocarbon values are expressed with respect to 0.95 NBS oxalic acid standard.

shell samples from this region obtained in reported stratigraphically associated contexts. The data suggest that oceanic circulation patterns, at least for this region, have been constant as far back as early Holocene times.

Over the last decade, it has become clear that the highly negative view of marine shell carbonates generated in the early years of radiocarbon research has been overly pessimistic. Undue weight was placed on a small number of samples from a region now known to suffer from severe upwelling effects. Carbonates from well preserved marine shells from open ocean environments where upwelling effects have been studied can yield radiocarbon values which are as reliable as standard terrestrial organics. For regions where contemporary marine shell values indicate significant variation over small distances, the cited error value should be increased to compensate for the added uncertainty introduced by the marine environment.

It is important to emphasize that a discussion of the problems associated with marine shell could be repeated for any sample type/geochemical environment where the contemporary radiocarbon values may not be in equilibrium with the atmosphere. In such cases, a specific contemporary standard must be used for each sample type or geochemical environment which can be related to the terrestrial biological radiocarbon standard. Special standards and/or correction values would, for example, be required for specifically defined oceanographic regions in the case of marine shell, standards for specific fresh water shell or gastropod geochemical environments, standards for Arctic and Antarctic specimens, and specific soil carbonate environments (75).

Variability in the Amount of Carbon in Reservoirs. In addition to variations in the production and distribution of radiocarbon over time and within portions of various carbon reservoirs, variations may result in situations where carbon not in equilibrium with the contemporary standard values is added or removed from any reservoir. Two instances of this are well documented since they occurred within the last century as a result of human intervention. The first is known as the industrial or Suess effect and is caused by the combustion of fossil fuels beginning about 1890, resulting in a depletion of atmospheric ^{14}C activities by about 3% (76). A more recent occurrence has been called the atomic bomb or Libby effect. The detonation of nuclear devices in the atmosphere beginning in 1945 produced large amounts of artificial ^{14}C, increasing the radiocarbon concentrations in the atmosphere by more than 100% in the Northern Hemisphere (77). Because of equilibration with the oceans, the levels have been diminishing steadily since the atmospheric testing was terminated by the major nuclear powers except France and the People's Republic of China. It remains to be seen what effect the testing of the neutron bomb in the atmosphere will do to ^{14}C activity in the contemporary world.

The significance of these effects to archaeologists are limited. Except when combined with variations in the production rate, they make it difficult to distinguish radiocarbon concentrations in organic samples over the past several hundred years and force laboratories to use 100 or 150 years as the minimum age which can be cited. This affects the accuracy of values for late prehistoric or protohistoric archaeological contexts for several areas of the world. It also explains why radiocarbon laboratories cannot use modern wood as a contemporary reference standard. Rather, artifical standards must be used to provide what the value of the ^{14}C activity in the atmosphere would have been without human intervention.

Variability in Carbon Isotope Ratios. For the radiocarbon method, the basic physical measurement used to index time is the ^{14}C/^{12}C ratio. Carbon has three naturally occurring isotopes which exist in the ratio of about 98.9% ^{12}C, 1.1% ^{13}C, and 10^{-10}% ^{14}C. Accurate estimates of the age of a sample using the ^{14}C method assumes that no change has occurred in the natural carbon isotope ratios except by the decay of ^{14}C and the resultant change in the ^{14}C/^{12}C ratio. Several physical effects other than decay, however, have been shown to alter the carbon isotope ratio and thus have introduced the need to correct ^{14}C values thus affected. One problem has to do with the effect of the fractionation of carbon isotopes under natural conditions. Another, much more often discussed, involves the problem introduced when carbon-containing compounds not indigenous to the original sample are physically or chemically introduced, resulting in the contamination of the original sample.

While all the isotopes of carbon follow the same chemical or physical pathway, the rate at which this occurs varies as a function of their difference in mass. The pioneering studies of Harmon Craig in the early 1950s first pointed to the need to consider variations in the stable isotope ratios (^{13}C/^{12}C or δ^{13}C) of samples if one wished to obtain precise and comparable ^{14}C/^{12}C (or δ^{14}C) values (78). He determined that the effect of any change in the δ^{13}C value was reflected in the δ^{14}C value by a factor of two. Thus a change of 1% (or as it is usually expressed, 10‰) in the stable isotope ratio would be reflected by a change of 2% (20‰) in the δ^{14}C value. Variations of up to several hundred years can result if the radiocarbon values on sample types with variable δ^{13}C ratios are not corrected for any deviation. Fortunately, most of the δ^{13}C values on standard sample materials such as charcoal and wood are such that normally little, if any, correction in δ^{14}C values are required. Problems arise, however, when it is desirable to compare radiocarbon values from a variety of sample types such as grasses, grains, seeds, succulents, and marine carbonates as well as standard terrestrial organics such as wood. In such cases, it is absolutely mandatory that $\delta^{13}C$ values be obtained and used to correct the $\delta^{14}C$ values onto a common scale. Unfortunately, this requirement is sometimes overlooked when radiocarbon values are quoted from sources in cases where stable isotope data have not been provided.

By contrast, the subject of contamination of radiocarbon samples by non-indigenous organics has enjoyed great vogue among archaeologists especially as an explanation of why their radiocarbon dates are wrong, i.e., why such radiocarbon values did not conform to their prior bias as to the age of the sample. Obviously, the sources and effects of the introduction of foreign organics are complex depending on the nature and condition of the sample material, the characteristic(s) of the geochemical environment(s) within which the sample has been embedded, and the time frame over which such action(s) occurred. Each situation may be unique, and precautions exercised to avoid contamination effects may have to be designed to fit each particular situation. While recognizing that exceptions to general characterizations occur, a few generalizations and accepted routine procedures in the pre-treatment of samples have been established by laboratories. These procedures are sample-type specific but are generally concerned with completely removing what is assumed to have not been present when the original sample was removed from the carbon reservoir. The high rating given to charcoal and wood from the beginning was based on the fact that such samples could be subjected to pretreatment by strong acids and bases, and thus the removal of absorbed carbonates and soil humic and fulvic acids and other soluble soil organic matter could be greatly facilitated. On the other hand, in addition to the problems mentioned previously, marine shell was highly suspect as a sample material in terms of its potential for contamination because it was thought that there would be no way to isolate the original shell matrix if there had been isotopic exchange through recrystalization with CO_2 dissolved in ground waters.

It is usually possible to infer the effect of known contamination effects on a given sample in terms of the direction the age change will take (i.e., making the resultant age too old or too young), but the magnitude of the anomaly can be calculated only if the true age of the original sample, the age of the contaminant, and the percentage contribution of the contaminant are all known (79). The effect of varying percentages of contamination of known activity on samples of differing age can be calculated to obtain an estimate of the order of magnitude of potential error that can be introduced. The curves in Figures 7 and 8 were prepared originally by Ingrid Olsson of the Uppsala Radiocarbon Laboratory in Sweden (80). It is clear from these curves and from other calculations that contamination with materials of even up to 1000 years older or younger than the original materials causes errors which are on the order of magnitude of the sta-

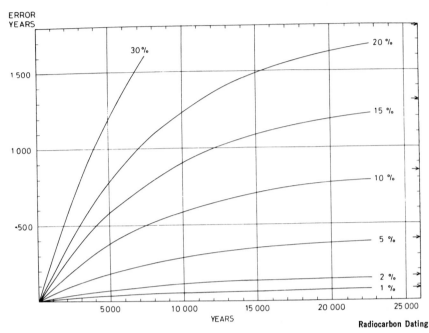

Figure 7. *Effect (in years along ordinate) of the introduction of 1–30% con-*
tamination older (lower in ¹⁴C activity) than sample to be dated. Values along
the abscissa indicate difference between the true age of sample to be dated and
the older contaminant (80). See discussion in Polach (75).

tistical error if the contamination is limited to a few percent. In the case
of Holocene charcoal and wood, it would be extremely rare for unob-
served and unremoved contamination to exceed a few percent unless the
sample derived from a highly unusual depositional environment such as
petroleum or tar deposits. The potential is somewhat greater with some
sample types such as peats and, under certain conditions, carbonates. The
problem does become significant with samples of Pleistocene age. Here
the potential for contamination becomes significant, and strict attention
to pretreatment methodologies is absolutely mandatory.

     For sample types typically encountered in most archaeological situ-
ations, one approach to evaluating the potential for contamination of
samples is to conduct fraction studies. This approach allows one to
identify empirically sources of contamination of samples by determining
the radiocarbon content of various chemical fractions of the same sample
matrix. If significant variations in ¹⁴C activity among the fractions are
noted, then the potential for contamination effects must be considered
seriously.

     Marine shells have been studied in this manner by obtaining ¹⁴C
analyses on the inner and outer portions of the same shell to study pos-

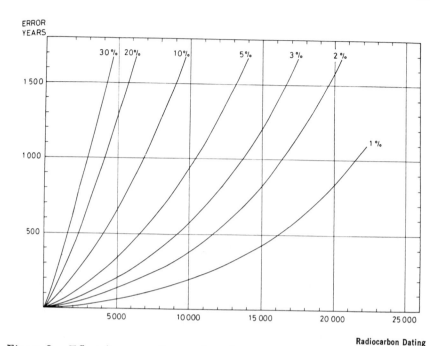

ERROR
YEARS

Radiocarbon Dating

*Figure 8. Effect (in years along ordinate) of the introduction of 1–30% con-
tamination younger (higher ^{14}C activity) than sample to be dated. Values along
the abscissa indicate difference between the true age of sample to be dated and
the younger contaminant (80). See also discussion in Polach (75).*

sible carbonate replacement effects. In cases where Pleistocene age shell
was being studied, some laboratories have reported variations of several
thousands of years between the surface and interior layors. However, in
over 90% of the Pleistocene age samples, the differences between the
inner and outer fractions did not exceed 5%. For Holocene age marine
shells, while there was a general pattern in which the outer shell fraction
yielded slightly younger values, with few exceptions, the deviation did
not exceed a few percent (81, 82).

Bone samples traditionally received an even lower reliability rating
than shell. Libby suggested originally that it was "barely conceivable"
that certain bone ^{14}C measurements under some circumstances might be
reliable (83). It was assumed that bone would constitute a poor sample
type basically because of its low organic carbon content and generally
porous structure. Early ^{14}C values using the inorganic or carbonate fraction of bone were in most cases clearly erroneous (84). For many sites,
the availability of usually reliable sample types such as wood or charcoal
obviated the need to use problematic samples such as bone. In some contexts however, especially where stratigraphic discontinuities may have
caused the association of skeletal material with, for example, charcoal to

be questioned, there was a concern to determine whether the organic or collagen fraction of bone structures had the isotopic integrity required to yield accurate ^{14}C values.

The first reported satisfactory agreement between bone and associated charcoal ^{14}C values for a suite of samples was obtained in 1964 using the organic or collagen fraction of the samples (85). Contamination by noncollagenous organics remained a possibility (86), and several investigators suggested methods to insure that a pure collagen fraction was being used (87, 88, 89, 90). The preparation of gelatin also was suggested as a means to avoid potential contamination of the collagenous samples (90, 91, 92). However, since the collagen fraction in many bones, especially of Pleistocene age, is insufficient for routine radiocarbon analysis, Haynes (93) investigated the reliability of using the bone apatite fraction in routine bone ^{14}C analyses. Hassan (94) has summarized the possible sources of anomalous results in ^{14}C values using both collagen and bone apatite. She has concluded that physical contamination is the principal source of error in collagen work but that erroneous apatite dates can result from the exchange of carbon in apatite structures during recrystallization and/or surface exchange reactions. Stable isotope studies suggested also the possibility of variability in the original ^{14}C activity in bone caused by differences in taxonomic order, age, diet, and/or geochemical environment of the animal.

Table II.   Radiocarbon Determinations on the Carbonate, Collagen, and Amino Acid Fractions of Two Human Bone Samples[a]

Sample Number	Sample Type	Organic Carbon Yield (%)	$\delta^{13}C$ (‰)	Corrected ^{14}C Age
UCR-449A	carbonate	—	−8.42	930 ± 140
UCR-449B	collagen	6.95	−19.89	2765 ± 155
UCR-449C	amino acids	1.26	−21.41	2930 ± 150
UCR-450A	carbonate	—	−9.43	830 ± 100
UCR-450B	collagen	6.49	−20.24	2835 ± 140
UCR-450C	amino acids	1.12	−21.29	2960 ± 140

[a] From Burial 36 at Site SJo-112 in Northern CA. See Ref. 67 for additional details.

Fraction studies have been applied to the study of the reliability of bone ^{14}C values by measuring the comparative ages of the carbonate, collagen, and amino acid fractions in single bone samples. Table II lists results obtained on two duplicate sample suites from a single burial. These data continue to support the early view that the carbonate fraction in bone yields highly unreliable results. In this case, the carbonate ^{14}C age was nearly 2000 years younger than the collagen and amino acid fractions. By contrast, the values obtained on all organic fractions are statistically identical and seem to reflect the actual age of the burial based on archaeological criteria including radiocarbon values on charcoal from other sites on the same time horizon.

Because of the large number of conditions and factors which can influence variations in the carbon isotope ratios of certain types of samples, this aspect of radiocarbon studies especially in archaeological contexts has been perhaps the least understood but the most often cited as an explanation for anomalous results by archaeologists. The fact is that, with few exceptions, problems of contamination for most Holocene age samples can be solved usually by applying standard pretreatment approaches or, in the case of more severe problems, by applying fraction studies.

Half-life of Radiocarbon. The fundamental constant which permits the conversions of the $\delta^{14}C$ value into an equivalent age is the decay constant or half-life of radiocarbon. We reviewed briefly the establishment of the Libby half-life of 5568 ± 30 years as the standard value used in the publication of ^{14}C age values. Redeterminations by several other researchers have led to a slight upward revision of the value. As we noted previously, the most widely accepted estimate is currently 5730 ± 40 years, which is approximately 3% more than the Libby half-life (*21*). Anyone wishing to recalculate published values need only multiply the published value by 1.03.

However, the issue of the correct half-life for radiocarbon has lost a considerable amount of its significance because of the discovery and documentation of the long- and short-term secular variation effects. The existence of dendrochronologically documented relationships between radiocarbon age and calendar age fortunately enables researchers to circumvent completely the problem of the real ^{14}C half-life. This will also hold true even for radiocarbon determinations on Pleistocene age materials where the use of the 5730 value increases ^{14}C values at, for example, 35,000 years B.P. by about 1000 years. This is somewhat more than the typical one sigma statistical error for his age range. However, it is likely that fluctuations in past ^{14}C production rates and the effects of other as yet undocumented geophysical variations make the correction of such values by only the Cambridge half-life figure somewhat questionable.

Measurement of Natural Radiocarbon Concentrations. By the middle of the 1950s, the original solid carbon method of assay of natural radiocarbon concentrations had been completely superseded by the development of either gas or liquid scintillation counting systems. More than 20 years of experimentation and development in the low level counting technology field has turned what was once a black art and analytical tour de force into a routine set of procedures which can be performed by any conscientious technician. This does not mean that the measurement of natural levels of radiocarbon in concentrations of one part in 10^{12} and below is not still a challenging analytical procedure; however, the problem now is the rigorous application of standardized techniques and statistical procedures on a consistant basis.

While the originally cited one-sigma counting errors for the solid carbon systems were on the order of 200–600 years, contemporary counting errors can be reduced routinely to around 0.5% (± 40 years) or below. However, as our previous discussions hopefully have emphasized, the statistical counting errors are by far the minor source of error in many radiocarbon values of interest to archaeologists. In fact, recent highly-precise radiocarbon determinations on contemporary (modern) tree ring samples indicate that the typical annual variation in radiocarbon concentration in wood is on the order of ± 80–100 years. Although by international convention of the active radiocarbon laboratories the published errors are only those involving counting statistics, geophysical reality suggests that the error values of less than ± 80 years may represent specious accuracy.

Most recently, the potential for a third generation of systems to detect radiocarbon and other low activity nuclides has been reported by several research groups. Their approach is to measure ^{14}C concentrations directly by counting the number of radiocarbon atoms contained in a sample rather than by monitoring decay events. At least three groups have attempted to use high energy particle accelerators as extremely sensitive mass spectrometers. The potential advantage of their approaches is the possibility of using extremely small samples (1–100 mg) and extending the maximum dating range from the present 50,000–70,000 years to over 100,000 years with precisions comparable with those with existing systems. A Berkeley group is using the 88-in. cyclotron to accelerate samples in gas form. Currently, the principal problem involves contamination of the beam by residual ^{14}N in the sample and in the ion source of the cyclotron (95). Two other teams at the University of Rochester and Simon Frazer University are using Tandem Van de Graaf accelerators with solid carbon samples. By counting only negative ions, they are attempting to circumvent the nitrogen contamination in the beam (96, 97). Further experience will be required to demonstrate if routine radiocarbon determinations can be carried out with either of these instrumental approaches at an appropriate level of precision and accuracy.

Radiocarbon Dating in Prospect

One of the core issues in contemporary paleoanthropology and prehistoric studies is the nature and causes of the evolution of hominid behavior over the last three to five million years as reflected in the physical residue of that behavior. Evolution, of course, means change over time, and it is precisely the ability to document and deal with human behavior within a temporal framework that allows archaeology to make its unique contribution to general behavioral science concerns focusing on the evolution of social and cultural systems within our species and its biological antecedents.

The concept of time is essentially a philosophical or cosmological issue irrelevant to this discussion. However, such a category proceeds from and is embedded in a perception of the process of change in the state of some specific physical entity. Those concerned with natural historical processes—whether geologists, geochemists, geophysicists, evolutionary biologists, paleontologists, or archaeologists—are concerned not with the concept of time but rather with the ability to increment or scale accurately the duration of changes in selected physical phenomena from one state to another. Each discipline, however, obviously requires that its units of scale fit the physical dimensions of the phenomena being studied or the questions being asked about the relationships between the phenomena. For most geological questions, the relevant time scales are usually sufficient if they have increments in units of 10^3–10^5 years over a time span on the order of 5.0×10^8 years. Unfortunately the hominid experiment has been of much more recent and compressed duration requiring the documentation of units on the order of 100–1000 years over a time span of only a few million years.

Until the introduction of the radiocarbon method, prehistorians were generally forced to use temporal incrementing techniques which were extremely difficult, if not impossible, to calibrate in terms of any real-time scale. Local and regional stratigraphic and seriational approaches, the development of local and regional artifact complexes, hopefully with chronological significance, and attempts to correlate climatic or even astronomical (i.e., the Milankovitch curve) processes with a given cultural expression all were used to develop chronologically meaningful frameworks. One of the results of the difficulty with which prehistoric chronology was derived was that, at least in New World studies, the creation of cultural historic frameworks tended to occupy an inordinate amount of the attention of pre-^{14}C era archaeologists, almost to the total exclusion of efforts to deal with broader issues such as those concerned with the nature of cultural evolutionary processes (98).

Only in Scandanavia and in the southwestern U.S. did archaeologists have a well established means to use some independent mechanism to document accurately real-time scales. In Scandanavia, varve chronologies, and in the southwestern U.S., dendrochronology, contributed significantly to the dating of regional archaeological materials. Both, unfortunately, were useful in only tightly defined geographic regions, and there were severe limitations in associating datable material in cultural contexts, especially in varve dating (99).

The significance of the radiocarbon method in prehistoric studies lies not only in its obvious immediate utility but perhaps even more importantly, in its longer-term impact that it is making in basic methodological and even theoretical issues which are being addressed especially in American archaeology. The immediate effect is easily appreciated and best known—the potential ability to state that a ceramic form, house type, stratigraphic unit, or the beginning of agriculture or urbanism occurred x number of years ago and y number of years before or after some other related event. What now could be documented was a rate of change, not merely a sequence of events. The promise and potential of this ability was, of course, immediately taken up by a generation of archaeologists and, dispite criticism from those whose most cherished views had to be significantly modified, a quarter of a century later, this ability continues to provide crucial insights into the human prehistoric past on a site-by-site and region-by-region basis.

It was fortunate that the basic set of propositions and assumptions on which the radiocarbon method rested turned out to be, in the main, very accurate. If a substantial number of the early values had subsequently been shown to be in serious error and had been discarded (as one writer in 1955 stated would be the case in a few years), there would have been a serious reaction, especially in New World studies (100). Fortunately, archaeologists did not have to face this on a wholesale basis. Radiocarbon values as they were produced were incorporated into the existing chronological frameworks established previously by seriation and artifact cross-ties. In some cases, a lengthening and shortening of numerous sequences was required, sometimes by thousands of years. But the general corpus of dates settled into place with surprisingly few major crises. Further studies resulted in individual values being revised, and the association of values with cultural expressions was rearranged from time to time. Also, the impact of the knowledge of the secular variation effect also has occasioned considerable discussions. The archaeological literature generated as a result of the documentation of secular variation effects has been significant. *Antiquity* introduced the custom of distinguishing between calibrated and uncalibrated ^{14}C values of affixing different identifiers (101). If uncalibrated, lower case notations would be used (bp, bc, ad); if calibrated, capital letters were to be used (BP, BC, AD). Several calibration approaches have been proposed involving various types of charts and tables in addition to the two noted in the text (102). It may, however, be premature to rely on any single scheme since a number of questions pertaining to the nature of the variations remains somewhat uncertain. The archaeological implications of secular variation effects have been exemplified most dramatically by the work of Renfrew in his revisions in the prehistory of Europe on the basis of calibrated radiocarbon values (*103*). Even semi-popular works on archaeology have noted the impact of secular variation values (*104*). It may be prudent to withhold judgement on some of the applications of the calibration values until some of the confusion in the data itself is clarified and until certain other geophysical parameters are more clearly understood (*105*). But, in general, the values fitted into the major sequences, and in sixth and seventh decades of the 20th century, the radiocarbon method is established as the principal dating technique in prehistoric studies.

In addition, however, to the easily observed results of the widespread utilization of radiocarbon values in prehistoric reconstructions, a more subtle influence has been slowly diffusing into the pursuit of prehistoric studies. As radiocarbon-based sequences began to be interwoven into the fabric of culture histories, archaeologists became conscious of the fact that such reconstructions were being established independently of the manipulations of artifact frequencies and seriation graphs or printouts. There now existed the ability to independently assign specific real-time temporal frameworks to archaeological materials by using a totally non-archaeological method. The validity of the results rested on mechanisms which could not be evaluated by one trained only as an archaeologist. Such values used to develop archaeologically based chronologies were supposedly now in the hands of a physicist or chemist. The issue and problem that had previously taken a major amount of his attention was now to be taken over by a machine—or so it would seem. There was thus joy and expectancy at having a dating method of such wide applicability and accuracy but also the uneasiness of a complete lack of background and experience to deal critically with the results so much desired.

One may see both positive and negative aspects in these developments. On one hand, for a short time, there was the temptation for some archaeologists to abdicate most of their responsibility in the critical evaluation of the ^{14}C values that were being given to them by the laboratories doing the measurements, sometimes even to the point of allowing themselves to become mere users of radiocarbon dates while some laboratories allowed themselves to become mere producers of dates. Some seemed to have lost sight of the fact that the archaeologist was the professional responsible for providing the basic documentation of the physical association of sample with event. A radiocarbon date is worthless if it lacks primary documentation of context and relationship with relevant geological and cultural features. While the radiocarbon specialist is responsible for the technical and analytical integrity of the physical measurements, it is the responsibility of the archaeologist to evaluate critically such data and to integrate it within the physical and cultural contexts from which the sample was derived. It may be necessary for the archaeologist to consult with those various other subject specialists, but in the end it is his responsibility to evaluate the dating evidence to reconstruct and interpret cultural processes.

Another generally negative aspect of this longer-range impact of radiocarbon on the pursuit of archaeology was the tendency to believe that it was possible to date a unit or site if one or at most two radiocarbon determinations were performed—usually one supposedly documenting the top of the sequence and one at the bottom. In the early development of the method, this methodology may have been justified in terms of cost and other practical consideration; it certainly can not be justified in terms of any sound research design. Today, it cannot be defended, in my judgement, on any grounds. A single or small number of radiocarbon determinations should not be used to date anything. Only with a suite of ^{14}C determinations, using, if relevant, different types of sample types collected with the full knowledge of the geochemical environments of the site and with the ability to compare the values obtained, can one date a site or feature.

The use of the radiocarbon method in archaeology today may be viewed as an excellent reminder and exemplar of the interdisciplinary focus of contemporary archaeological studies. Archaeologists, by the very nature of their subject matter, must look beyond the confines of a tight disciplinary framework to carry out the mandate of their science. The contribution of physical chemistry and nuclear physics to the pursuit of archaeology has been highlighted dramatically by the success of the radiocarbon method. The solution of a number of the problems that confronted those developing the method required both interdisciplinary collaboration and cooperation on a long-term basis that transcended the tendency of the interaction to be that of a producer–user relation. This same model of close long-term multidisciplinary cooperation and collaboration is being extended to include many types of relations between archaeologists and physical scientists of whatever disciplinary specialization. A new professional society has even emerged hopefully to facilitate this increasingly important aspect of modern prehistoric studies. We can look forward to this focus becoming an even more significant and productive part of modern archaeological and archaeometric research in the coming decades.

Literature Cited

1. Nobel Foundation, "Nobel Lectures, Chemistry 1942–1962," pp. 587–592, Elsevier, Amsterdam, 1964.

2. Daniel, G., "The Origins and Growth of Archaeology," p. 266, Crowell, New York, 1967.
3. Chang, K. C., "Rethinking Archaeology," p. 24, Random House, New York, 1967.
4. Michael, H. N., Ralph, E. K., "Dating Techniques for the Archaeologist," Chapter 1, M.I.T., Cambridge, 1971.
5. Fleming, S., "Dating in Archaeology: A Guide to Scientific Techniques," Chapter 3, St. Martins, New York, 1977.
6. Zuckerman, H., "Scientific Elite: Nobel Laureates in the United States," Chapters 1–3, Free Press, New York, 1977.
7. Kamen, M. D., "Early History of Carbon-14," *Science* (1963) **140**, 584.
8. Kurie, F. N. D., "A New Mode of Disintegration Induced by Neutrons," *Phys. Rev.* (1934) **45**, 904.
9. Libby, W. F., "History of Radiocarbon Dating," *in* "Radioactive Dating and Methods of Low Level Counting," pp. 3–26, International Atomic Energy Agency, Vienna, 1967.
10. Libby, W. F., "Ruminations on Radiocarbon Dating," *in* "Radiocarbon Variations and Absolute Chronology," pp. 630–640, Almquist and Wiksell, Stockholm, 1970.
11. Libby, W. F., "Radiocarbon Dating, Memories and Hopes," *in* "Proceedings of the 8th International Conference on Radiocarbon Dating," pp. XXVII–XLIII, Royal Society of New Zealand, Wellington, 1973.
12. Kamen, M. D., "Early History of Carbon-14," *Science* (1963) **140**, 586.
13. Libby, W. F., "Radioactivity of Neodymium and Samarium," *Phys. Rev.* (1934) **46**, 196.
14. Libby, W. F., Lee, D. D., "Energies of the Soft Beta-Radiations of Rubidium and Other Bodies. Method for Their Determination," *Phys. Rev.* (1939) **55**, 245.
15. Kamen, M. D., "Early History of Carbon-14," *Science* (1963) **140**, 590.
16. Korff, S. A., Danforth, W. E., "Neutron Measurements with Boron-Trifluoride Counters," *Phys. Rev.* (1939) **55**, 980.
17. Groueff, S., "Manhattan Project: the Untold Story of the Making of the Atomic Bomb," p. 20, Little, Brown, and Company, Boston, 1967.
18. Ibid., p. 92.
19. Libby, W. F., "Radiocarbon Dating," p. 35, University of Chicago, Chicago, 1955.
20. Engelkemeir, A. G., Hamill, W. H., Inghram, G., Libby, W. F., "The Half-life of Radiocarbon(C14)," *Phys. Rev.* (1949) **75**, 1825.
21. Godwin, H., "Half-life of Radiocarbon," *Nature* (1962) **195**, 984.
22. Editorial statements in *Radiocarbon* (1963–   ).
23. Libby, W. F., "Radiocarbon Dating," p. 7, University of Chicago, Chicago, 1955.
24. Ibid., p. 30.
25. Rubey, W. W., "Geological History of Sea Water," *Bull. Geol. Soc. Am.* (1951) **62**, 1111.
26. Libby, W. F., "Radiocarbon Dating," p. 30, University of Chicago, Chicago, 1955.
27. Ibid., pp. 66–68.
28. Ibid., p. 12, footnote 1.
29. Libby, W. F., "History of Radiocarbon Dating," *in* "Radioactive Dating and Methods of Low Level Counting," p. 17, International Atomic Energy Agency, Vienna, 1967.
30. Libby, W. F., Anderson, E. C., Arnold, J. R., "Age Determination by Radiocarbon Content: World Wide Assay of Natural Radiocarbons," *Science* (1949) **109**, 227.
31. Editorial statements in *Radiocarbon* (1962–   ).

32. Johnson, F., "Reflections upon the Significance of Radiocarbon Dates," in "Radiocarbon Dating," pp. 141–161, University of Chicago, Chicago, 1955.
33. Grootes, P. M., Mook, W. G., Vogel, J. C., de Bries, A. E., Haring, A., Kistmaker, J., "Enrichment of Radiocarbon Dating Samples Up to 75,000 Years," Z. Naturforsch. (1975) 30a, 1.
34. Taylor, R. E., "Chronological Problems in West Mexican Archaeology: A Dating Systems Approach to Archaeological Research," Chap. 2, unpublished doctoral dissertation, University of California, Los Angeles (1970).
35. Olsson, I. U., Ed., "Radiocarbon Variations and Absolute Chronology," Almqvist and Wilsell, Stockholm, 1970.
36. Rafter, T. A., Grant-Taylor, T., Eds., "Proceedings of the 8th International Conference on Radiocarbon Dating," Royal Society, Wellington, 1973.
37. Berger, R., Seuss, H. E., Sds., "Advances in Radiocarbon Dating: Proceedings of the IX International Radiocarbon Conference," University of California, Berkeley, in press.
38. Meighan, C. W., "Responsibilities of the Archaeologist in Using the Radiocarbon Method," (University of Utah) Anthropological Papers (1956) 26, 48.
39. Michael, H. N., Ralph, E. K., "Dating Techniques for the Archaeologist," Chapter 1, M.I.T., Cambridge, 1971.
40. Spaulding, A. C., "The Significance of Differnces Between Radiocarbon Dates," Am. Antiq. (1958) 23, 309.
41. Long, A., Rippeteau, B., "Testing Contemporaneity and Averaging Radiocarbon Dates," Am. Antiq. (1974) 39, 205–215.
42. Wise, E. N., "The C-14 Age Determination Method," in "Geochronology," University of Arizona, Tucson, 1955.
43. Libby, W. F., "Accuracy of Radiocarbon Dates," Science (1963) 140, 278.
44. Willis, E. G., Tauber, H., Munnich, K. O., "Variations in the Atmospheric Radiocarbon Concentration Over the Past 1300 Years," Radiocarbon (1960) 2, 1.
45. de Vries, H., "Variation in Concentration of Radiocarbon with Time and Location on Earth," Proc. Ned. Akad. Wet., Ser. B (1958) 61, 1.
46. Fergusson, C. W., "Dendrochronology of Bristlecone Pine Prior to 4000 B.C., in "Proceedings of the 8th International Conference on Radiocarbon Dating," pp. A2–A27, Royal Society of New Zeland, Wellington, 1973.
47. Fergusson, C. W., "Dendrochronology of Bristlecone Pine, Pinus aristata. Establishment of a 7384-year Chronology in the White Mountains of Eastern-central California, U.S.A.," in "Radiocarbon Variations and Absolute Chronology," pp. 237–259, Almqvist and Wiksell, Stockholm, 1970.
48. Fergusson, C. W., personal communication.
49. LaMarche, V. C., Harlan, T. P., "Accuracy of Tree Ring Dating of Bristlecone Pine for Calibration of Radiocarbon Time Scale," J. Geophys. Res. (1973) 78, 8849.
50. Damon, P. E., Long, A., Gray, A. C., "Arizona Radiocarbon Dates for Dendrochronologically Dated Samples," in "Radiocarbon Variations and Absolute Chronology," pp. 615–618, Almqvist and Wiksell, Stockholm, 1970.
51. Damon, P. E., Long, A., Wallick, E. I., "Dendrochronologic Calibration of the Carbon-14 Time Scale," in "Proceedings of the 8th International Conference on Radiocarbon Dating," pp. A28–A41, Royal Society of New Zealand, Wellington, 1973.
52. Suess, H. E., "Bristlecone-pine Calibration of Radiocarbon Time 5200 B.C. to the Present," in "Radiocarbon Variations and Absolute Chronology," pp. 303–312, Almqvist and Wiksell, Stockholm, 1970.

53. Ralph, E. K., Michael, H. N., "MASCA Radiocarbon Dates for Sequoia and Bristlecone-pine Samples," *in* "Radiocarbon Variations and Absolute Chronology," pp. 619–624, Almqvist and Wiksell, Stockholm, 1970.

54. Ralph, E. K., Michael, H. N., Hann, M. C., "Radiocarbon Dates and Reality," *Mus. Appl. Sci. Cent. Archaeol. Newsl.* (1973) **9**, 1.

55. Elsasser, W. E., Ney, P., Winckler, J. R., "Cosmic-ray Intensity and Geomagneticism," *Nature* (1956) **178**, 1226.

56. Bucha, V., Neustupny, E., "Changes of the Earth's Magnetic Field and Radiocarbon Dating," *Nature* (1967) **215**, 261.

57. Bucha, V., "Changes of the Earth's Magnetic Moment and Radiocarbon Dating," *Nature* (1969) **224**, 681.

58. Barbetti, M., "Geomagnetic Field Behavior Between 25,000 and 35,000 yr B.P. and Its Effect on Atmospheric Radiocarbon Concentration: A Current Report," *in* "Proceedings of the 8th International Conference on Radiocarbon Dating," pp. A104–A113, Royal Society of New Zealand, Wellington, 1973.

59. Denham, C. R., Cox, A., "Evidence that the Laschamp Polarity Event Did Not Occur 13,300–30,400 Years Ago," *Earth Planet. Sci. Lett.* (1971) **13**, 181.

60. Yang, A. I. C., Fairhall, A. W., "Variations of Natural Radiocarbon During the Last 11 Millenia and Geophysical Mechanisms for Producing Them," *in* "Proceedings of the 8th International Conference on Radiocarbon Dating," pp. A44–A57, Royal Society of New Zeland, Wellington, 1973.

61. Farmer, J. G., Baxter, M. S., "Short-term Trends in Natural Radiocarbon," *in* "Proceedings of the 8th International Conference on Radiocarbon Dating," pp. A58–A71, Royal Society of New Zealand, Wellington, 1973.

62. Berger, R., "The Potential and Limitations of Radiocarbon Dating in the Middle Ages: the Radiochronologist's View," *in* "Scientific Methods in Medieval Archaeology," pp. 89–140, University of California, Berkeley, 1970.

63. Lerman, J. C., Mook, W. G., Vogel, J. C., "C-14 in Tree Rings from Different Localities," *in* "Radiocarbon Variations and Absolute Chronology," pp. 273–301, Almkvist and Wiksell, Stockholm, 1970.

64. Deevey, E. S., Jr., Gross, M. S., Hutchinson, G. E., Kraybill, H. L., "The Natural C^{14} Contents of Materials from Hard-Water Lakes," *Proc. Nat. Acad. Sci. U.S.A.* (1954) **40**, 285.

65. Chatters, R. M., Crosby, J. W., III, Engstrand, L. G., "Fumarole Gaseous Emanations: Their Influence on Carbon-14 Dates," *Washington State University, College of Engineering, Res. Div. Circular* (1969) **32**, 1.

66. Libby, W. F., "Radiocarbon Dating," pp. 44, University of Chicago, Chicago, 1955.

67. Taylor, R. E., Slota, P. S., "Fraction Studies on Marine Shell and Bone for Radiocarbon Dating," *in* "Advances in Radiocarbon Dating," University of California, Berkeley, in press.

68. Broeker, W., "Radiocarbon Dating: Fictitious Results with Mollusk Shells," *Science* (1963) **141**, 634.

69. Rubin, M., Taylor, D. W., "Radiocarbon Activity of Shells from Living Clams and Snails," *Science* (1963) **141**, 637.

70. Johnson, F., "Reflections Upon the Significance of Radiocarbon Dates," *in* "Radiocarbon Dating," p. 150, University of Chicago, Chicago, 1955.

71. Kulp, J. L., Tryon, L. E., Eckelman, W. R., Snail, W. A., "Lamont Natural Radiocarbon Measurements II," *Science* (1952) **116**, 409.

72. Taylor, R. E., "Chronological Problems in West Mexican Archaeology: An Application of a Dating Systems Approach to Archaeological Research," Chapter 2, unpublished doctoral dissertation, University of California, Los Angeles (1970).

73. Berger, R., Taylor, R. E., Libby, W. F., "Radiocarbon Content of Marine Shells from the California and Mexican West Coast," *Science* (1966) **153**, 864.

74. Taylor, R. E., Berger, R., "Radiocarbon Content of Marine Shells from the Pacific Coasts of Central and South America," *Science* (1967) **158**, 1180.

75. Polach, H. A., "Radiocarbon Dating as a Research Tool in Archaeology— Hopes and Limitations," *in* "The Procedings of a Symposium on Scientific Methods of Research in the Study of Ancient Chinese Bronzes and Southeast Asian Metal and Other Archaeological Artifacts," pp. 255–298, National Gallery of Victoria, Melbourne, 1975.

76. Suess, H. E., "Radiocarbon Concentration in Modern Wood," *Science* (1955) **122**, 415.

77. Berger, R., Fergusson, G. F., Libby, W. F., "UCLA Radiocarbon Dates IV," *Radiocarbon* (1965) **7**, 336.

78. Craig, H., "The Geochemistry of Stable Carbon Isotopes," *Geochim. Cosmochim. Acta* (1953) **3**, 53.

79. Olson, E., "The Problem of Sample Contamination in Radiocarbon Dating," unpublished doctoral dissertation, Columbia University (1963).

80. Olsson, I. U., "Some Problems in Connection with the Evaluation of C¹⁴ Dates," *Geol. Foeren. Stockholm Foerh.* (1974) **96**, 311.

81. Dyck, W., Fyles, J. G., "Geological Survey of Canada Radiocarbon Dates III," *Radiocarbon* (1964) **6**, 167.

82. Lowden, J. A., Fyles, J. G., Blake, W., Jr., "Geological Survey of Canada Radiocarbon Dates VI," *Radiocarbon* (1967) **9**, 156.

83. Libby, W. F., "Radiocarbon Dating," p. 45, University of Chicago, Chicago, 1955.

84. Tamers, M. A., Pearson, F. J., "Validity of Radiocarbon Dates on Bone," *Nature* (1965) **208**, 1053.

85. Berger, R., Horney, A. G., Libby, W. F., "Radiocarbon Dating of Bone and Shell from Their Organic Components," *Science* (1964) **144**, 999.

86. Haynes, C. V., "Carbon-14 Dating and Early Man in the New World," *in* "Pleistocene Extinctions: the Search for a Cause," pp. 267–286, Yale University, New Haven, 1967.

87. Krueger, H. W., "The Preservation and Dating of Collagen in Ancient Bones," *Proc. Int. Conf. Radiocarbon and Tritium Dating, 6th* (CONF-650652) United States Atomic Energy Commission, Oak Ridge, 1965.

88. Sellstedt, H., Engstrand, L., Gejvall, N.-G., "New Application of Radiocarbon Dating to Collagen Residue in Bones," *Nature* (1966) **212**, 572.

89. Berger, R., Protsch, R., Reynolds, R., Rozaire, C., Sackett, J. R., "New Radiocarbon Dates Based on Bone Collagen in California Paleoindians," *Contrib. Univ. Calif. Archaeol. Res. Fac.* (1971) **12**, 43.

90. Longin, R., "New Method of Collagen Extraction for Radiocarbon Dating," *Nature* (1971) **230**, 241.

91. Protsch, R. R., "The Dating of Upper Pleistocene Subsaharan Fossil Hominids and Their Place in Human Evolution: With Morphological and Archaeological Implications," unpublished doctoral dissertation, University of California, Los Angeles (1973).

92. Sinex, F. B., Faris, B., "Isolation of Gelatin from Ancient Bones," *Science* (1959) **129**, 969.

93. Haynes, C. V., "Radiocarbon Analysis of Inorganic Carbon of Fossil Bones and Enamel," *Science* (1968) **161**, 687.

94. Hassan, A. A., "Geochemical and Mineralogical Studies on Bone Material and Their Implications for Radiocarbon Dating," unpublished doctoral dissertation, Southern Methodist University (1975).
95. Muller, R. A., "Radioisotope Dating with a Cyclotron," *Science* (1977) **196**, 489.
96. Bennett, C. L., Beukens, R. P., Clover, M. R., Gove, H. E., Liebert, R. B., Litherland, A. E., Purser, K. H., Sondheim, W. E., "Radiocarbon Dating Using Electrostatic Accelerators: Negative Ions Provide the Key," *Science* (1977) **198**, 508.
97. Nelson, D. E., Korteling, R. G., Stott, W. R., "Carbon-14: Direct Detection at Natural Concentrations," *Science* (1977) **198**, 507.
98. Taylor, R. E., "Dating Methods in New World Archaeology," *in* "Chronologies in New World Archaeology," pp. 1–27, Academic, New York, 1978.
99. Butzer, K. W., "Environment and Archaeology: an Ecological Approach to Prehistory," 2nd ed., Aldine, New York, 1971.
100. Wise, E. N., "The C-14 Age Determination Method," *in* "Geochronology," p. 586, University of Arizona, Tucson, 1955.
101. Daniel, G., *Antiquity* (1972) **46**, 265.
102. Clark, R. M., "A Calibration Curve for Radiocarbon Dates," *Antiquity* (1975) **49**, 251–266.
103. Renfrew, C., "Before Civilization: The Radiocarbon Revolution and Prehistoric Europe," Jonathan Cape, London, 1973.
104. Wilson, D., "The New Archaeology," Knopf, New York, 1974.
105. Taylor, R. E., "Science in Contemporary Archaeology," *in* "Advances in Obsidian Glass Studies: Archaeological and Geochemical Perspectives," pp. 1–21, Noyes, Park Ridge, New Jersey, 1976.

RECEIVED September 19, 1977.

4

Spark Source Mass Spectrometry in Archaeological Chemistry

A. M. FRIEDMAN and J. LERNER

Chemistry Division, Argonne National Laboratory, Argonne, IL 60439

Techniques for analysis and sample preparation have been developed for using spark source mass spectrometry (SSMS) to study archaeological samples. Comparative studies of neutron activation and SSMS on identical samples have been made. The technique is used to determine the ores of origin of two series of early Peruvian artifacts.

Analysis of metal artifacts has long been a major tool in archaeological studies. Since by their very nature metal samples are electrically conducting, they are also well suited for analysis by spark source mass spectroscopy (SSMS). For several years we investigated the use of this technique and have applied it to archaeologically interesting samples.

Figure 1 is a diagram of the spectrograph. The sample is formed into two small electrodes ($2 \times 2 \times 10$ mm each), and a pulsed radiofrequency spark is induced between them, volatilizing and ionizing some of the material. The resulting ion beam is accelerated, defined, mass-analyzed, and monitored. The individual masses then are deposited on a photographic emulsion. A series of exposures are made in which the intensity varies over a factor of 10^6; the plate then is developed and scanned by a microdensitometer. This data is fed into a computer which is programmed to take into account elemental isotopic ratios, sensitivity factors, and machine parameters and which then generates an analysis of the electrode material. In general we have determined at least 11 elements in each sample with sensitivities of the order of 0.1 ppm or better.

The SSMS is particularly well suited for the analysis of archaeological samples. Relatively small amounts of material are required for the electrodes (ca. 100 mg) yet the photographic plate records simultaneously all elements, including those not specifically sought. Thus it is

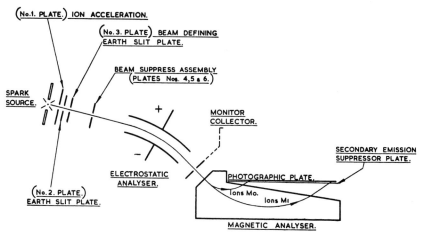

Figure 1. Mass spectrograph

possible to re-examine the plate at a later time to check for unexpected abundances or to verify that a specific element is indeed absent.

Since a pulsed RF source is used for sparking rather than a continuous arc, the electrodes remain relatively cool. Tests have shown that the concentration of even volatile elements (e.g., mercury) is unaffected by the sparking. Furthermore, the radiofrequency interruptions (ca.

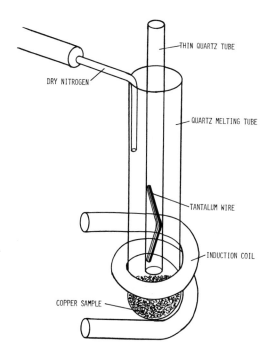

Figure 2. Sample electrode preparation apparatus

Table I. Results of Analysis of

	CA0			CA3		
Element	ppma actual	ppma exp.	actual exp. ×100	ppma actual	ppma exp.	actual exp. ×100
Sb	261	282.6	92.4	22.4	59.7	37.5
Bi	152	98.1	154.9	24.3	22.4	108.5
Cr	269	495.2	54.3	36.7	142.3	25.8
Ga	319	799.6	39.9	36.4	187.0	19.7
Pb	123	114.7	107.2	15.4	26.3	58.6
Ag	277	312.8	88.6	35.3	52.6	67.1
Sn	257	292.1	88.0	26.8	57.3	46.8

500 kHz) cause the spark to move erratically over the surface of the electrodes—a desirable feature for samples that are rarely uniformly homogeneous. While the analyses may be relatively imprecise, in general high orders of accuracy are not required for most archaeological applications of SSMS. The entire process of electrode preparation, spectrometry, plate development, scanning, and data card punching takes about one day per sample.

Table II.

Abundance (ppm)	Error (%)
100–1000	15
10–100	20
4–10	50
1–3	80

Table III. Comparison of SSMA and

Mass Spectrometry[a]

Sample	Ag	Co	Cr	Fe	Sb	Sc	Se
34-21-35AW	77	0.38	0.35	200	0.66	0.14	0.89
34-21-35L	150	ND	0.35	320	170	0.26	3.1
34-21-35Z	42	ND	0.12	19	ND	170	1.4
43-21-16D	3.3	1.1	5.3	470	0.36	1.5	10.0
66-21-00	45	0.63	0.32	55	1.8	0.30	2.4
30-05-10As	44	0.60	0.50	32	1.3	0.42	4.1
90-20-01-S	0.26	14	1.4	26	0.64	0.61	2.0
16-11-50Es	84	7.6	0.41	7200	0.28	0.22	0.78
16-12-58Hs	98	21	0.54	27000	320	0.21	7.2

[a] ND = Not Detected.

CA Series of Copper Standards

	CA6			CA7			CA8	
ppma actual	*ppma exp.*	*actual exp. ×100*	*ppma actual*	*ppma exp.*	*actual exp. ×100*	*ppma actual*	*ppma exp.*	*actual exp. ×100*
2.61	5.28	49.4	1.57	4.70	33.4	0.52	0.86	60.5
2.74	1.95	140.5	0.91	1.47	61.9	0.61	0.29	210.3
3.67	8.10	45.3	2.44	7.58	32.2	1.22	3.28	37.2
4.56	45.98	9.9	2.73	25.93	10.5	2.73	14.19	19.2
2.15	2.65	81.1	1.23	2.56	48.1	0.92	1.02	90.2
3.53	5.80	60.9	2.36	5.30	44.5	0.12	3.82	3.1
3.21	4.40	73.0	1.61	3.11	51.8	—	1.08	—

Sample Preparation

It is clearly desirable to minimize the amount of material withdrawn from an archaeological sample for analysis. In some cases it has been possible mechanically to shear off slivers of metal that can serve directly as electrodes. In other cases the material is somewhat granular but can be compressed in a die to form cohesive electrodes. However, under many circumstances there is no alternative to melting and casting, and for this the apparatus shown diagrammatically in Figure 2 was developed.

The sample is maintained under a nitrogen atmosphere at the base of the quartz melting tube. Application of an RF current to the induction coil melts the material. Suction is immediately applied to the top of the inner quartz tube to draw a layer of metal over the tantalum wire core; contact with the cold quartz walls solidifies the sample immediately. Active stirring of the molten bead by the RF field and subsequent rapid chilling to produce a fine-grained solid help to promote homogeneity within the electrodes. The entire melting and solidification process requires only a few seconds and consumes roughly 25–100 mg of sample.

After further cooling the inner quartz tube is removed and broken to release an electrode of appropriate size and shape. Only a small

NAA (concentrations in PPMA)

			Neutron Activation[a]			
Ag	Co	Cr	Fe	Sb	Sc	Se
120	ND	ND	ND	1.5	0.034	ND
140	0.55	ND	1400	ND	0.51	0.56
180	0.52	ND	960	ND	0.22	0.16
7.2	0.14	1.5	570	0.24	0.38	ND
45	ND	0.38	100	0.44	ND	ND
78	0.47	1.9	720	3.6	0.06	0.86
ND	0.26	ND	250	0.71	0.30	0.15
8.8	3.2	ND	240	0.52	0.022	ND
540	24	ND	32000	1300	ND	0.25

amount of etching and pre-sparking is required to remove surface contamination. Although the use of a tantalum core precludes analysis for this element in the sample, the determination would be suspect in any case because virtually the entire ion source is constructed of tantalum.

Data Analysis

While the sensitivities for each of the elements are relatively uniform, there are differences caused by variations in ionization efficiency, volatility, and overlap of isotopic mass lines. These effects require the calibration of the technique with a variety of standards covering a range of concentrations, matrix materials, and other trace elements. Table I contains a set of analyses of copper matrix standards and compares the spark source mass spectrograph results to the true values. A series of experiments was performed on standard electrodes containing a variety of elements and yielded the average errors shown in Table II. The errors are a function of elemental abundance, and in the ppm range the analyses are in general only roughly correct.

A set of samples was analyzed by neutron activation analysis (NAA) and spark source mass spectrometry. Some of these results are shown in Table III. The differences correspond to the range of accuracies shown in Table II except possibly for some of the iron results. In those cases it was felt that the NAA samples were contaminated.

Application to Archaeology

In previous reports (1, 2) the origins of copper samples have been traced on the basis of NAA-determined impurities to three kinds of ore: native copper (Cu metal) ores, oxidized (CuO or $CuCO_3$) ores, and reduced (Cu_2S or CuMS) ores. This data has been used as the basic parameter of a computer program (1) which calculates the probability of a sample originating in a given ore type. The program generates this probability as a linear combination of uncorrelated individual impurity distributions. This calculation has been described in Refs. 1 and 2, and the Fortran listing is available on request.

We have been investigating a similar scheme set up on the basis of SSMS-determined impurities; presumably the classification might be even more definitive since more trace elements can be estimated. The mass spectrometer data for 44 type I ores is presented in Figures 3 and 4; for comparison the NAA data for 214 type I ores is shown in Figure 5. In general, both the location of the peaks and the width of the distributions for corresponding elements are quite similar.

Since there is reasonable agreement between the two methods of analysis, the SSMS data was processed using the program described

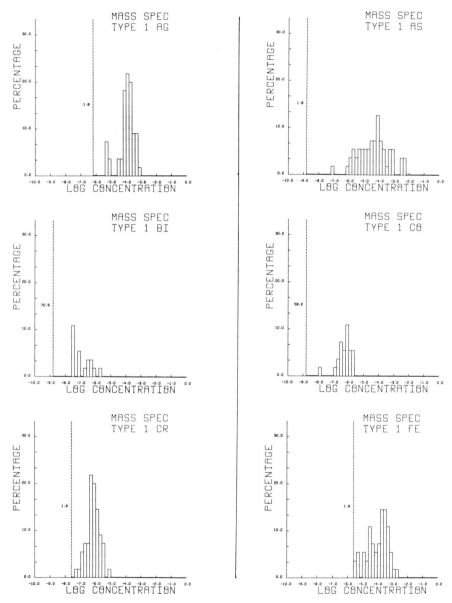

Figure 3.   Impurity distribution in type I ores—SSMS data

above. Of the 44 type I ore samples only two (5%) were classified as other types. Establishment of SSMS probability tables for the three ore types and the use of 11 trace elements should result in improved identification of the sources of copper.

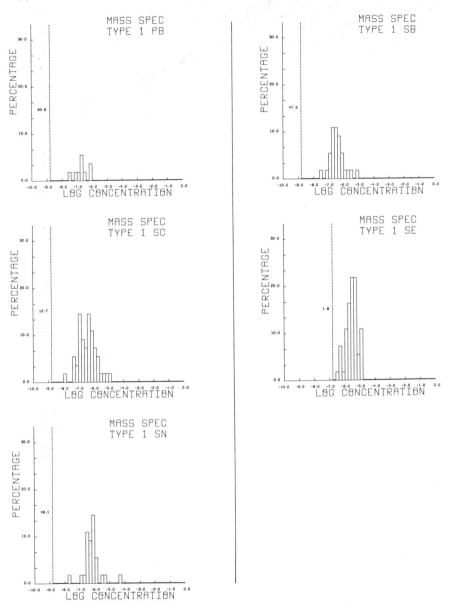

Figure 4. Impurity distribution in type I ores—SSMS data

In addition, a series of early Peruvian artifacts were analyzed and assigned to ore types by the program. The results are shown in Table IV. The early samples have a high probability of being simple type I ores, and the latter are apparently more complex mixtures.

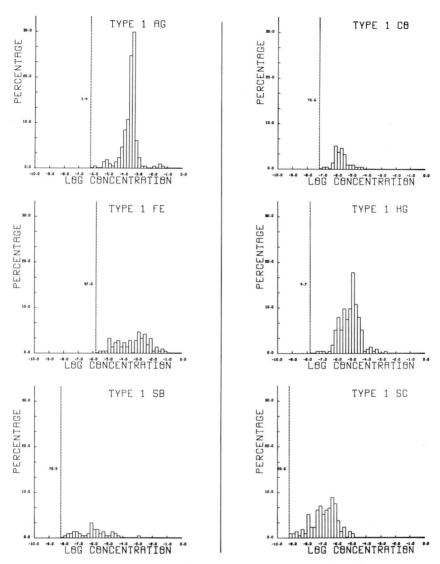

Figure 5. Impurity distribution in type I ores—NAA data

These results imply a sharp change in metallurgical patterns between the late intermediate and late horizon periods, and we have now started to study a series of Moche artifacts obtained from Gary Vesculius in order to investigate this transition further. When the artifacts in this set are grouped into broad time periods, our preliminary results indicate that these also show a transition between the use of type I and type II copper ores. Of the seven artifacts believed to originate before 500 AD, 57% had a high probability ($> .95$) of being made from type I ores, and 43%

Table IV. Selected Moche Results and Localized Chronological Table

Sample No.	Era	Type 1	Type 2	Type 3
4645	early intermediate (Moche)	91.01	4.56	4.43
4891-32	early intermediate (Moche)	80.69	10.49	8.83
169897	early intermediate (Moche)	71.30	27.77	0.94
5712-16	early intermediate (Moche)	93.95	5.74	0.31
A1936-32	late horizon or late intermediate	23.24	38.72	38.07
2200	late horizon (Inca Bronze)	12.07	24.59	63.35
170082	?	25.77	33.52	40.71

Dates	Era	Style
1450–1532	late horizon	Inca
1200–1450	late intermediate	Chiamu, late Chancay, Ica
800–1200	middle horizon	Coast Tialuanco, Hicara
0–800	early intermediate	Mochica, early Lima, Nazca

had a high probability ($>.95$) of being made from type II ores. Of the 15 artifacts believed to have been made between 500 and 1000 AD, 40% had a high probability(> 0.9) of being made from type I ores, and 60% had a high probability (0.7–.95) of being made from type II ores. Of the seven artifacts studied which were made after 1000 AD, only one (or 13%) had a high probability (0.9) of being made from type I ores, and the remainder (87%) had high probabilities (0.6–.95) of being made from type II ores. Since the artifacts were chosen at random from the set, this variation in ore types should indicate the general trend, but at present the sampling is still too small for a definite conclusion. However, these two examples do indicate the general utility of the method.

Acknowledgments

Without the efforts of George Lamich and Argonne Undergraduate Honors Program students Cathy Richardson, Charles Shuey, Robert Bowman, and Gary Sanford this report would not have been possible. We also wish to thank the Field Museum for making available the Moche samples. Work performed under the auspices of the USERDA.

Literature Cited

1. Friedman, A. M., Conway, M., Kastner, J., Milsted, J., Metta, D., Fields, P., Olsen, E., *Science* (1966) **152**, 1504.
2. Bowman, R., Friedman, A. M., Lerner, J., Milsted, J., *Archaeometry* (1975) **17**, 157.

RECEIVED September 19, 1977.

Applications of X-Ray Radiography in the Study of Archaeological Objects

PIETER MEYERS

Research Laboratory, The Metropolitan Museum of Art, New York NY 10028

Although the use of X-ray radiography in the study of paintings is well known, its application to other museum objects has been more limited. However, X-ray radiography of several museum objects provided new, useful information: a ceramic bowl is composed of unrelated fragments and plaster; there is a clay core in a bronze figurine from ca. 2500 B.C.; and the famous Bronze Horse was cast by the lost wax process and not, as had been suggested, by a modern piece-mold sand casting process. In addition, X-ray radiography helped to provide details on the methods of manufactuer of Sasanian silver and a Chinese bronze vessel.

The use of x-ray radiography to study art objects began when German scientists produced the first x-ray radiographs of paintings soon after the discovery of x-rays by Roentgen in 1895. During the next half century this became a standard technique for examining paintings in many museum laboratories (1), although its further development in Germany was apparently hampered by a patent on the radiography of paintings (2). Since then the usefulness and limitations of the technique have been clearly established. Systematic studies of x-ray radiographs provided much important information on the painting techniques of individuals and schools of artists. Today most of the important paintings in museum collections have been radiographed, and the published literature contains hundreds of articles dealing with radiographic studies of paintings. Conservators as well as museum curators and art historians make frequent use of x-ray radiographs, which often can be interpreted straightforwardly by the trained eye. Furthermore, the necessary equipment is readily available and relatively simple to use.

In contrast, the application of x-ray radiography to archaeological materials and other three-dimensional objects has not reached the same level of sophistication. The reasons seem obvious: the more powerful

0-8412-0397-0/78/33-171-**079**$05.00/1

equipment necessary for radiography of most three-dimensional objects is not readily available; the greater complexity of the actual process and of the subsequent interpretation of the radiographic images may have discouraged potential investigators; fewer scholars have realized the utility of such investigations for studies of ancient technologies. The wider application of x-ray radiography to archaeological materials also is perhaps being prevented by the scarcity in the literature of comparative material. Proper descriptions of radiographic studies are needed to provide detailed information on specific aspects of the development of ancient technology. This chapter was written in the hope that more extensive discussion of the techniques and results of applying x-ray radiography to archaeological artifacts will increase interest in and contribute to the success of current and future projects.

Methods

Although x-ray radiography equipment used for paintings is sometimes suitable for the study of objects made of wood, ivory, bone, or other organic materials and for some ceramics, most three-dimensional objects require more powerful equipment. Industrial x-ray radiography units with a maximum x-ray energy of 200–300 kV are required for most metal and stone artifacts and for large ceramics.

The principles of x-ray radiography need not be elaborated in detail; the theory is well known, and the practical aspects are fully described in the literature (3). However, in museum laboratories radiography is frequently performed by personnel not necessarily familiar with the various practical aspects. Even though the available equipment restricts the choice of conditions for radiography, the operator must make a number of decisions that can seriously affect the success of the project. It may be of use, therefore, to list some of the variables encountered in x-ray radiography.

High Voltage or Maximum X-Ray Energy. Optimum conditions are determined by the thickness of the object to be radiographed and by the atomic numbers of the elements of which it is composed. The kilovoltage must be high enough to allow a sufficient amount of x-ray radiation to penetrate the object. Since the contrast in the radiographic image decreases with increasing x-ray energy, the kilovoltage must not be too high.

X-Ray Film. The best x-ray films have very fine grain, such as Kodak type M. These give the highest resolution, but they require longer exposure time than the larger grained films. A full description of x-ray films is given in the product literature or in Bridgman's discussion of x-ray films (4). To obtain high quality radiographs it is essential to follow accurately the manufacturer's film processing directions.

Intensifying Screens and Lead Screens. Chemical intensifying screens may shorten exposure times by factors up to 200; however their use is not recommended because they decrease resolution. Lead screens, on the other hand, serve dual purposes: they not only absorb a large amount of unwanted secondary radiation, but they also act as intensifying screens because secondary electrons are emitted by the lead upon impact of x-rays. Such screens, 0.005 in. thick, should be placed on either side of the film when exposures greater than 150 kV are used.

Filters. Metal filters placed near the exit window of the x-ray tube reduce the low energy x-rays emitted at the anode. Their use will reduce the contrast but will improve the resolution.

Image Registration. The standard method of observing x-ray images is through the use of x-ray film. Other methods of registering radiographic impressions include fluoroscopy, xeroradiography, and color radiography. Fluoroscopy involves the projection of the radiographic image on a fluorescent screen which can be viewed directly or can be photographed. The resolution is inferior to that of the standard film radiography. In xeroradiography the radiographic image is recorded on an electrically charged plate. Resolution and contrast in prints obtained from these plates compare favorably with those in standard film radiographs. Xeroradiography has not been widely used yet in studies of archaeological objects (5). Color radiography appeared to be a promising technique approximately 10 years ago (6, 7), but since the literature describes no recent applications, it may be assumed that this technique did not live up to its initial promise.

Microradiography has not yet been used extensively to study archaeological materials. This technique, which allows study of details too fine to be seen with the naked eye by enlarging the x-ray radiograph, may find useful applications in the future.

Stereoradiography permits the study of a very informative three-dimensional radiographic image by a simple but effective method. Two radiographs are obtained under identical conditions except that the x-ray target is moved 4 in. (the distance between the human eyes) parallel to the film between the first and second exposures. Then, by a simple mirror system the radiographs are viewed in such a way that each eye sees only one film. The superimposed images give the impression of a three-dimensional radiograph.

A good radiograph obtained by one of the methods described provides an accurate and lasting record of the object under study. When duplicates or prints are produced much information is lost, especially in the processing of prints. Most photographic films cannot accommodate the large range of densities typical for x-ray radiographs. Methods of reproducing radiographs have been summarized by White (8).

Applications

Among the objects most susceptible to successful radiographic study are those with well-defined internal structures such as cast bronzes, jewelry, furniture, musical instruments, and ceramics. The variety of materials that have been studied is described in the literature. For example Bridgman reported on the radiography of museum objects (9), Gettens examined Chinese bronzes (10), Gorelick studied cylinder seals (11), Bertrand et al. and Gugel et al. described applications to ceramics (12, 13), and Harris and Weeks radiographed mummies of ancient Egyptian royalty (14).

Most publications discuss x-ray radiography of only a few three-dimensional objects and deal with specific aspects of these such as method of manufacture or repairs and hidden decorations. The few radiographic studies where methods of manufacture have been studied on a systematic basis using large groups of objects have been very successful. The radiographic study of Chinese bronzes by Gettens (10) contributed significantly to a better understanding of Chinese bronze casting techniques. An investigation at The Metropolitan Museum of Art of Renaissance bronzes enabled R. E. Stone to correlate casting characteristics observed in x-ray radiographs with workshops of master sculptors such as Antico,

Figure 1.   Glazed Minai ware bowl, 12–13th century A.D., Persian. The Metropolitan Museum of Art, Acc. # 57.36.11.

Riccio, and Severo (*15*). In addition, the x-ray evidence made possible a detailed reconstruction of the casting processes and allowed proper attribution of several Renaissance bronzes previously unattributed or erroneously ascribed.

The following sections present some examples of x-ray radiographic studies. All the x-ray radiographs illustrating this discussion were made with a Norelco 300 MG industrial x-ray unit on Kodak industrial x-ray film type M encased in 0.005-in. lead screens, unless otherwise indicated.

Islamic Ceramic Bowl. X-ray investigation can help solve the common problem of how much of an apparently complete ceramic vessel is original and how much is restoration. Several Islamic glazed Minai bowls of the 12th and 13th centuries A.D. presented such a problem; examination with UV light indicated that large areas were overpainted and apparently restored. When x-ray radiographs were obtained, it became evident that among the 15 objects examined, more than half were composed of unrelated ceramic fragments with plaster fills. The bowl shown in Figure 1 is an example of this group. The x-ray radiograph of that bowl (Figure 2) shows clearly that the bottom part bears no relationship to the fragments making up most of the rest of the bowl. Areas between these fragments, slightly darker in color, are plaster fills. Another bowl from this group has been described by Pease (*16*).

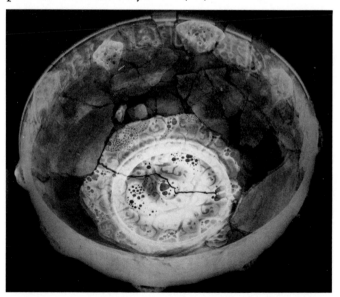

Figure 2. X-ray radiograph of bowl shown in Figure 1. The bowl is assembled from fragments of two unrelated vessels as indicated by the differences in density and in structure. Exposure: 50 kV, 5 mA, recorded on Kodak No-Screen film.

Sasanian Silver. The subject of Sasanian silver was well represented in a previous symposium (*17, 18, 19*). Articles presented at this meeting dealt predominantly with elemental compositions. In a general technical study on Sasanian silver objects at the Research Laboratory of the Metropolitan Museum of Art, considerable attention has been given to the methods of manufacture. Careful observations using a binocular microscope, measurements of the thickness of the metal walls, and especially x-ray radiography of a large number of Sasanian silver vessels resulted in a well defined theory of the practices of the Sasanian silversmith. The findings of this study, which are entirely consistent with other evidence, will be reported in detail elsewhere, but some conclusions are listed here. As shown below, the investigation produced disagreement with the widely accepted theory based on statements by Orbeli and Trevers (*20*) that the Sasanian silversmith used three basically different techniques to produce plates and bowls.

HAMMERING. The plate or bowl was shaped by hammering from a cast blank. The interior of the plates or exterior of the bowls was decorated by chasing or engraving. When decoration in relief was required, the background was carved away around the figures (Figure 4); to produce high-relief decoration pre-cast and pre-hammered pieces were added to the plate, as shown in the plate illustrated in Ref. *18*.

CASTING.

DOUBLE SHELL TECHNIQUE. Plates were constructed of a hammered undecorated exterior plate and a hammered interior plate with a repoussé decoration. The two were secured together either by soldering or by bending the edges of the exterior plate inward and hammering them down over the interior plate.

In general x-ray radiographs consistently show characteristics indicating the method of manufacture. For example cast plates and bowls invariably contain trapped gas or voids in the metal, which appear in radiographs as dark spots (Figure 3); the radiographic image of a cast plate or bowl is often mottled and grainy. Frequently the wall thickness of cast objects increases in areas of greatest curvature, supposedly to avoid contact between the inner and outer investment walls during the casting operation. Hammered plates and bowls, on the other hand, do not contain trapped gas; their radiographic images are smooth and sharp with characteristic density changes caused by the hammering. Wall thickness decreases as a result of the more extensive hammering needed to produce curved surfaces. Plates or bowls made by the double shell technique are distinguished readily from cast or hammered objects. Their x-ray radiographs in most cases show small density variations that do not conform with variations in apparent metal thickness. Radiographs

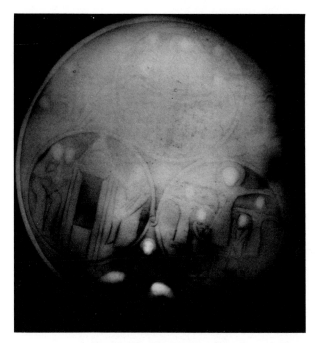

Figure 3. X-ray radiograph of a cast plate. Irregularly distributed dark spots indicate gas trapped in the metal, a characteristic found in cast silver plates and not in hammered plates. Radiograph obtained with Philips PG 200 x-ray unit, 200 kV, 3 mA, no lead screens.

of vessels produced by the double shell method should show the use of solder and the three-ring image of the bent-over rim: two images of the outside plate with the impression of the inner plate between them.

Radiographic studies on more than 100 Sasanian silver artifacts and an equal number of related silver objects strongly indicate that the Sasanian silversmith used hammering exclusively as the major shaping technique for all his objects. Among vessels accepted as genuine Sasanian, none were found that appeared to have been made by the double shell technique or that could be positively identified as cast.

The use of x-ray images to establish the method of manufacture may be illustrated by referring to a silver gilt bowl (Figures 4 and 5) originally thought to have been made by the double shell technique because many small hemispheres were visible on the inside of the bowl. These hemispheres are generally produced by punches applied on the outside; however no traces of punches appear on the outside. The craftsman might most simply have constructed the bowl by using two plates, an inside one with the punched hemispheres and an outside one with a repoussé decoration of an eagle holding a gazelle. An x-ray radiograph (Figure

6) provides most of the evidence that this bowl was made in an entirely different way. Silver metal was first shaped into a bowl, probably by hammering. Punches on the outside produced the hemispheres on the inside. The punch holes on the outside were then almost obliterated by shaving away the surrounding metal. The x-ray radiograph shows that the hemispheres in the background areas are entirely solid. Small cavities

Figure 4. (top) Sasanian silver gilt bowl with eagle and gazelle. The State Hermitage Museum, Leningrad, U.S.S.R. Figure 5. (bottom) Inside view of the silver gilt bowl shown in Figure 4.

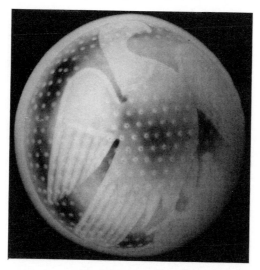

*Figure 6.   X-ray radiograph of the silver gilt
bowl shown in Figure 4.*

*The opacity of the hemispheres (white) indicates
that the bowl is solid. The black dots visible in
the center of the white circles on the eagle's
raised body but not visible in the white circles in
the background indicate that the hemispheres on
the inside of the bowl were obtained by punches
on the outside. When the outside surface was
carved down to show the eagle and gazelle in low
relief, polished, and decorated, the punched holes
in the background disappeared completely, leaving microscopic cavities on the eagle. Radiograph
obtained with Philips PG 200 x-ray unit, 200 kV,
3mA, no lead screens.*

still exist in the slightly thicker areas of the eagle but are visible only
through a microscope. The low-relief decoration of an eagle holding a
gazelle was created by carving away more silver from the background.
The decoration was completed by polishing, chasing, and partial gilding.

The incorrect description by Orbeli and Trever of Sasanian silver-
smithing techniques did not misinform the serious scholar alone; the
modern forger was fooled also. Many "Sasanian" silver objects exist that
were manufactured either by the double shell technique or by casting.
Since our study indicated that these techniques were not used by the
Sasanian silversmith, the authenticity of such objects should be considered very questionable. (Combined anomalies in style, iconography,
elemental analysis, method of manufacture, and corrosion should provide,
of course, the definitive evidence that these objects are not made by
a Sasanian silversmith.)

An authentic silver gilt plate is shown in Figure 7; its x-ray radiograph (Figure 8) is consistent with a plate shaped by hammering and

Figure 7. (top) *Sasanian silver gilt plate, The State Hermitage Museum, Leningrad, U.S.S.R.*
Figure 8. (bottom) *X-ray radiograph of Sasanian silver plate shown in Figure 7. Decorated sections appear as lighter areas indicating solid metal. This radiograph is typical for hammered Sasanian silver plates with a carved relief decoration; the light circle in the center is the radiographic impression of the foot ring. Radiograph obtained with Philips PG 200 x-ray unit, 200 kV, 3 mA, no screens.*

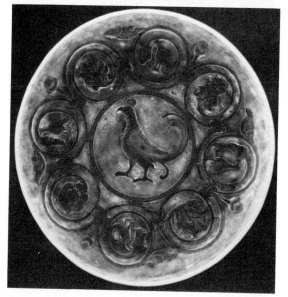

Figure 9. (top) Forgery of Sasanian silver gilt plate, private collection. Figure 10. (bottom) X-ray radiograph of forgery shown in Figure 9. Decorations appear as darker arecs indicating hollow area between two surfaces. This plate is manufactured by the double shell technique, a method not used by Sasanian silversmiths. Exposure: 200 kV, 5 mA, lead screens.

decorating by carving away the background around the design. A modern forgery of this plate is shown in Figure 9. The dark appearance of the decoration on the radiograph (Figure 10) indicates a hollow area between two surfaces and provides strong evidence for the double shell technique.

Cast Bronzes. The greatest potential of x-ray radiography for the study of archaeological materials probably lies in its application to cast bronzes. The following examples illustrate how certain details of bronze-casting procedures can be reconstructed based on x-ray images.

CHINESE BRONZE VESSEL. Scholars had never fully understood how the copper figures that decorate the surface of a Chinese vessel of the late Chou dynasty (Figure 11) were applied; "inlaid" hardly appeared to be a satisfactory description. (The questions regarding the method of manufacture of this object were brought to the attention of the Metropolitan Museum of Art Research Laboratory by W. T. Chase.)

Radiographs of sections of the vessel (Figure 12) provided reliable evidence for reconstructing the method of manufacture. The figures and decorations, visible on the surface of the bronze, were probably cut out of copper sheet. Attached to them are numerous square and rectangular supports. White spots on or near these supporting rods seem to indicate that they were once soldered to the copper figures. The figures with their supports were carefully placed between the inner and outer molds, with the copper figures inserted into or resting against the outer mold while the rods rested against the inner mold When the molten bronze was poured into the mold, the solder melted and formed little round globules. The temperature was not high enough to obtain fusion between molten bronze of copper figures or rods. The many voids, seen as dark areas in the radiograph, were caused by rapid cooling and solidification of the bronze around the relatively cool figures and supporting rods. The vessel was then undoubtedly polished to produce a smooth surface on which the "inlays" were level with the vessel wall. A more extensive report on this method of manufacture will be published elsewhere.

SUMERIAN BRONZE IBEX. The lost-wax process dates back to the early Bronze Age in the ancient Near East (fourth millenium B.C.), but the earliest use of ceramic or clay cores in bronze casting remains a mystery. This development in bronze casting technology is of interest because it not only saved expensive metal, but it also allowed the foundry master to produce better and more sophisticated castings. One of the earliest examples of a bronze figure cast around a ceramic core is a Sumerian statuette of an ibex (Figure 13). Its x-ray radiograph (Figure 14) indicates the sophistication of bronze casting during the middle of the third millenium B.C. The ibex was cast around a sausage-shaped ceramic core. During casting the core was held in place by two copper rods which can still be observed in the shoulder and in the haunch. The stem of the

Figure 11. Chinese Hu (ritual vessel), late Chou Dynasty, bronze with copper decoration. The Metropolitan Museum of Art, Acc. #29.100.545.

Figure 12. X-ray radiograph of section of the wall of the Hu shown in Figure 11. From the evidence seen in this radiograph—square and rectangular supports behind each figure, the white spots on or near each rod (solder), and dark areas (voids in metal) around each figure—the casting procedure described in the text can be deduced. Exposure: 275 kV, 4 mA, lead screens.

Figure 13. Ibex on stand, arseni-cal copper, Sumerian, ca. 2500 B.C. The Metropolitan Museum of Art, Acc. #1974.190.

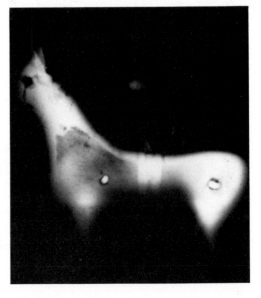

Figure 14. X-ray radiograph of ibex shown in Figure 13. The ibex is cast around a ceramic core. Two supporting metal rods are clearly visble. The stem of the superstructure ex-tends down through the core. The head was cast separately, attached at the neck by a tongue-in-groove method and secured through a metal rod. Exposure: 300 kV, 4 mA, lead screens.

structure above the animal's head extends through the entire core down to the outside wall of the belly. This may indicate that the ibex was cast around an already prepared superstructure. The head was cast separately because of failure to cast the head integrally with the ibex and was attached to the neck by an ingenious tongue-in-groove method. A metal pin, faintly visible in the radiograph, secured the neck to the vertical tongue of the head.

BRONZE HORSE. X-ray radiography clarified the technique used to produce the much-publicized Bronze Horse (Figure 15) in the collection of The Metropolitan Museum of Art and thus helped to authenticate the sculpture. This object had been considered a prime example of classic Greek bronze sculpture until, in 1967, Joseph V. Noble, then Vice-Director for Administration at the Museum, declared it to be a modern forgery (22) primarily because it had been fabricated by a modern sand casting technique. This conclusion was based on a study of gamma-ray radiographs of the horse (Figure 16) and observation of a casting fin caused by a piece mold. This casting fin was described as "a line, running from the tip of his nose up through his forelock, down the mane and back, up under the belly, and all the way around." As evidence of manufacture by the sand piece-mold process, visible in the radiograph, Noble mentioned the sand core, the iron wire running through the horse, and the ends of the transverse wires that also held the core.

An extensive technical study undertaken by the Museum's conservation department immediately after the announcement provided overwhelming evidence that the horse was not a modern forgery but was manufactured in antiquity (23, 24, 25). (The examination was conducted by Kate C. Lefferts, Lawrence J. Majewski, Edward V. Sayre, and the author.) The most recent thermoluminescence dating tests indicate a date of manufacture in the last five centuries B.C. (26). The early stages of the technical study revealed the horse to be covered with a thick layer of black and green pigmented wax, a common treatment for bronze objects in the 1920s and 1930s. When the wax was removed, the casting fin disappeared leaving a depressed line on the face extending from a spot between the eyes to the nose. All experts consulted have considered this to be not a mold mark but a deliberately sculptured line. The casting fin that disappeared with the wax coating was presumably caused by piece molds made for forming a reproduction cast of the horse.

The radiographic evidence strongly supports a lost-wax casting procedure and does not indicate a sand piece-mold process as demonstrated by the following observations. The iron armature is not an iron wire, as used in modern processes, but is an irregular band, approximately 12 mm wide and 2 mm thick (Figure 17). Similar armatures and also iron chaplets have been observed in many bronzes manufactured in

Figure 15 (above); *Figure 16* (below). *Description on opposite page.*

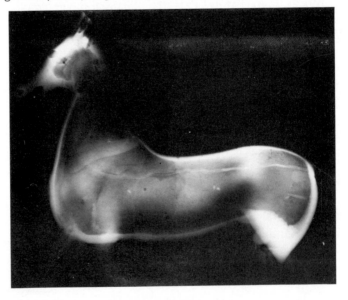

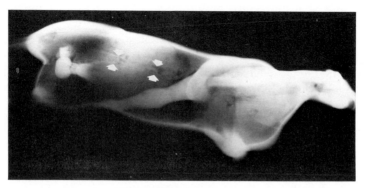

*Figure 15.   (opposite top) Bronze Horse, Hellenistic, Roman (?).
The Metropolitan Museum of Art, Acc. #23.69.   Figure 16.   (opposite, bottom) Gamma-ray radiograph of the Bronze Horse using
^{192}Ir.   The modeled ceramic core, the iron armature (see also Figure 17), and the iron chaplets all indicate lost-wax casting. Density
variations particularly visible in the neck area correspond to thickness variations in the metal and indicate that the artist shaved
away the clay core while working on the wax model to avoid making the metal wall of the bronze figure too thin.   Figure 17.
(above) Gamma-ray radiograph, using ^{137}Cs, of the Bronze Horse
obtained by placing the film behind the horse's back. The armature, previously described as a wire, is shown to be an irregularly
shaped band approximately 12 mm wide (arrows).*

classical antiquity. Moreover, in modern casting procedures, armatures,
chaplets, and core material are withdrawn immediately after casting. The
use of iron armatures extending into three legs (the fourth leg is a cast-on
ancient repair) is inconsistent with the sand casting technique; in the
wax process the legs need support, but in the piece-mold sand casting
technique, there is no need for an armature. The density variations in the
radiograph (Figure 16), particularly visible in the area of the horse's
neck, correspond to thickness variations in the metal. They are hard to
explain in a piece-mold sand casting. In the lost-wax casting process,
however, the artist may have shaved away section of the clay core to
sustain reasonable wall thicknesses of his wax model. This would prevent difficulties in the casting because of thin walls and would result in
uneven wall thicknesses. The observed density variations in the x-ray
radiograph are entirely consistent with such thickness variations.

The radiographic study eliminated the most important arguments
used against the authenticity of the horse. However the fact that the
horse was proven to be cast with the lost-wax process did not auto-
matically reestablish its authenticity. Careful visual examinations,
elemental analyses, studies on corrosion products, metallurgical investigations, and various other technical studies were necessary to demonstrate
that all physical characteristics of the bronze were entirely consistent

with a date of manufacture in antiquity. Sophisticated thermoluminescence tests on samples from the core provided the definitive evidence
that the horse was produced in classical antiquity.

There is, however, still some ambiguity about the exact date of
manufacture. Some scholars believe that the horse was not made during
the fifth century B.C., the originally suggested date of manufacture, but
during the Hellenistic or Roman periods. At present, thermoluminescence
dating cannot distinguish between these periods, and unless these dating
techniques improve, one must rely on the art historian or archaeologist
for a more exact attribution.

Literature Cited

1. Burroughs, Alan, "Art Criticism from a Laboratory," Little, Brown and Co.,
   Boston, 1938.
2. Bridgman, Charles F., *Stud. Conserv.* (1964) **9**, 135–139.
3. Anon, "Radiography in Modern Industry," 2nd ed., Eastman Kodak Co.,
   X-Ray Division, Rochester, 1957.
4. Bridgman, Charles F., *IIC-AG Bull.* (1968) **9**(1), 23–94.
5. Heinemann, S., *Am. Antiq.* (1976) **41**(1), 106–111.
6. Ryan, M. C., *Mater. Eval.* (1968) **26**(8), 159–162.
7. Beyer, N. S., Staroba, J. S., *Mater. Eval.* (1968) **26**(8), 167–172.
8. White, Mary Lou, *ICOM Comm. For Cons.*, 4th Meeting, pp. 75/9/2-1
   to -4 (1975).
9. Bridgman, Charles F., *Expedition* (1973) **15**(3), 2–14.
10. Gettens, Rutherford John, "The Freer Chinese Bronzes: Volume II, Tech-
    nical Studies," Oriental Studies, No. 7, Smithsonian Institution, Freer
    Gallery of Art, Washington, 1969.
11. Gorelick, L., *Dent. Radiogr. Photogr.* (1975) **48**(1), 17–21.
12. Bertrand, A., Legrand, C., Farges, P., *Bull. Soc. Fr. Ceram.* (1962) **55**,
    45–54.
13. Gugel, E., Dichtl, H., Mitsche, R., Kupzog, E., *Ber. Dtsch. Keram. Ges.*
    (1969) **5**(46), 254–259.
14. Harris, James E., Weeks, Kent R., "X-Raying the Pharaohs," Scribners,
    New York, 1973.
15. Stone, Richard E., private communication, unpublished data.
16. Pease, Murray, *Metrop. Mus. Art Bull.* (1958) **16**(8), 236–240.
17. Beck, Curt W., ed., "Archaeological Chemistry," ADV. CHEM. SER. (1974)
    138, 11–21.
18. Ibid., 22–33.
19. Ibid., 124–147.
20. Orbeli, G., Trevers, C., "Orfèverie Sasanide," Academie, Moscou-Leningrad,
    1935.
21. Frantz, J. H., Meyers, P., unpublished data.
22. Noble, Joseph V., *Metrop. Mus. Art Bull.* (1968) **26**(6), 253–256.
23. Lefferts, K. C., Majewski, L. J., Sayre, E. V., Meyers, P., *Metrop. Mus. J.*,
    to be published.
24. Zimmerman, D., Yuhas, M., Meyers, P., *Archaeometry* (1974) **16**(1),
    19–30.
25. Shirey, David L., *The New York Times* (Sunday, Dec. 24, 1972) B37.
26. Zimmerman, D., private communication.

RECEIVED September 26, 1977.

Organic Materials

6

Trace Element Analysis in the Characterization of Archaeological Bone

G. WESSEN, F. H. RUDDY,[1] C. E. GUSTAFSON, and H. IRWIN[2]

Departments of Chemistry and Anthropology, Washington State University, Pullman, WA 99163

Trace element analysis using neutron activation has been used to characterize archaeological bone. The alkaline earth elements strontium and barium appear to be reliable indicators of bone origin. Studies of recently killed specimens suggest that these elements are homogeneous throughout the skeletal matrix so that small samples may be regarded as representative of the entire organism. Alterations of elemental concentrations resulting from interactions of the sample with the depositional environment have been examined empirically by analyzing various samples in contact with contrasting depositional environments for different time periods. The results of the analysis of over 350 morphologically distinct specimens have provided identification criteria for archaeological artifacts made from bone of unknown origin.

Trace element analysis of selected archaeological mateials has been an expanding of enquiry since the early 1950s (*1*). The ability to characterize materials in terms of distinctive trace element content has provided the opportunity for detailed reconstruction of prehistoric economies and technologies in many parts of the world. In particular, such studies have been especially informative in the examination of trade and exchange networks and the growth of associated economic centers.

A number of inorganic materials have provided useful trace element data including obsidian (*2, 3, 4, 5, 6*), ceramics (*7, 8*), coins (*9*), and

[1] Present address: Hanford Engineering Development Laboratory (Westinghouse Hanford), P.O. Box 1970, Richland, WA 99352.
[2] Deceased April 1978.

other metal objects (1). However, the application of trace element studies to organic materials such as bone has been particularly limited (10, 11, 12, 13). The analysis of archaeological bone presents several problems which are encountered to a much lesser degree in the analysis of inorganic substances where often a discrete source is involved and where the matrix is not as open to contamination by the depositional environment. Nevertheless, bone is one of the most commonly found archaeological materials, and any inferences made from its trace elemental composition would certainly be useful. We relate herein our experiences with the trace element analysis of archaeological bone, the problems encountered in these analyses, and some of the conclusions that we have reached as the result of our measurements.

Experimental Considerations

Trace element data may be useful for characterization purposes provided that the following criteria are met:

(1) The concentrations measured are homogeneous throughout the entire matrix and are characteristic of the organism studied.

(2) The concentrations measured are not altered by contamination.

(3) Trace elements have not been lost through interaction with the depositional environment.

(4) The variations of trace element content may be reasonably and reliably interpreted.

In order to satisfy the first condition, trace element concentrations were measured in samples from various portions of single bones and throughout the skeletons of recently killed specimens. Conditions two and three seemed to be met best by choosing elements that were chemically similar to the major elements in the matrix. Calcium is a major constituent of bone, and elements from the alkaline earth group were chosen for study. Attempts to measure magnesium were unsuccessful, and attention was focused on barium and strontium which were measureable in most samples. Neither barium nor strontium has a demonstrated nutritional or metabolic role so they both may be regarded as biologically inert. Both are calcium analogs and are are found mainly in bone. The fact that environmental availabilities of strontium and barium are variable (14, 15), lead to optimism that the choice of these elements might satisfy condition four.

The method of analysis used is neutron activation. The samples are irradiated for 10 min in a thermal neutron flux of 6×10^{12} n/(cm^2 sec) in the Washington State University TRIGA Mark III reactor. After 10 min have been allowed for decay of short-lived activities, the samples are assayed for gamma-ray activity using a Ge(Li) detector whose signals are monitored by a 1024-channel ND-160 analyzer. The analyzer is interfaced to a PDP-15 computer which is used to find the peaks in the gamma ray spectra and to determine the intensities by numerical peak integrations. The amounts of the elements present are determined by comparison with known standards containing a measured amount of the elements. A typical standard consisted of 1.114 mg of calcium, 802 μg of strontium, and 650 μg of barium as the nitrate salts evaporated to dryness from standard solutions in the bottom of an irradiation vial. Seven samples and one standard are irradiated simultaneously. Immediately after cooling the samples and standards are counted for 2 min each to determine the activity of 8.80-min ^{49}Ca ($E\gamma = 3084.4$ keV). The calcium concentrations are roughly constant in the range 15–25% by weight and indicate metabolic imbalance or other abnormalities in a sample. After the calcium counts a longer count (8–16 min) at a higher amplifier gain setting determines the activities caused by 170-min ^{87m}Sr ($E\gamma = 388.5$ keV) and 82.8-min ^{139}Ba ($E\gamma = 165.8$ keV).

Special consideration has been paid to sample preparation to minimize problems resulting from loss of trace elements to or contamination by the depositional environment. The isotopes ^{56}Mn and ^{24}Na have been particularly troublesome as background activities. According to the x-ray microprobe work of Parker and Toots (16) sodium appears to be associated with the apatite part of the bone (probably substituting for calcium in apatite), whereas manganese tends to occur in voids and fractures. It was hoped that sample pretreatment might lower the concentration of elements such as manganese that are probably present from contamination by the depositional environment. We found that all treatments short of washing with strong mineral acids did not reduce substantially the manganese content of the samples. Although mineral acids did reduce the manganese content, the bone matrix deteriorated so that this procedure could not be used. We adopted a procedure that involves repeated ultrasonic agitation of the bone samples in demineralized water to remove contamination that might adhere to the surface of the bone and rejected any harsher chemical pretreatments.

The inadvertent presence of manganese in our samples prevented accurate measurement of the magnesium content of the bone since the 846.7-keV gamma ray of 2.58 hr ^{65}Mn cannot be resolved from the 843.8-keV gamma ray from 9.46 min ^{27}Mg. The 1014.5-keV gamma ray from ^{27}Mg is subject to a large background correction owing to the intense Compton continuum from the 1811.2-keV γ of ^{56}Mn. ^{27}Mg can only be measured in the presence of ^{56}Mn when the concentration of magnesium is many times greater than the concentration of manganese in the matrix. This was not the case in most of our samples. Compton interference from the 846.7-keV ^{56}Mn gamma ray provides background interference at the ^{87m}Sr and ^{139}Ba peak positions, occasionally completely obscuring the peak. [24]Na is also a source of background interference. Chemical separation of the elements of interest would eliminate these interference problems, but a major advantage of the method chosen is the large number of samples that can be processed in a short time. We are able to obtain calcium, barium, and strontium concentrations on seven samples per irradiation. Complete reduction of the data can be accomplished within hours of the irradiation, and several irradiations may be performed in a day. The use of chemical separations would lead to delay and would reduce the rate of sample analysis considerably.

The Poisson statistics of radioactive decay provide the major source of error in these analyses. Where the background interference was least, the standard deviation of the measured quantity of alkaline earth element is approximately 15%. In many cases, particularly when the alkaline earth metal concentration is low and the background is high, the standard deviation can be as much as 40%. Concentrations with greater uncertainties were reported as upper limits.

A 10-min irradiation followed by two counts of each sample appears to be the optimum choice for obtaining sensitivity for the elements of interest (Ba, Ca, Sr) while keeping the activity of the interfering elements (Mn, Na) as low as possible. Although strontium and barium are often determined by neutron activation analysis in much longer irradiations (8 hr or more), in the present application a longer irradiation could lead to more drastic interferences from other elements such as sodium and phosphorus because of the nature of the matrix being analyzed. Relatively small sample sizes are used (10–20 mg) which can be an advantage in the analysis of archaeological specimens.

Results and Discussion

In order to check the uniformity in the skeletal matrix for the elements studied, samples of a recently killed deer taken at Potlach, ID and a recently killed fur seal from Cape Alava, WA were analyzed. A total of 34 samples from different portions of the two skeletons yielded the results shown in Figures 1 and 2 (13, 17). The average concentrations and ranges for calcium, strontium, and barium are shown in Table I. The concentrations of calcium and strontium showed little variation within the two skeletons. However, the barium concentrations were an average of 30 times higher in the deer than in the fur seal. This variation was evident (but to a lesser degree) in samples analyzed at three different Northwest coast archaeological sites (13). On the basis of these results, we conclude that a single bone specimen is representative of the entire skeleton in trace element concentrations and that, at least in the case of barium, characteristic variations in concentrations do exist.

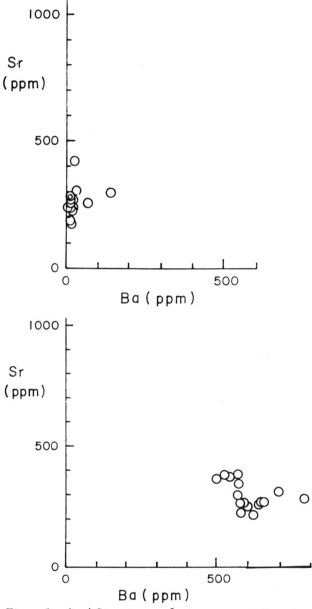

Figure 1. (top) *Strontium vs. barium concentrations for a recently killed Cape Alava, WA fur seal. All concentrations are reported on a weight ppm (µg/g) basis.*

Figure 2. (bottom) *Strontium vs. barium concentrations for a recently killed Potlatch, ID deer*

Table I. Barium, Calcium, and Strontium Concentrations from Recently Killed Specimens of Fur Seal and Deer

Species	Element	Average Concentration[a]	Low	High
Fur seal	Ca	19.3%	14.7%	25.8%
(*Callorhinus*	Sr	258 ppm	174 ppm	433 ppm
ursinus)	Ba	20.5 ppm	8.6 ppm	139 ppm
Deer	Ca	15.9%	13.8%	21.9%
(*Odocoileus*	Sr	302 ppm	217 ppm	392 ppm
hemionus)	Ba	601 ppm	494 ppm	780 ppm

[a] Elemental concentrations are reported as percent by weight or weight ppm ($\mu g/g$).

No matter how carefully prior contamination from the depositional environment is removed or contamination of the samples during analysis is avoided, altered concentrations could still be observed. Therefore, experiments were conducted to determine empirically the nature and extent of variation in concentrations that could be attributed to post-mortem conditions related to the depositional environment.

In the first experiment samples of elk and bison bones from two ar-chaeological sites with different ages were studied. The first site was Wawawii (45-Ga-17), a late prehistoric site near Central Ferry, WA with associated radiocarbon dates of approximately 2000 B.P. The second site examined was Lind Coulee (45-Gr-92), an early postglacial site near Warden, WA with associated radiocarbon dates of approximately 9000 B.P. The results of the analyses for strontium and barium are shown in Figures 3 and 4. These two populations represent the same types of organisms in similar environments, the main difference between the two sites being age. The Wawawii results show a variation in both strontium and barium concentrations that is consistent with data from several archaeological sites (*13, 16, 17*). The Lind Coulee results, on the other hand, show extreme internal variations in both strontium and barium concentrations. At present neither elk nor bison occur in the immediate vicinity of these sites. The Wawawii elk are consistent with modern elk from nearby Northern Idaho (*13, 16*). However, only the lower end of the Lind Coulee distribution is similarly consistent.

In order to examine the possibility of a time-related depositional effect, we analyzed a series of samples from throughout a stratigraphic column at Marmes Rockshelter (45-Fr-50), a site located approximately halfway between Wawawii and Lind Coulee. At Marmes we were able to sample deer and pronghorn antelope throughout a period of more than 7000 years. The results appear in Figures 5 and 6. No significant time-related variation was indicated for either strontium or barium.

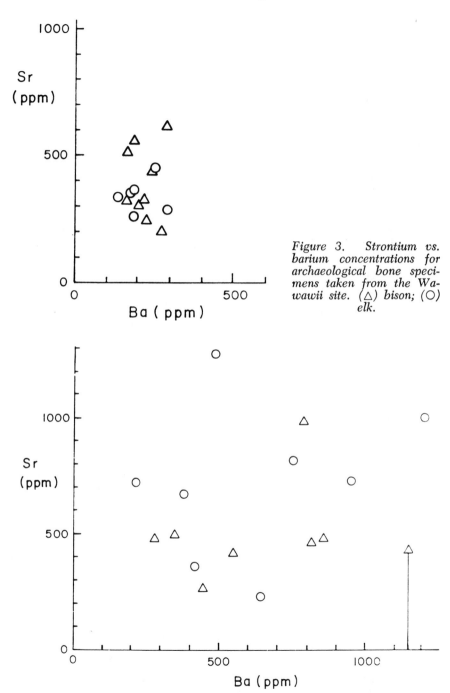

Figure 3. *Strontium vs. barium concentrations for archaeological bone specimens taken from the Wawawii site. (△) bison; (○) elk.*

Figure 4. *Strontium vs. barium concentrations for archaeological bone specimens taken from the Lind Coulee site. (△), bison; (○) elk.*

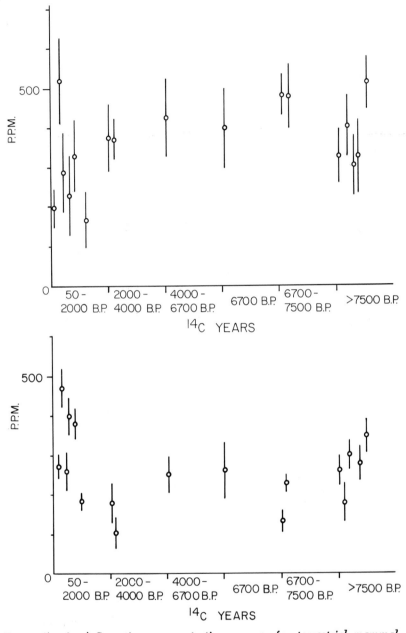

Figure 5. (top) *Strontium concentration vs. age for terrestrial mammal bone samples from the Marmes Rockshelter. Each point represents a single sample. The error bars represent the standard deviation based on the statistics of radioactive decay.*

Figure 6. (bottom) *Barium concentration vs. age for terrestrial mammal bone sample from the Marmes Rockshelter*

Nonetheless, we still find the differences between the Wawawii and Lind Coulee distributions disturbing. The fact that depositional alteration does not appear to occur at one or even many localities does not preclude its occurrence elsewhere. It appears to us that depositional alteration, if it occurs, is more likely to be a factor with older materials and/or when represented material covers an extended time interval. We do believe, however, that this matter warrants further study, and we are continuing to examine it. A basic precaution in our archaeological studies has been to work within single stratigraphic units or at least within single localities. Differences can exist in the range of concentrations for the same species at different localities (*13, 17, 18*). Caution must be used when comparing distributions between localities.

Summary and Conclusions

Regardless of the seemingly insurmountable difficulties associated with extracting meaningful trace element data from samples taken from archaeological sites, useful results have been obtained. We have determined that barium is a useful indicator of environment and possibly diet but that strontium seems to be much less useful in this regard (*18*). The characteristic differences in barium concentration have already proven useful in an archaeological context. Artifactual bone material consisting of harpoon valves from the Ozette archaeological site (45-Ca-24) on the northwest Washington coast have been analyzed for strontium and barium content. It was shown (*13*) on the basis of the observed barium concentrations that seven of the 12 harpoon valves analyzed represented terrestrial mammal bone (deer or elk), but only two had concentrations consistent with marine mammal bone (seal, sea lion, walrus, whale). Since more than 90% of the bone at Ozette represents sea mammal (*19*), a clear selection process is evident. Reasons for such selection will doubtless require a broader range of experiments, but we believe that trace element studies have a real potential in this area.

The present method of analysis offers several distinct advantages. The first of these is speed; calcium, barium, and strontium concentrations are determined simultaneously in a short irradiation, and analysis of up to seven samples can be completed within hours. Neutron activation analysis is highly sensitive to the elements of interest compared with other methods such as x-ray fluorescence and atomic absorption techniques. Additionally, the required sample size is small (10–20 mg), thus resulting in little alteration of archaeological specimens.

We have analyzed over 350 samples of recent and archaeological bone and have found patterned relationships in the barium and strontium concentrations. Provided that suitable caution is exercised to exclude effects leading to portmortem alterations in the concentrations, such relationships may be used to investigate problems concerning prehistoric economy and technology.

Literature Cited

1. Perlman, I., Asaro, F., Michel, H. V., *Ann. Rev. Nucl. Sci.* (1972) **22**, 283.
2. Renfrew, C., Cann, J. R., Dixon, J. E., *Ann. Br. School Archaeol. Athens* (1965) **60**, 225.
3. Weaver, J. R., Stross, F. H., *Contr. Univ. Calif. Archaeol. Res. Facil.* (1965) **1**, 89.
4. Gordus, A. A., Wright, G. A., Griffen, J. B., *Science* (1968) **161**, 382.
5. Bowman, H. R., Asaro, F., Perlman, I., *Archaeometry* (1973) **15**, 1.
6. Nelson, D. E., D'Auria, J. M., Bennett, R. B., *Archaeometry* (1975) **17**, 1.
7. Perlman, I., Asaro, F., *Archaeometry* (1969) **11**, 21.
8. Harbottle, G., *Archaeometry* (1970) **12**, 23.
9. Ambrisino, G., Pindrus, P., *Rev. Mett.* (1953) **50**, 136.
10. Brown, A. B., *Contr. Geol.* (1974) **13**, 47.
11. Brown, A. B., Thesis, University of Michigan, Ann Arbor, MI, 1973.
12. Gilbert, R., Thesis, University of Massachusetts, Amherst, MA, 1975.
13. Wessen, G., Ruddy, F. H., Gustafson, C. E., Irwin H., *Archaeometry* (1977) **19**, 2.
14. Hill, M. N., "The Sea," Wiley-Interscience, New York, 1963.
15. Mitchel, R. L., "Trace Elements in Soil" in "Chemistry of the Soil," F. E. Bear, Ed., Reinhold, New York, 1964.
16. Parker, R. B., Toots, H., *Geological Soc. Am. Bull.* (1970) **81**, 925.
17. Wessen, G., Thesis, Washington State University, Pullman, WA, June 1975.
18. Wessen, G., Ruddy, F. H., Gustafson, C. E., Irwin, H., unpublished data.
19. Gustafson, C. E., *Science* (1968) **161**, 49.

RECEIVED September 15, 1977.

Amino Acid Analysis in Radiocarbon Dating of Bone Collagen

AFIFA A. HASSAN

Department of Geology, Washington State University, Pullman, WA 99164

P. E. HARE

Geophysical Laboratory, Carnegie Institution of Washington, Washington, DC 20008

Erroneous radiocarbon dates on bone collagen may result from impurities remaining after sample processing. Amino acid analysis and a determination of the nitrogen/carbon ratio can quickly test the purity of the extracted collagen to indicate if further purification is required. Collagen is separated from bone by dialysis in dilute HCl and in distilled water. The collagen is then converted to gelatin. The results indicate the presence of non-collagen organic materials in fossil and in modern bones. When these materials are indigenous to the bone and not contaminants, they do not affect the radiocarbon dates. The amino acid composition of "collagen" residue from some fossil bones indicates that the residue is not collagen. Radiocarbon dates on non-collagen material extracted from fossil bone should be interpreted with caution.

Bone tissue is composed of inorganic and organic material. The main constituent of the organic fraction is the protein collagen, and the inorganic fraction is composed of the bone mineral apatite. Collagen, in contrast with the bone mineral, does not exchange carbon with the bone diagenetic environment. Thus, collagen is a suitable dating material if it is uncontaminated. Many investigators have devised methods to separate collagen (*1–11*). In most of these methods, however, the purity of the separated collagen is questionable. The method of selective liquid chromatography of amino acids (*2, 9*) seems to eliminate the problem of

0-8412-0397-0/78/33-171-**109**$05.00/1

impurities; however, the separated amino acids could be from other sources besides collagen. It would be safe to date the amino acids that exist only in collagen, but that is not practical since it requires an enormous amount of bone sample. Also the amount of collagen left after sample preparation is often insufficient for radiocarbon dating because of the breakdown and loss of collagen as HCl-soluble material. If the HCl-soluble fraction of collagen could be recovered, it would be easier to obtain sufficient quantity for radiocarbon dating.

Dialysis of bone in dilute HCl retains the HCl-soluble fraction of collagen inside the dialysis tube, thus improving the collagen yield (11). Conversion to gelatin (8) removes the water-insoluble non-collagenous material, but it is still necessary to check the separated collagen for the presence of water-soluble non-collagenous material of high molecular weight (11).

In this study, the modern and fossil bone collagen separated by dialysis is analyzed quantitatively for amino acids and the nitrogen/carbon ratio to determine if the separated material is pure collagen or if it contains impurities which might affect radiocarbon dating. The fossil bone specimens with associated ages are: *Bison antiquus* (10,000–11,000 B.P.) Folsom site, NM; *Mammuthus columb.* femur and rib (mammoth III) (11,170 ± 360 B.P.) Blackwater Draw, Clovis level, NM; *Mammuthus imperator* (11,045 ± 647 B.P.) Domebo site, Clovis level, OK; *Mammut americanum* (16,540 ± 170) Boney Springs, MO; *Mammut americanum* (34,300 ± 1,200 B.P.) Trolinger bog, MO; and whale bone ($\sim$ 100,000 B.P.) Baffin Island, Canada, Early Wisconsin. Fossil bones and their expected ages, except the whale bone, were obtained from C. Vance Haynes, Jr. of the University of Arizona at Tucson. The whale bone was obtained from Gifford Miller of the University of Colorado. The modern bones belong to an elephant that died in 1964 and was buried for six years prior to re-excavation. The elephant bones were obtained from Charles McNulty of the University of Texas at Arlington.

Experimental

Nitrogen content of the different bone samples was determined by a carbon–nitrogen–hydrogen analyzer (Hewlett–Packard model 185). Nitrogen percentages which reflect the amino acid content of the samples are used to calculate the weight of the different samples required to obtain a sufficient amount of collagen. The collagen is separated by a method developed by Hassan (11) in which the crushed, sized, and calcite-free bone samples are placed in dialysis tubes of 12,000 mol wt, cut off, and filled with distilled water. The dialysis tubes, with their top end untied to permit the generated CO_2 to escape, are placed in long cylinders filled with 0.3*N* HCl. The HCl solution in the cylinders is changed frequently until the bone apatite is dissolved, after which the HCl is replaced with distilled water. The distilled water is changed frequently until all the Ca^{2+} from the apatite is removed from the dialysis tube. The high-molecular-weight fractions of collagen (HCl-insoluble and HCl-soluble) are retained. Small organic molecules also are removed from the dialysis tube while higher-molecular-weight materials such as humates, proteins, and some polysaccharides are retained. The two fractions, HCl-soluble and HCl-insoluble, are separated by centrifugation, are converted separately to gelatin (100°C, pH 4, 2 hr), and are filtered through 0.20-μ cellulose filter paper. Conversion to gelatin and filtration removes solid contaminants such as water-insoluble humates.

After evaporation to dryness in a vacuum desiccator over NaOH the gelatin samples are weighed and the nitrogen and carbon content of the gelatin determined. An aliquot of the gelatin sample is hydrolyzed to free amino acids [6N HCl for 15 min at 155°C (*12*)]. Standard amino acid mixtures also are carried through the same procedure.

A sensitive amino acid analyzer based on the ion-exchange method of Spackman, Moore, and Stein (*13, 14*) was used to determine quantitatively the amounts of amino acids. The system is highly stable, sensitive, and capable of 1% reproducibility at the nanomole level on consecutive runs (*12*). The basic amino acids can be analyzed separately if desired on a single ion-exchange column using pH 4.4 citrate buffer at 50°C.

Table I.   Collagen,[a] Expressed as Weight % of Dry Bone and Its Nitrogen/Carbon Ratio

	% Collagen		Nitrogen/Carbon	
Bone Sample	*HCl-insol.*	*HCl-sol.*	*HCl-insol.*	*HCl-sol.*
Modern elephant	27.0	2.7	0.37	0.34
Folsom bison	4.5	2.4	0.35	0.36
Blackwater Draw mammoth	1.9	0.5	0.05	0.11
Blackwater Draw mammoth III	0.1	0.4	0.13	0.10
Domebo mammoth	1.7	2.6	0.34	0.17
Boney Springs mastodon	0.2	0.2	0.13	0.06
Trolinger bog mastadon	0.6	1.7	0.29	0.22
Whale bone	22.5	5.0	0.37	0.36

[a] Collagen is separated from bone by dialysis in HCl then is converted to gelatine.

Results

The weight of the separated collagen and its nitrogen/carbon ratios are recorded in Table I. These results indicate an increase in the percentage of the HCl-soluble fraction of collagen with fossilization. The results also indicate lower values of nitrogen/carbon ratios in the HCl-soluble fraction of collagen compared with those of the HCl-insoluble fraction. The nitrogen/carbon ratios are also lower in the fossil collagen compared with those of the modern collagen. Table I shows that the nitrogen/carbon ratio of the separated collagen generally is proportional to the percentage of the collagen in the bone samples; however, there are exceptions. For example the amount of the HCl-insoluble "collagen" separated from the Blackwater Draw mammoth bone sample is more than expected based on the nitrogen/carbon ratio which may indicate the presence of non-collagen contaminants. This sample was covered in the field with Alvar preservative which was scraped off during sample pretreatment.

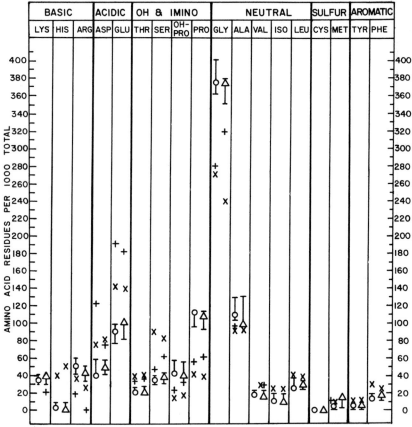

Fig. 2. A plot of the relative proportions of the amino acids in bone collagen.
HCL-INSOLUBLE COLLAGEN: O=MODERN, I=ALL FOSSILS, +=BLACKWATER DRAW,
X=BONEY SPRINGS
HCL – SOLUBLE COLLAGEN: △=MODERN, I=ALL FOSSILS, +=BLACKWATER DRAW,
X=BONEY SPRINGS

Organic Geochemistry

Figure 1.   Relative proportions of the amino acids in bone collagen. HCl-insoluble collagen: (O) modern; (I) all fossils; (+) Blackwater Draw; (×) Boney Springs. HCl-soluble collagen: (△) modern; (I) all fossils; (+) Blackwater Draw; (×) Boney Springs (12).

The relative proportions of the amino acids in the fossil and modern bone collagen are plotted in Figure 1. The proportions of the amino acids of most of the fossil bones are similar to those of the modern bones and indicate the presence of collagen. The samples from Blackwater Draw and Boney Springs show the least amount of collagen. These two samples show relatively high proportions of aspartic, threonine, serine, glutamic, and lower proportions of OH-proline, arginine, proline, and glycine compared with the modern bone. To determine quantitatively the amount of organic impurities, the total amino acid compositions of the separated "collagens" were calculated as weight percentages. The quantitative amino acid analyses for the separated collagens and their radiocarbon dates (*11*) and/or their expected ages are recorded in Table II. In most cases the recovery of the amino acids is less than 100%. The proportion of the non-proteinaceous "contaminants" are higher in the fossils than in modern bone and higher in the HCl-soluble fraction than in the HCl-insoluble fraction of collagen.

These contaminants could be inorganic materials such as iron compounds complexed with non-collagen organic matter. To determine the percentages of non-collagen impurities, the amino acid contents of the separated collagens were calculated. The calculated percentages are derived from the nitrogen/carbon ratios of the collagen of the different samples (Table I) and the value 0.37 (the nitrogen/carbon ratios of the HCl-insoluble fraction of modern collagen which yield approximately 100% amino acids). The calculated percentages of amino acids and non-nitrogeneous organic materials are recorded in Table II. These organic materials coexist with both fractions of fossil collagen as well as with the HCl-soluble fraction of modern collagen. Proportionately, however, they make up a larger fraction of the total organic matter in the fossils.

Problems and Conclusions

The presence of organic impurities in fossil as well as in the HCl-soluble fraction of modern bone collagen indicates that part of these non-collagen organics may be a component of bone, possibly carbohydrates. It is possible that upon fossilization under certain environmental conditions of burial, the relative amount of these materials increases independently of age. This may result from bacterial reworking of the bone organic matter.

The small amount of organic impurities coexisting with the separated collagen from Folsom bison and whale bone indicates that the method used is successful in these two cases. Both of the collagen fractions, HCl-insoluble and HCl-soluble, are pure and could be used safely for radiocarbon dating. This is reflected by the good agreement between the

Table II.   Quantitative Amino Acid

Amino Acids as % Wt of Collagen	Modern Elephant		Folsom Bison		Blackwater Draw Mammoth	
	HCl-insol	HCl-sol	HCl-insol	HCl-sol	HCl-insol	HCl-sol
OH–pro	5.1	1.8	7.1	5.1	<0.1	0.1
Aspartic	4.3	2.4	5.6	3.9	0.2	0.3
Threonine	2.3	0.8	2.3	1.8	<0.1	0.1
Serine	3.4	1.4	2.9	2.0	<0.1	0.2
Glutamic	12.0	5.2	11.0	8.4	0.3	0.8
Proline	12.0	4.3	10.8	8.2	0.1	0.2
Glycine	26.4	10.0	27.6	20.3	0.2	0.8
Alanine	9.3	3.2	4.6	7.3	0.1	0.3
Cystine	0	0	0	0	0	0
Valine	2.3	0.2	2.7	2.2	<0.1	0.1
Methionine	1.0	1.2	0.5	0.3	<0.1	<0.1
Isoleucine	1.7	0.6	2.3	1.7	<0.1	0.1
Leucine	3.0	1.4	3.8	3.1	<0.1	0.2
Tyrosine	1.2	0.4	0.4	0.2	<0.1	<0.1
Phenylalanine	2.1	1.0	1.2	1.4	<0.1	0.1
Histidine	0.5	0.1	0.5	0.5	<0.1	0.1
OH–lysine	n.d.	n.d.	n.d.	n.d.	n.d.	n.d.
Lysine	4.8	1.9	3.8	2.7	0.1	0.1
Ammonia	n.d.	n.d.	n.d.	n.d.	n.d.	n.d.
Arginine	8.5	2.7	9.1	6.0	<0.1	0
Total (measured)	99.9	38.6	97.1	75.1	1.2	3.4
% Amino acids (calculated)	100	92	95	97	14	30
% Other organics	0	8	5	3	86	70
Collagen ^{14}C date (B.P.) (11)	modern		10,314 ± 104[a]		1,674 ± 57 (?)[b]	
Expected age B.P. based on stratigraphy and/or ^{14}C dates on associated material	modern		10,000–11,000		11,170 ± 360[e]	

Radio carbon lab numbers: [a](SMU-179), [b](SMU-174), [c](SMU-176), [d](SMU-170), [e](A-481), [f](SM-695), [g](TX0-1478), [h](A-1080A).

radiocarbon date of the Folsom bison collagen and the expected age based on the association with known culture.

The erroneous carbon-14 age obtained from Blackwater Draw collagen could have been anticipated from the amino acid analyses. The relative proportions of the amino acids of this sample indicate that little or no collagen is present. The sample also contains a high percentage of organic impurities in the HCl-insoluble and HCl-soluble fractions. Part of these impurities could be Alvar, a commercial preservative used to cover the sample in the field which also dissolves in acidic boiling water. Thus, a purification method other than conversion to gelatin should have been used. There is no radiocarbon date on the Boney Springs collagen. However, the high percentage of impurities and the difference in the amino acid proportions of this sample compared with those of the modern collagen indicate that the radiocarbon dates will not be accurate without further purification.

In the sample from the Domebo site only the HCl-soluble fraction contains a relatively high proportion of non-collagenous organics. If

Analysis of Bone Collagen

Blackwater Dr. Mammoth III		Domebo Mammoth		Boney Springs Mastodon		Trolinger Bog Mastodon		Whale Bone	
HCl-insol	HCl-sol	HCl-insol	HCl-sol	HCl-insol	HCl-sol	HCl-insol	HCl-sol	HCl-insol	HCl-sol
0.8	0.3	6.0	0.7	<0.1	<0.1	4.7	0.4	5.9	4.6
1.1	0.7	4.8	0.8	0.1	0.2	3.4	0.5	4.8	5.5
0.5	0.2	2.3	0.3	0.1	0.1	1.9	0.2	2.9	2.8
0.7	0.3	2.7	0.4	0.2	0.2	2.4	0.2	3.8	3.8
2.0	1.4	10.0	1.6	0.4	0.3	7.6	1.0	10.9	9.6
1.7	0.7	11.1	1.3	0.1	0.1	7.5	0.7	12.0	10.9
4.3	1.6	26.1	3.1	0.4	0.3	18.8	1.7	25.1	23.2
1.6	0.8	9.9	1.2	0.1	0.2	7.9	0.7	8.6	8.0
0	0	0	0	0	0	0	0	0	0
4.0	0.2	2.3	0.3	<0.1	0.1	1.5	0.2	2.4	1.4
<0.1	0.1	0.2	<0.1	<0.1	<0.1	0.2	<0.1	0.6	0.5
0.3	0.1	1.7	0.2	0.1	0.1	1.7	0.1	2.4	1.9
0.5	0.2	3.0	0.4	0.1	0.1	3.4	0.3	3.1	3.1
0.1	<0.1	0.4	<0.1	<0.1	<0.1	0.2	<0.1	0.6	0.5
0.3	0.1	1.8	0.2	0.1	0.1	1.6	0.1	2.4	1.9
0.2	0.1	0.5	0.1	0.1	0.1	0.6	0.1	0.5	1.0
n.d.	n.d.	n.d.	n.d.	n.d.	n.d.	n.d.	n.d.	n.d.	n.d.
0.7	0.2	4.3	0.6	0.1	0.1	4.1	0.3	4.4	3.4
n.d.	n.d.	n.d.	n.d.	n.d.	n.d.	n.d.	n.d.	n.d.	n.d.
1.0	0.4	7.4	0.8	0.1	0.1	5.7	0.4	9.4	6.2
19.8	7.4	94.5	12.2	2.2	2.2	73.2	6.9	99.8	88.3
35.1	27	92	46	35.1	16.2	78	60	100	97.3
64.9	73	8	54	64.9	83.8	22	40	0	2.7
n.d.		9,295 ± 329[c]		n.d.		21,935 ± 2,334[d]		n.d.	
11,170 ± 360[e]		11,045 ± 647[f]		16,540 ± 170[g]		34,300 ± 1,200[h] (?)		100,000 (?)	

these analyses were obtained before the dating, it would have been advisable to date only the HCl-insoluble fraction or to date the two fractions separately for comparison. This may explain the difference between the date obtained from the two combined fractions and the date on bone organic matter obtained from 10% HCl treatment of bone in which no purification steps were taken. Thus, contamination with older organic material is possible.

The Trolinger bog sample is another example which shows the presence of these non-collagen organic materials in both fractions, although they are more abundant in the HCl-soluble fraction. These organic materials could possibly affect the radiocarbon date. However, there is no other date for comparison except that of an older peat unit in which the mastodon was buried.

Another aspect of this study is the racemization of the amino acid residues in the separated collagen fractions (15). Contamination is often cited if the extent of racemization is lower than expected. We have found that the non-collagen fraction is often more highly racemized than is the collagen fraction, and furthermore there is evidence of vastly different rates of racemization in the initial stages of bone diagenesis compared with later stages when intact collagen changes to soluble collagen and gelatin. Accumulated data on racemization of amino acids residues in collagen and non-collagen protein components will help to distinguish contamination from ground water and microbiological alteration from in situ bone protein material.

In conclusion, nitrogen/carbon ratios combined with quantitative amino acid analyses could determine the level of impurities that may co-exist with fossil bone collagen and could help in selecting the optimum method of collagen separation. An extraction method may be successful in some cases but could fail to remove the impurities from bone collagen in other samples. Chemical analysis of the impurities and their radiocarbon dates also should be obtained.

Acknowledgments

The authors would like to thank C. Vance Haynes, Jr., Gifford Miller, and Charles McNulty for providing the samples; Scott Cornelius for drafting the figures; and B. J. Arrasmith for typing the manuscript.

Literature Cited

1. Berger, R., Horney, A. G., Libby, W. F., *Science* (1964) **144**, 999.
2. Ho, T. Y., Marcus, L. F., Berger, R., *Science* (1969) **164**, 1051.
3. Tamers, M. A., Pearson, F. J., Jr., *Nature* (1965) **208**, 1053.
4. Krueger, H. A., *6th Int. Conf. Radiocarbon and Tritium Dating Proc.*, Pullman, Washington (1965) 332.
5. Sellstedt, H., Engstrand, L., Gejvall, H. G., *Nature* (1966) **212**, 572.
6. Haynes, C. V., Jr., "Radioactive Dating and Methods of Low-Level Counting," *IAEA* (1967) 163.
7. Berger, R., Protsch, R., Reynolds, R., Rozaire, C., Sackett, J. R., Contribution of the Univ. of California Archaeol. Research Facility, Berkeley (1971) **12**, 43.
8. Longin, R., *Nature* (1971) **230**, 241.
9. Protsch, R., Berger, R., *Science* (1973) **179**, 235.
10. Olsson, I. N., El-Daoushy, M. F. A. F., Abd-El-Mageed, A. I., Klasson, M., *Geol. Foere. Stockholm Foerh.* (1974) **96**, 171.
11. Hassan, A. A., "Mineralogy and Geochemistry of Bone Material and Their Implications for Radiocarbon Dating," University Microfilm, Ann Arbor, 1976.
12. Hare, P. E., "Organic Geochemistry," G. Eglinton, M. Murphy, Eds., p. 438, Springer-Verlag, New York, 1969.
13. Hare, P. E., "Protein Sequence Determination," S. B. Needleman, Ed., 2nd ed., Springer-Verlag, Berlin and New York, 1975.
14. Spackman, D. H., Stein, W., Moore, S., *Anal. Chem.* (1958) **30**, 1190.
15. Unpublished data.

RECEIVED September 19, 1977.

Amino Acid Racemization Dating of Bone and Shell

PATRICIA M. MASTERS

Scripps Institution of Oceanography, University of California, San Diego, La Jolla, CA 92093

JEFFREY L. BADA

Scripps Institution of Oceanography and Institute of Marine Resources, University of California, San Diego, La Jolla, CA 92093

Amino acid racemization dating is a promising new technique for dating fossil materials of biological origin which are about 1000 to several hundred thousand years old. The analytical procedures used in racemization dating are described. Bone racemization dates are compared with independently deduced ages. The racemization rates derived from well dated fossil bones correlate strongly with the estimated temperature exposure of the various samples. The reliability of racemization dates on bone is compared with those on mollusc shell. In another application of racemization dating, D/L aspartic acid ratios in teeth from a medieval cemetery population in Czechoslovakia are used to determine the age at death. Six criteria can now be applied in judging the reliability of a racemization-deduced age.

During the past few years, a new method of dating fossil materials based on the racemization reaction of amino acids has been developed (*see* Refs. *1, 2, 3, 4* for reviews). This geochronological technique is the first one to be based on an organic diagenetic reaction. The racemization dating method uses the fact that the amino acids present in the proteins of living organisms, especially higher organisms, consist only of the L-enantiomers. However, over geologic time, these L-amino acids undergo slow racemization, producing the corresponding D-amino acids. Both L- and D-amino acids have been found in fossil materials, and the

0-8412-0397-0/78/33-171-**117**$05.50/1
© 1978 American Chemical Society

D/L amino acid ratio increases with the age of the sample until a racemic mixture is eventually produced. The length of time required for the different amino acids to reach equilibrium varies because each amino acid has a characteristic racemization rate which can be predicted from the racemization mechanism. The extent of racemization of amino acids has been used to estimate the ages of various fossil materials. Although the racemization dating technique has been applied primarily to fossil bones and shells, it has also been used to estimate the ages of fossil wood (5), coprolites (6), and deep-sea sediments (7, 8, 9).

Recently, racemization also has been found to take place in several metabolically stable proteins in living mammals (10, 11, 12). The extent of racemization in these proteins provides a means for estimating the ages of natural populations of mammals and may also have implications concerning the aging process (13).

Since racemization is a chemical reaction, the reaction rate is temperature dependent. Thus, in order to date a fossil using racemization, it is necessary to evaluate the average temperature to which the sample has been exposed. This temperature can be evaluated using a procedure in which the in situ rate of amino acid racemization for a particular site is calculated by measuring the extent of racemization in a fossil which has been previously dated either by radiocarbon or by some othe independent dating technique (14, 15, 16, 17, 18). After this calibration has been carried out, other fossils in this same general area can be dated based on their extent of amino acid racemization. Ages of fossil bones determined using this calibration procedure have agreed closely with ages deduced using carbon-14 or other chronological information (15, 17, 18). This calibration procedure has also been used to date fossil woods (5) and, recently, shells (19, 20).

Amino acid racemization dating has been particularly useful in estimating the ages of fossil bones (18). The method has an effective dating range beginning at a few thousand years B.P. (before present) and extending to several hundred thousand years B.P., the actual range depending upon the temperature of the region where the bone was found. Only a few grams of bone are required for amino acid racemization analysis. By comparison, the radiocarbon dating limit is $\sim 40,000$ years, and quantities of up to a kilogram of bone are needed for a collagen-based carbon-14 analysis.

Of the various amino acids, aspartic acid has been the most widely used in fossil bone dating. This amino acid has one of the fastest racemization rates of the stable amino acids (21, 22). At 20°C in bone, the half-life for aspartic acid racemization is $\sim 15,000$ years. Thus, for most mid- or low-latitude sites the racemization rate of aspartic acid is much slower than the decay rate of radiocarbon and therefore can be used to date bones which are too old to be dated by carbon-14.

Although the first demonstration that amino acid racemization took place in fossils used mollusc shells (*23*), the application of this reaction in dating these materials has been extensively investigated only recently (*19, 20*). Work on *Mercenaria* (*19*), *Chione* (*20*), and other species (*24*) has tested the application of racemization dating to fossil mollusc shells from geological contexts and Indian shell middens. These and other studies have shown that there are problems with amino acid racemization dating of carbonaceous fossils which are not encountered with bone. Reversible first-order racemization kinetics which are observed in bone (*25*) are not found in *Chione* shell (*20*), foraminifera (*8, 26*), or calcareous sediments (*7, 9*). Moreover, species differences in racemization rates are found both in molluscs (*20, 24*) and in foraminiferal tests (*26*).

Experimental Methods

Fossil bone (usually ~ 5 g) is prepared for analysis by cleaning the fragment in dilute HCl solutions with ultrasonication. Organic glues and preservatives are removed by Soxhlet extraction in diethyl ether for 2–4 weeks. After cleaning, the sample is hydrolyzed in excess double-distilled 6M HCl for 24 hr. The HCl solution is then evaporated to dryness and the residue redissolved in double-distilled water. The sample then is desalted on Dowex 50W-X8 (100-200 mesh) resin which has been cleaned with NaOH and protonated with double-distilled HCl (*14*). Amino acids are eluted from the column with 1.5M NH$_4$OH. The effluent is brought to dryness in a rotary evaporator at ~ 50°C under partial vacuum. Aspartic acid is separated by chromatography on a calibrated column of BioRad AG1-X8, 100–200 mesh, anion exchange resin. The resin is regenerated with four column volumes of 1M sodium acetate followed by two column volumes of 1M acetic acid. One half of the sample is applied to the column, and elution is carried out with 1M acetic acid (*27*). The aspartic acid fraction is evaporated using the rotary evaporator. The L-leucyl–DL-aspartic acid dipeptides are synthesized according to the procedure of Manning and Moore (*28*), and the D/L aspartic acid ratio is determined on a Beckman model 118 automatic amino acid analyzer (*14*).

The D/L enantiomeric ratios for alanine, glutamic acid, and leucine are obtained by a gas chromatographic method (*29*), using the remaining half of the desalted amino acid fraction. The N-trifluoroacetyl-L-prolyl-DL-amino acid methyl esters are synthesized, and separation is performed on a Hewlett-Packard model 5711A gas chromatograph with flame ionization detector and a 20-ft column using 8% SP 2250 coated on Chromosorb W-AW-DMCS solid support.

Amino acid racemization analyses on shell are carried out in two ways: sampling all the amino acids extracted from the shell (called total) and sampling only the protein-bound amino acids (referred to as protein). The first method is essentially identical to the processing procedure developed to analyze fossil bone. In this case, all the amino acids within the shell—whether they occur as free amino acids, in small peptides, or in proteins—are included in the final fraction which is analyzed. The second method separates the higher-molecular-weight proteins from the small peptides and free amino acids, thus resulting in a more homogeneous fraction which is then broken down by 6M HCl hydrolysis into amino acids for analysis.

First, ~ 20 g of shell are cleaned thoroughly by sonication in double-distilled water. The outer layers of the shell are dissolved in 2M HCl, and then the remainder of the shell is dissolved completely in 6M HCl. Because of the amount of calcium carbonate present in the shell, the pH of the solution is raised during this step. It is thus necessary to evaporate the samples to ~ 20 mL in a rotary evaporator. (Complete dryness is not possible, because of the hydroscopic property of the $CaCl_2$ formed.) The sample is resuspended in 250 mL of 6M HCl and is hydrolyzed for 24 hr. The sample is brought nearly to dryness again in a rotary evaporator, is diluted with double-distilled water, and is desalted as described above.

The alloisoleucine/isoleucine (alleu/iso) ratio is determined directly on the automatic amino acid analyzer, and D/L enantiomeric ratios for additional amino acids are obtained by gas chromatography (above).

For an analysis of the protein fraction, the shell is cleaned as described above. It is then dissolved in 6M HCl at 4°C to avoid hydrolyzing the peptide bonds. This solution is reduced to a small volume with the rotary evaporator at low temperature ($< 50°C$), is diluted 1:1 with double-distilled water, and is dialyzed exhaustively against water. The dialysate is brought to dryness in the rotary evaporator and resuspended in 6M HCl. Hydrolysis and subsequent processing of the sample follows the procedures described above.

Amino Acid Racemization Dating of Fossil Bone

Aspartic acid has been the amino acid most extensively used in racemization dating of fossil bone. The reaction can be written as:

$$L\text{-aspartic acid} \underset{k_{asp}}{\overset{k_{asp}}{\rightleftharpoons}} D\text{-aspartic acid,} \tag{1}$$

where k_{asp} is the first-order rate constant for interconversion of the L- and D-enantiomers. Racemization ages are calculated from the equation [see Refs. 9 and 25 for derivation]:

$$\ln\left[\frac{1 + D/L}{1 - D/L}\right] - \ln\left[\frac{1 + D/L}{1 - D/L}\right]_{t=0} = 2k_{asp}t, \tag{2}$$

where t = time in years and D/L is the aspartic acid enantiomeric ratio in the bone sample. The $t = 0$ term is 0.14 in modern bone hydrolyzed for 24 hr.

Comparisons of Racemization and Independently Derived Dates. Amino acid racemization dates on bone have been reported for 36 sites throughout the world (*14–18, 30–32*). At 25 of these localities, ages derived from independent evidence such as radiocarbon dating, geological interpretation, and historical records were available for comparison and agreed well with the racemization ages. Some of these comparisons are listed in Table I.

In only two instances was there a large disagreement between racemization-deduced ages and those derived from other information. One of these disparate results came from Olduvai Gorge in East Africa. In this particular case, the racemization age is compatible with the age estimated from archaeological and geological evidence, but the radiocarbon age deduced on the same sample is much too young [*see* discussion in Ref. *18*]. In some instances where bones are so badly contaminated that the radiocarbon dates are essentially meaningless, the racemization ages may still provide a reasonable indication of the actual age of the sample.

A sample from Sakkara, Egypt has also yielded an anomalous racemization age (*30*). In this case, the racemization age is much too young compared with the actual historical age and is thus erroneous. Analyses of other Egyptian samples, however, have yielded racemization ages which agree closely with the historical ages of the samples (*30*). The reason for the anomalous Sakkara racemization age is not known, but this result indicates that when a racemization age is erroneous, it will be too young rather than too old. In other words, contamination of the indigenous amino acids is more likely to introduce younger (less racemized) amino acids as was also the case with *Australopithecus* fossils from Swartkrans (*22*).

Influence of Environmental Variables on Racemization. The accuracy of the racemization dating method depends to a large extent on the reliability of the calibration. The calibration procedure presupposes that physical and chemical variables in the environment are evaluated when k_{asp} is determined. This assumption has been tested by investigating the effects of pH, rainfall, and temperature on k_{asp}.

The effects of pH on racemization rates of free amino acids in aqueous solution are complex (*21*). However, when bone fragments were heated at 100°C in solutions of pH 2–9, the rate constant for aspartic acid racemization was essentially independent of pH (*37*). It was proposed that the phosphate from the bone's inorganic phase acts as a buffer. This same buffering action would take place in a bone under natural conditions of burial as well.

If the age of a fossil bone is known from other dating information, then the temperature can be deduced from the extent of racemization.

Table I. Comparisons between Aspartic Acid Racemization and

Site	D/L Aspartic Acid
Europe	
Muleta Cave, Mallorca, Spain (15)	0.273
	0.293
	0.455
Abri Pataud, Les Eyzies, France (15)	0.148
	0.178
Stránská skála, near Brno, Czechoslovakia	
layer no. 7, 5.0–5.1 m deep	0.512 }
layer no. 5, 5.5–5.6 m deep	0.574 }
Külna Cave, near Brno, Czechoslovakia	
layer no. 7a, 3.8–4.0 m deep	0.072[b]
layer no. 11, 7.0–7.2 m deep	0.092[b]
Middle East	
Cayönu, Turkey (18)	0.169
Sarab, Iran	0.173
Asiab, Iran (near Sarab)	0.155
	0.167
Tarkhan, near Cairo	
F 1516	0.450[e,f] }
F 1691	0.475[e,g], 0.465[e,h] }
F 1556	0.453 }
Sakkara, near Cairo	0.137[b]
Tura, near Cairo	0.416
South African Coast	
Nelson Bay Cave (17)	0.151
	0.167
Klasies River Mouth Cave 1 (17)	
level 13[i]	0.370 }
level 16	0.467 }
level 18	0.474 }
level 19	0.548 }
Swartklip I (33)	0.55
North America	
Laguna Woman, Laguna Beach, CA (16)	0.25
Los Angeles Man, near Baldwin Hills,	0.35
Los Angeles, CA (16)	
Murray Springs, AZ (15)	0.33
	0.52
Double Adobe, AZ (18)	0.50

[a] Except as indicated, samples hydrolyzed 24 hr. [b] Sample hydrolyzed 6 hr. D/L aspartic acid ratio for modern bone hydrolyzed 6 hr = 0.026 (33); thus, $t = 0$ term used in Equation 2 for these samples is 0.052. [e] Sample hydrolyzed 4 hr. D/L aspartic acid ratio for modern bone hydrolyzed 4 hr = 0.023 (33); thus, $t = 0$ term used in Equation 2 for these samples is 0.046.

Either Radiocarbon, Historical, or Geologically-Inferred Ages [a]

Radiocarbon, Historical, or Geologically Inferred Age (yr)	Aspartic Acid Age (yr)
16,850 ± 200 UCLA 1704D	$k_{asp} = 1.25 \times 10^{-5}$ yr^{-1}
18,980 ± 200 UCLA 1704E	18,600
28,600 ± 600 UCLA 1704A	33,700
23,010 ± 170 GrN 4721	$k_{asp} = 3.41 \times 10^{-6}$ yr^{-1}
33,260 ± 425 GrN 4719	32,100
690,000 [d]	$k_{asp} = 7.7 \times 10^{-7}$ yr^{-1}
45,600 + 2850 or −2200 GrN 6060	50,000
Last Interglacial (?) [e]	80,000
8,340 UCLA 1703B	$k_{asp} = 1.21 \times 10^{-5}$ yr^{-1}
7,620 UCLA 1703C	8,700
7,620 ± 70 UCLA 1714A	$k_{asp} = 1.14 \times 10^{-5}$ yr^{-1}
8,700 ± 100 UCLA 1714C	8,600
1st and 2nd Dynasty	$k_{asp} = 9.31 \times 10^{-5}$ yr^{-1}
3200–2700 B.C.	3180 B.C.
	2910 B.C.
4th Dynasty	760 A.D. (?)
2500–2400 B.C.	
Middle Kingdom	2030 B.C.
2100–1600 B.C.	
16,700 ± 240 I-6516	$k_{asp} = 4.92 \times 10^{-6}$ yr^{-1}
18,100 ± 500 UW-185 }	
18,660 ± 110 GrN-5889 }	20,000
	65,000
	89,000
∼ 80,000 to 120,000 [f]	90,000
	110,000
	∼ 110,000
beginning of last glacial?	
(i.e., ∼ 80,000) [k]	
17,150 ± 1,470 UCLA 1233A	$k_{asp} = 1.08 \times 10^{-5}$ yr^{-1}
23,600 UCLA 1430	26,000
5,640 ± 60 A-905A,B	$k_{asp} = 4.84 \times 10^{-5}$ yr^{-1}
11,230 ± 340 A-805	10,500
10,420 ± 100 A-1152	9,900

[d] Brunhes/Matuyama paleomagnetic reversal (690,000 years) occurs between layers 5 and 7 (*34*). [e] Ref. *34*. [f] Aspartic acid content ≃ 0.005–0.01 mg/g bone. [g] Aspartic acid content ≃ 0.05–0.1 mg/g bone. [h] Aspartic acid content ≃ 0.01–0.2 mg/g bone. [i] The larger the level number, the lower the sample stratigraphically. [j] See Ref. *35*. [k] See Ref. *36*.

The rate constant calculated for a particular site using a carbon-14-dated bone should provide an estimate of the average temperature over the entire time interval since the bone was deposited. Because the earth's climate has been relatively constant since the end of the Pleistocene 10–12,000 years ago, k_{asp} values determined for Holocene bones should be roughly proportional to present-day mean annual air temperatures at the sites. We tested this proposal by determining D/L aspartic acid (D/L asp) ratios in radiocarbon dated bones older than 12,000 years from several localities around the world (18, 30). These sites represented diverse environments ranging from desert through savanna, grassland, and forest to steppe and tundra. The ratios then were used to calculate the average temperature for each site, and these temperatures were compared with present-day mean annual air temperatures. In all cases, the temperature deduced from the k_{asp} value agreed reasonably with the actual present-day temperature. Given the short period over which present-day temperature data have been recorded, the diversity of environments, and the lack of ground temperature information (which would be more directly comparable to the calculated temperatures), we feel that this agreement is remarkable. The k_{asp} values were nearly identical for samples from areas which have the same present-day mean annual air temperatures but widely differing annual rainfalls (37). If environmental factors other than temperature were influencing racemization rates, then the temperatures determined from the k_{asp} values would show a poor correlation with the actual present-day temperatures. Based on our temperature comparisons, it was concluded that temperature is the major variable affecting the rate of racemization and that the calibration procedure compensated adequately for this effect. Consequently, we feel the dates calculated from the calibrated k_{asp} values are accurate and reliable.

Accuracy of Racemization Dates beyond the Range of Radiocarbon. A series of cave deposits in South Africa provided the opportunity to check the accuracy of racemization dates beyond the range of radiocarbon (17). The Klasies River Mouth Cave I deposits begin on a raised beach which corresponds in age with the high eustatic sea stand 120,000 years ago. The Middle Stone Age sequence overlying the beach sands contains abundant marine food refuse, and all carbon-14 dates from MSA occupations were greater than 30,000 years. Racemization dates ranged between 65,000 years for an upper level and 110,000 years in the lowest unit (Table I). This time interval corresponds with periods when sea level was close to but slightly lower than the present level and when the cave occupants were exploiting marine food resources. Once sea level dropped, 60–70,000 years ago during glacial advances, the cave was abandoned by MSA peoples and was not reoccupied until 6,000 years ago. Correspondence of the racemization ages and well dated climatic events provides sound evidence that amino acid racemization can be used to deduce accurate ages for fossil bones which are beyond the dating limit of radiocarbon.

Means of Calibrating Racemization Rates. Theoretically, the racemization technique has an effective dating range of ~ 500,000 years in cool temperate environments. One problem in working with these older samples is finding an appropriate calibration sample. Radiocarbon-dated bones 17–20,000 years old provide suitable calibrations for the last ~ 100,000 years (*17*), but for older fossils other dating methods must be used to calibrate the k_{asp}. The Middle Pleistocene ante-Neandertals from Arago cave in southern France were assigned racemization ages of ~ 330,000 years by using uranium-series dates on cave travertines to calculate the racemization rate (*31, 32*). The paleomagnetic reversal 690,000 years ago in the Stránská skála cave sequence in Czechoslovakia provided a calibration rate constant for dating the deposits in Külna cave (*30*). Cultural horizons may also be used for the calibration procedure. The Upper Paleolithic level at Sefunim, a cave on Mt. Carmel in Israel, supplied a date for calculating the k_{asp} used to determine racemization ages for hominid levels in Tabūn and Skhūl (*31*).

Racemization Ages Deduced Using Other Amino Acids. It should be possible to determine racemization ages using several different amino acids. The reason that aspartic acid has been used extensively so far is that this amino acid has the fastest racemization rate, and thus easily detectable racemization of aspartic acid takes place in a time interval datable by radiocarbon. Alanine and glutamic acids have racemization rates which are about half that of aspartic acid, but the amino acids valine and leucine undergo racemization considerably more slowly. Recently we have determined glutamic acid racemization ages for some samples which also had been dated by aspartic acid racemization. These preliminary results are listed in Table II. In general, glutamic acid racemization ages are compatible with those deduced from aspartic acid. However, the glutamic acid racemization ages are not as accurate as those deduced from aspartic acid racemization because the D/L glutamic acid ratios determined by gas chromatography are less precise than the D/L aspartic acid ratios determined on the amino acid analyzer. In the future, we plan to do extensive studies on using the extent of racemization of several amino acids to deduce a series of racemization ages for a single fossil bone sample.

The extent of racemization of several amino acids in a fossil bone can also be used to determine the degree of secondary amino acid contamination (*22*). Elevated temperature kinetics experiments have shown that the extent of amino acid racemization in a bone should have the

Table II.   Racemization Ages Deduced Using Aspartic and Glutamic
Acids: Murray Springs, AZ[a]

	D/L Ratio		Racemization Age (yr)[b]	
Radiocarbon Age (yr)	Aspartic Acid	Glutamic Acid	Aspartic Acid	Glutamic Acid
5,640 ± 60, A-905A,B[c]	0.33	0.15	$k_{asp} = 4.86$ $\times 10^{-5}$ yr^{-1}	$k_{glu} = 1.83$ $\times 10^{-5}$ yr^{-1}
11,230 ± 340, A-805	0.52	0.21	10,500	~ 9,000
Double Adobe[d] 10,420 ± 600, A-1152	0.50	0.21	9,900	~ 9,000

[a] The aspartic acid results are from Refs. 15 and 18.
[b] Calculated from Equation 2. The $(D/L)_{t=0}$ term for glutamic acid = 0.05 for a
modern bone hydrolyzed 24 hr (38).
[c] Sample used to calculate k values.
[d] Site is located close to Murray Springs and has the same general climate.

pattern D/L aspartic acid > D/L alanine ≃ D/L glutamic acid > D/L leucine. The presence of this pattern in a fossil bone sample provides
evidence that the sample was not contaminated.

Even though a sample may be contaminated with secondary amino
acids, amino acid racemization still provides a minimum age (22).
Contamination introduces amino acids that are more modern (that is,
ones that consist mainly of the L-enantiomers) than the indigenous col-
lagen-bound amino acids. Contamination thus lowers the actual D/L
amino acid ratio which should be present in the bone.

California Paleoindian Dates. The principal controversy involving
the validity of racemization dates concerns the ages determined for some
California paleoindian skeletons. Since 1973 this laboratory has identified
four different sites along the coast of San Diego County and one in
northern California where human skeletal remains have amino acid dates
of 40–50,000 years (16, 18). These ages would place the entry of peoples
into the New World at ~ 70,000 years B.P. (16), much earlier than
generally believed. It has been suggested that few, if any, of the published racemization dates are accurate (39, 40). Although the evidence
discussed above contradicts this assertion, we have continued to look for
ways to cross-check the validity of the high D/L asp ratios (~ 0.5)
observed in the Del Mar Man remains and in bones from the other three
sites.

Numerous physical, chemical, and biological processes will have
acted on fossil bones during their taphonomic history. If one or more of
these processes could have produced the high D/L ratio found in the Del
Mar Man fossils, for example, then these remains may actually be Holocene in age. If this were the case, then other California bones of known
Holocene age ought to be affected similarly. To test this, we analyzed

Holocene radiocarbon-dated Indian skeletons from sites throughout California.

These results are listed in Table III. In none of the cases do the D/L asp ratios determined for these skeletons approach the ratios of ~ 0.4–0.5 determined for the skeletons such as Del Mar Man and Sunnyvale (*16, 18*). The Holocene samples which we investigated came from a variety of localities which differed greatly in their general environmental characteristics. If variations in environmental parameters could produce anomalously high racemization, this should have been detected in at least some of the samples which we analyzed. Since the extent of racemization in the Holocene samples from throughout California is consistently much less than in the Del Mar Man skeleton, etc., this indicates that the most reasonable explanation for the highly racemized amino acids in these latter skeletons is that they are considerably older than the Holocene skeletons.

Reliability of the Laguna Calibration. The Del Mar Man and other Paleoindian racemization ages could be erroneous if the Laguna skull radiocarbon age of 17,150 years (UCLA 1233A) is incorrect. It has also been suggested that the Laguna skeletal material that was dated by radiocarbon was not from the original skeleton recovered in 1933. The Laguna skull had a rather bizarre post-excavation history. The skeleton was sent to Europe and even to Africa for examination by several noted anthropologists including L. S. B. Leakey. There has been some question as to whether the sample that was returned and dated was, indeed, the

Table III. Racemization of Aspartic Acid in Holocene Radiocarbon-Dated Aboriginal Skeletons from California

Site	Radiocarbon Age	D/L Aspartic Acid[a]
Northern California		
Marin 152	3,270 ± 70 (UCLA 1891A)	0.112
SDA-66	9,040 ± 210 (UCLA 1795B)	0.150[b]
	7,750 ± 400 (UCLA 1995C)	0.172[b]
Stanford Man	5,130 ± 70 (UCLA 1861)	0.14
Central California		
Tranquility	2,550 ± 60 (UCLA 1623B)	0.127
Southern California		
W-9, SDi-4660, SDM skeleton 19241	6,700 ± 150 (LJ-79)	0.154
W-12, SDi-4669, SDM skeleton 16709	8,360 ± 75 (Pta-1725)	0.189
Ora 64, 56-60 cm level	6,960 ±140 (GAK-4136)	0.170

[a] All samples were hydrolyzed 24 hr except as indicated.
[b] Sample hydrolyzed 4 hr. A 24-hr hydrolysis would yield a ratio of ~ 0.25.

original material. However, the radiocarbon-dated material closely resembled the skeletal remains shown in photographs taken at the time the skeleton was excavated (41). Nevertheless, this uncertainty still exists.

One way to test the reliability of the Laguna calibration and the authenticity of the skeleton itself would be to use the Laguna-deduced k_{asp} value to date other bones which, in turn, have also been dated by radiocarbon. We have carried out this kind of comparison for several Holocene samples found along the southern California coast, and these results are presented in Table IV. (The Holocene k_{asp} derived from the Laguna calibration cannot be applied to the other samples listed in Table III because they have not been exposed to the same temperatures as the southern California bones.) As can be seen, the ages deduced using the corrected Laguna calibration rate constant agree closely with the radiocarbon ages determined for the various bones. If the radiocarbon date on the Laguna skull were incorrect or if the dated specimen were not the original skull, these correlations would be expected to be much poorer than they are.

Table IV. Racemization Ages of Holocene Aboriginal Skeletons from Southern California Coastal Sites[a]

Site No.	SDM No.	D/L Aspartic Acid	Aspartic Acid Age (yr)[b]	C-14 Age (yr)
W-9, SDi-4660	19241	0.154	5,700	6,700 ± 150[c] (LJ-79)
W 12, SDi-4669	16709[d]			
4-hr hydrolysis		0.142	7,900	8,360 ± 75
24-hr hydrolysis		0.189	8,100	(Pta-1725)
Ora 64, 50–60 cm	—	0.170	6,800	6,960 ± 140[c] (GAK-4136)

[a] Except as indicated, all the samples were hydrolyzed 24 hr. The $t = 0$ term for a bone sample hydrolyzed 4 hr is ~ 0.05 (33).
[b] Dated using $k_{asp} = 1.5 \times 10^{-5}$ yr^{-1} (given in Ref. 16).
[c] Radiocarbon date on associated shell.
[d] This sample was also analyzed by G. Dungworth and H. Kessels, University of Nijmegen (42) and had a D/L aspartic acid ratio of 0.208. This D/L ratio yields an age of 9,400 years.

Another indication of the validity of the Laguna calibration is the fact that the racemization age for the Los Angeles Man skeleton deduced using the Laguna-based k_{asp} value is 26,000 years (16), which is consistent with the radiocarbon age of > 23,600 years (UCLA 1430) determined for this skeleton. Concordant racemization and radiocarbon ages have also been obtained for a mammoth from Santa Rosa Island (16). However this result should be considered somewhat less significant because of adjustments required in the Laguna k_{asp} value to account for a lower temperature environment at the Santa Rosa Island site.

A result which provides additional evidence that the Laguna k_{asp} value can be used for dating Upper Pleistocene bones from sites along the southern California coast involves an event which took place on the Scripps Institution of Oceanography campus. In October 1975, a portion of the sea cliff approximately 60–80 m north of Scripps Pier collapsed, exposing several fossil bones. Excavation of the freshly exposed cliff face and the collapsed cliff rubble resulted in the recovery of the nearly complete leg bone and the scapula of an extinct horse, *Equus occidentalis*. Since this horse became extinct in southern California near the end of the Pleistocene (43), the horse bones are more than 10,000 years old. Racemization analysis of these horse bones yielded a D/L aspartic acid ratio of 0.54, which is essentially the same as that of Del Mar Man. Using the Laguna k_{asp} value, the racemization age of the horse skeleton is therefore ~ 50,000 years. If the age of the horse skeleton can be ascertained by other techniques, this not only will provide another test of the validity of the Laguna calibration but will also provide an independent reference sample to use for determining the racemization age of Del Mar Man, etc.

The geology of the Scripps cliff has been discussed previously by Carter (44) and by Karrow and Emerson (45). The cliff is composed of Pleistocene alluvial and colluvial sediments overlying the Ardath Shale and Scripps Formation which are Eocene in age. There is a marine fossiliferous terrace deposit at the contact between the Pleistocene sediments and the Eocene conglomerate. Shells obtained from this terrace, ~ 40–50 m north of the area where the cliff collapsed, have been analyzed by amino acid racemization and corresponded in age to a period of high sea level during the last interglacial, i.e., ~ 120,000 years B.P. (46, 47). This age thus provides a maximum age for the horse skeleton.

Also exposed on the cliff face, both above and below the horse bones, were small flakes of charcoal. The origin of this charcoal is uncertain, but it is thought to result from a natural brush fire, rather than from human campfires, although this interpretation is speculative. Charcoal chunks in sufficient amounts for radiocarbon dating were recovered from two levels, one ~ 30–40 cm directly below and the other ~ 3 m above and ~ 10 m south of where the horse bones were recovered. These charcoal samples were submitted to the Mount Soledad Radiocarbon Laboratory and were found to have ages of 36,800 ± 2,000 years (LJ 3470) and 38,000 ± 3,000 years (LJ 3530), respectively. Both of the ages are very close to the upper dating limit of the Mount Soledad Radiocarbon Laboratory and should probably be considered minimum ages. They are consistent with some earlier radiocarbon dates (48, 49)

for the Scripps cliff of 21,500 ± 700 years (W-142) and > 34,000 years
(LJ 217) for charcoal recovered from horizons stratigraphically above
and below, respectively, the horse skeleton. The location of the various
dated samples and their corresponding racemization and radiocarbon
ages are shown in Figure 1.

The charcoal radiocarbon dates and the age of the marine terrace
which underlies the horse skeleton indicate that the horse bones are
> 30,000–< 120,000 years old. This age range is consistent with the
~ 50,000-year racemization age estimated for the horse. Since the Scripps
cliff horse bones have essentially the same D/L aspartic acid ratio as Del
Mar Man, this implies that the age of this skeleton also falls in the range
> 30,000–< 120,000 years.

*Figure 1. Photograph taken about two months after the collapse of the cliff
north of the Scripps Institution of Oceanography pier.*

The locations and ages of the various samples are as follows: (A) horse scapula,
~ 50,000 years (aspartic acid racemization); (B) horse leg bone, ~ 50,000 years
(aspartic acid racemization); (C) charcoal, 38,000 ± 3,000 years (LJ-3530); (D) charcoal, 36,800 ± 2,000 years (LJ-3470); and (E) contact between marine terrace deposit
(~ 120,000 years, amino acid racemization on shell) and Eocene. A charcoal sample
dated at 21,500 ± 700 years (W-142) was obtained in 1960 from a level which was
stratigraphically above the horse but was ~ 25–30 m south of the area of the recent
cliff collapse. Another charcoal sample which yielded an age of > 34,000 years (LJ-
217) was collected in 1953–54 (?) from a stratigraphic layer which underlies the horse
bones ~ 50 m north of the collapse. Charcoal was also obtained from a large cliff
chunk which had fallen onto the beach. This sample was dated at > 39,000 years
(LJ-3469) and was thought to have come from the same stratigraphic level as (D).

These various comparisons demonstrate that the Laguna skull radiocarbon date of 17,150 years is correct, that the dated Laguna material is indeed the original skull, and that the k_{asp} value derived from the racemization and radiocarbon analyses of the Laguna skull can be used to calibrate the amino acid racemization reaction for the southern California coast.

Amino Acid Racemization Dating of Fossil Mollusc Shell

Amino acid racemization dating of mollusc shell has not been investigated as extensively as that of fossil bone. Recently several studies have used D/L enantiomeric ratios to resolve stratigraphic relationships (24) and to date geological deposits by the calibration procedure (19, 20). Racemization dating has also been applied to midden shell from archaeological sites with limited success (20).

Nonlinear Kinetics. Among the limitations of amino acid racemization dating of carbonaceous fossils is the fact that reversible first-order kinetics are not observed in foraminifera (8, 26), in calcareous deep-sea sediments (7, 9), or in *Chione* shell (20). Racemization dates can be assigned only within the linear range of the reaction. Beyond the linear range, minimum ages can be calculated, but they tend to be less accurate. We have proposed (20) that nonlinear kinetics result from the presence of many distinct proteins which vary in amino acid composition. Amino acids in various proteins undergo diagenetic changes (hydrolysis, decomposition, racemization) at differing rates, so complex kinetics may result from a mixture of proteins and higher-molecular-weight peptides in the protein hydrolysates, as well as from free amino acids in the total hydrolysate. In contrast to the complicated kinetics in carbonaceous fossils, reversible first-order kinetics are observed in bone. This results from a more homogeneous protein composition since 90% of the organic material in modern bones is collagen.

Species Effects. Another problem with racemization analyses of mollusc shell is that different mollusc species taken from the same deposit can vary considerably in the extent of racemization (20, 21). The variation in alleu/iso ratios again may be attributed to different types of structural proteins and to differing rates of diagenetic hydrolysis of these proteins, resulting in variable proportions of free amino acids, small peptides, and protein-bound amino acids (20).

Racemization Dating of Shell from Interglacial Marine Terraces in San Diego County. Numerous fossiliferous terrace deposits occur along the coast of San Diego County. Only one reliable date has been reported for these terraces: a uranium-disequilibrium age of 120,000 ± 10,000 years on coral for a locality on Point Loma in San Diego (50). This and other units which are thought to correspond to the last interglacial sea level maximum are designated the Nestor Terrace. With the coral date for calibration, D-alloisoleucine/L-isoleucine (alleu/iso) ratios were determined and dates calculated for *Chione* shell from six additional terraces north of San Diego (*20*). The alleu/iso ratios and deduced ages are listed in Table V. Reproducibility of the ratios as shown in duplicate analyses is excellent.

Table V.   Isoleucine Racemization Dates on Total Hydrolysates of
Fossil *Chione* from Interglacial Marine Terraces in
San Diego County (*20*)

Terrace	Reference	Total Alleu/Iso	Calculated[b] Age ($\times 10^3$ yr)
14	a	0.304	98
130	a	0.595, 0.593[c]	220
1	a	0.570, 0.522	210, 180
SDSU 318	51	0.352, 0.359[d]	120
UCLA 3458	52	0.338	110
SDSU 1854	50	0.389	130
SDSU 2577	50	0.365, 0.361	calibration

[a] Samples supplied by G. Kuhn and P. Remeika.
[b] Ages deduced by using as a calibration the mean alleu/iso ratio (0.363) and a uranium disequilibrium date of 120,000 ± 10,000 years (*50*) at SDSU 2577 to calculate $k_{Iso} = 3.0 \times 10^{-6}$ yr^{-1}.
[c] Duplicate analyses on separate shells.
[d] Analyses performed on *Chione gnidia*.

Two terraces, UCLA 3458 and SDSU 318, are of particular interest because they have been designated exposures of the late Pleistocene Nestor Terrace on geomorphic and paleontological grounds (*51, 52*). The alleu/iso ratios for these two localities yield dates consistent with the Nestor Terrace age of 120,000 ± 10,000 years determined by Ku and Kern (*50*).

Radiometric evidence for glacial and interglacial events generally corroborate the racemization dates in Table V. Oxygen isotope data (*53*) suggest a low ice volume (i.e., sea level maximum) at 100,000 years as well as at 120,000 years. Locality 14 may represent the 100,000-year event. Localities 1 and 130 yielded racemization dates of 180,000–220,000 years. These ages correspond to the boundary of oxygen isotope stages 6 and 7 which has an estimated date of 195,000 years (*53*). This boundary marks the end of the second-to-last interglacial high sea stand.

The alleu/iso ratios for these latter two terrace exposures are close to the nonlinear portion of the kinetics plot (*30*). Therefore, it may not be possible to date absolutely older terraces in the San Diego area by the extent of isoleucine racemization in total hydrolysates of *Chione*. Because of the slower kinetics of isoleucine racemization in the protein fraction of *Chione*, dating of the next earlier interglacial episode may yet be achieved using this fraction.

Racemization Dating of *Chione* from Indian Middens. The prehistoric inhabitants of coastal southern California left abundant kitchen midden deposits marking their living sites. Mollusc shells are the major constituents found in these middens. We attempted to deduce racemization dates for shell from six archaeological sites along the coast of southern California (20). All of these sites had carbon-14 dates on shell carbonate available for comparison.

The results are summarized in Table VI. Using the carbon-14 ages for each site or stratigraphic level and a value of 0.95 for K' (20), the rate constants (k_{iso}) were calculated for the various sites. Since the present-day average annual air temperatures, i.e., 16°C, are essentially the same for the various coastal localities (54), the rate constants should be very similar for the six sites. The seven rate constants listed in the first half of Table VI vary by more than a factor of two. This degree of variation occurs even within the same site between samples having the same radiocarbon age (*see* Fairview and Great Western). The deviation is even greater for four of the nine Del Mar k_{iso} values. The average difference between the racemization dates and the carbon-14 dates in the first half of Table VI is 29%. The discrepancies between the radiocarbon and racemization ages again are much larger for the Del Mar samples; for these comparisons, the average difference is 88%.

The discrepancies may result from errors in the radiocarbon dates on shell carbonate, the alleu/iso ratios, or possibly both. None of the shells from sites listed in the first half of Table VI could be obtained from the exact unit or level as the shell used for the carbon-14 analysis. Only with the shells from the Del Mar middens were radiocarbon and racemization analyses carried out on samples having the same provenance. Mixing within the midden from rodent or human disturbance may account for some of the differences. Heating or cooking the shell during food preparation would also contribute to differences between the dates by increasing the alleu/iso ratio. Either reworking of the deposit or differences in heating may be responsible for the lack of reproducibility (and stratigraphic consistency) in duplicate analyses from the Del Mar site listed in Table VI.

Radiocarbon Dating of Amino Acids Extracted from *Chione* Shell. We attempted to resolve the question of which dating method produced more accurate shell dates by submitting amino acids extracted from *Chione* shell for a radiocarbon analysis (20). Because of the possibility of exchange with groundwater carbonates, the inorganic carbon in mollusc shell can be contaminated easily. We reasoned that the carbon in the amino acids would provide a more reliable age estimate than the carbon-

Table VI.   Isoleucine Racemization Results on *Chione*

Site	Laboratory Number	Shell Carbonate C-14 Age (yr BP)
Ora 370	LJ 3515	1,370 ± 40
Las Flores	LJ 3173	2,070 ± 50
Fairview		
E9	UCLA 1496B	3,685 ± 100
G2		3,685 ± 100 (?)
Ora 64	GAK 4137	6,790 ± 140
Great Western		
J-15-6	LJ 3245	8,120 ± 90
J-6-8	LJ 3160	8,270 ± 80
Del Mar W-34		
Dm 3	LJ 3175	4,590 ± 60
Dm 7	LJ 3176	5,440 ± 70
Dm 10–11	LJ 3507	7,380 ± 220 [a]
Dm 15	LJ 3177	9,260 ± 100
Del Mar W-34A		
Dm 3	LJ 3221	8,040 ± 110
Dm 6	LJ 3220	8,960 ± 120
Dm 8	LJ 3262	9,080 ± 120
Dm 10	LJ 3263	6,130 ± 100
Dm 11	LJ 3219	6,800 ± 100
Dm 12	—	—

[a] k_{iso} values calculated from Equation 2 using $K' = 0.95$ and carbon-14 ages for each sample.

[b] Ages calculated from Equation 2 using $k_{iso} = 2.0 \times 10^{-5}$ yr^{-1} from Ora 64.

ates. The carbon-14 date on amino acids isolated from *Chione* shell combined from Dm 10 and Dm 11 at site W-34 was 12,000 ± 1,100 (LJ 3631). The carbonate fraction combined from two shells representing Dm 10 and Dm 11 yielded an age of 7,380 ± 220 (LJ 3507). The 12,000-year date results in a k_{iso} value of 1.8×10^{-5} years^{-1} which agrees well with the Ora 64 calibration (*see* Table VI).

We have concluded (20) that racemization dates on fossil shell are reasonably accurate when samples are derived from geological deposits such as interglacial terraces. The results of racemization analyses of Indian midden shell appear to be less reliable since the reproducibility is poorer and since radiocarbon comparisons are widely divergent in 6 out of 15 pairs of dates. However, the carbon-14 analysis of amino acids extracted from shell indicates that the radiocarbon ages on shell carbonate may not provide accurate comparative data. Concerning the reliability of recemization dates on bone vs. mollusc shell, there are several advantages to bone analyses: linear kinetics, longer effective dating range, no species differences, good reproducibility, and concordance with collagen-based radiocarbon dates.

from Archaeological Sites in Southern California (20)

Total Alleu/Iso	k_{iso} [a] ($\times 10^{-5}$ yr^{-1})	Calculated Age (yr BP) Ora 64 k_{iso} [b]
.050	2.4	1,670
.046	1.4	1,460
.133	3.2	5,800
.076	1.6	2,950
.149	2.0	—
.155	1.7	6,900
.230	2.6	10,800
.291	6.1	14,000
.325 (.226) [d]	5.8 (3.9)	15,800 (10,600)
.231	2.9	10,800
.295 (.378)	3.1 (4.1)	14,200 (18,800)
.170	1.9	7,700
.222	2.3	10,400
.164	1.6	7,400
.328 (.246)	5.2 (3.8)	16,000 (11,600)
.281	4.0	13,500
.417	—	21,000

[c] Radiocarbon date is 12,000 ± 1100 (LJ 3631) on amino acids extracted from *Chione* in Dm 10 and Dm 11. The k_{iso} calculated from this date is 1.8×10^{-5} yr^{-1}.

[d] Values in parentheses represent analyses on separate shells from the same decimeter level.

Age at Death Estimates Based on Aspartic Acid Racemization in Teeth

Racemization has recently been found to be an in vivo as well as a diagenetic reaction. We have demonstrated that D-aspartyl residues are accumulating in stable proteins during the human lifetime because of the high mammalian body temperature. Nonturned-over proteins which undergo racemization are tooth enamel (10), dentine (11), and proteins from the nucleus of the lens (12). The extent of aspartic acid racemization in these proteins could be used to calculate the age of the individual.

The extent of aspartic acid racemization in the teeth of a skeleton can also be used, with certain limitations, to estimate the age of the individual at death. When the burial environment is relatively cold, the amount of postmortem racemization is negligible, so the D/L asp ratio in a tooth from a skeleton is a measure of the amount of in vivo racemization.

A medieval cemetery population in Czechoslovakia provided an ideal situation for demonstrating this application of the racemization technique.

The burials are being studied by M. Stloukal, a physical anthropologist at the National Museum in Prague, who has assigned ages from morphological evidence. Historical records indicate that the cemetery dates from the 7th or 8th century. The mean annual temperature in this region is ~ 9°C, and thus the amount of racemization occurring over the ~ 1,000-year burial period should be very small. Furthermore, we were able to correct for the small amount of postmortem racemization by determining the extent of aspartic acid racemization in a tooth from a skeleton whose age had been well established by Stloukal. After this correction was made, the ages of other skeletons from the cemetery were determined and were compared with Stloukal's estimates; these results are shown in Table VII. The racemization-deduced ages are in excellent agreement with those estimated by osteological criteria. An advantage of the racemization method is that ages should be more reliable with older individuals because the D/L ratios are higher and more accurately measured. Morphological age estimates, on the other hand, are less accurate beyond ~ 45 years.

Table VII.   Ages at Death Estimated for Skeletons from Medieval Cemetery in Czechoslovakia

Sample No.	Age Estimated by Stloukal (yr)	Racemization Age (yr)
79	18–20	calibration
80	30–40	33–35
77	40–50	48–50
94	60+	~ 80

Criteria for Evaluating the Reliability of an Amino Acid Racemization Date

With the increasing use of amino acid racemization dating, it is important to establish a set of criteria for assessing the reliability of a racemization-deduced age. Below, we have listed information which should be given when a racemization age is reported.

(1) The D/L aspartic acid ratio should be explicitly stated and the technique used to determine this ratio described.

(2) The calibration constant (k) used to calculate an age should be reported along with information on how this value was obtained.

(3) The temperature calculated (T_{calc}) from the k calibration value should compare favorably with the estimated temperature history of the sample.

(4) The extent of amino acid racemization should follow the pattern D/L aspartic acid > D/L alanine ≃ D/L glutamic acid > D/L leucine for an uncontaminated sample. If this pattern is not observed, the calculated age should be regarded as a minimum estimate.

(5) If sufficient material is available, the racemization analysis of a sample should be repeated several times to determine the reproducibility of the D/L ratio.

(6) The comparison of the racemization ages deduced from several amino acids should provide an estimate of the uncertainty of the racemization date.

The first two criteria should be considered a minimum requirement and should always be reported; otherwise, the racemization date should be considered meaningless. The second criterion is by far the most important. The D/L amino acid ratio for a particular bone can generally be unequivocally determined (criterion 1). However, the conversion of this ratio into an age estimate can be done only if a suitable calibration value has been determined for the site or locality where the bone was found. The reliability of the date increases with the number of criteria which are fulfilled. In the future, it should eventually be possible to satisfy an additional criterion and report an average racemization age for a sample deduced using several different amino acids.

Acknowledgments

We thank the numerous individuals who supplied the samples used for racemization analyses for their generosity, and we thank D. Darling and E. Hoopes for technical assistance. This work was supported by NSF grants EAR73-00320-A01 and EAR77-14490.

Literature Cited

1. Bada, J. L., Schroeder, R. A., *Naturwissenschaften* (1975) **62**, 71.
2. Schroeder, R. A., Bada, J. L., *Earth–Sci. Rev.* (1976) **12**, 347.
3. Dungworth, G., *Chem. Geol.* (1976) **17**, 135.
4. Vičar, J., *Chem. Listy* (1977) **71**, 160.
5. Lee, C., Bada, J. L., Peterson, E., *Nature* (1976) **259**, 183.
6. Petit, M. G., *Quat. Res. (N.Y.)* (1974) **4**, 340.
7. Bada, J. L., Luyendyk, B. P., Maynard, J. B., *Science* (1970) **170**, 730.
8. Wehmiller, J. F., Hare, P. E., *Science* (1971) **173**, 907.
9. Bada, J. L., Schroeder, R. A., *Earth Planet. Sci. Lett.* (1972) **15**, 1.
10. Helfman, P. M., Bada, J. L., *Proc. Nat. Acad. Sci. U.S.A.* (1975) **72**, 2891.
11. Helfman, P. M., Bada, J. L., *Nature* (1976) **262**, 279.
12. Masters, P. M., Bada, J. L., Zigler, J. S., Jr., *Nature* (1977) **268**, 71.
13. Helfman, P. M., Bada, J. L., Shou, M.-Y., *Gerontology* (1977) **23**, 419.
14. Bada, J. L., Protsch, R., *Proc. Nat. Acad. Sci. U.S.A.* (1973) **70**, 1331.
15. Bada, J. L., Schroeder, R. A., Protsch, R., Berger, R., *Proc. Nat. Acad. Sci. U.S.A.* (1974) **71**, 914.
16. Bada, J. L., Schroeder, R. A., Carter, G. F., *Science* (1974) **184**, 791.
17. Bada, J. L., Deems, L., *Nature* (1975) **255**, 218.
18. Bada, J. L., Helfman, P. M., *World Arch.* (1975) **7**, 160.
19. Mitterer, R. M., *Earth Planet. Sci. Lett.* (1975) **28**, 275.
20. Masters, P. M., Bada, J. L., *Earth Planet Sci. Lett.* (1977) **37**, 173.
21. Bada, J. L., *J. Am. Chem. Soc.* (1972) **95**, 1371.
22. Bada, J. L., Kvenvolden, K. A., Peterson, E., *Nature* (1973) **245**, 308.

23. Hare, P. E., Mitterer, R. M., *Carnegie Inst. Washington Yearb.* (1967) **65**, 362.
24. Miller, G. H., Hare, P. E., *Carnegie Inst. Washington, Yearb.* (1976) **74**, 612.
25. Bada, J. L., *Earth Planet. Sci. Lett.* (1972) **15**, 223.
26. King, K., Jr., Neville, C., *Science* (1977) **195**, 1333.
27. Hirs, C. H. W., Moore, S., Stein, W. H., *J. Am. Chem. Soc.* (1954) **76**, 6063.
28. Manning, J. M., Moore, S., *J. Biol. Chem.* (1968) **243**, 5591.
29. Hoopes, E. A., Peltzer, E. T., Bada, J. L., unpublished data.
30. Bada, J. L., Helfman, P. M., Hoopes, E. A., Darling, D., *Proc. Int. Radiocarbon Conf., IX,* in press.
31. Bada, J. L., Helfman, P. M., "Colloque I: Datations Absolues et Analyses Isotopiques en Préhistoire; Methodes et Limites," Union Int. Sci. Préhistoriques, IX Congrès, CNRS, Paris, 1976.
32. de Lumley, H., de Lumley, M.-A., Bada, J. L., Turekian, K. K., *J. Hum. Evol.* (1977) **6**, 223.
33. Bada, J. L., McCurdy, A., Roy, O., unpublished data.
34. Valoch, K., Moravske Museum, Brno, personal communication.
35. Klein, R., *World Arch.* (1974) **5**, 249.
36. Klein, R., *Quat. Res.* (N.Y.) (1975) **5**, 275.
37. Bada, J. L., Shou, M.-Y., *Abstracts of Annual Meeting,* Geological Society of America (1976) **8**, 762.
38. Hoopes, E. A., McCurdy, A., Bada, P. L., unpublished data.
39. Hare, P. E., *MASCA Newsletter* (1974) **10**, 4.
40. Hare, P. E., *Carnegie Inst. Washington, Yearb.* (1974) **73**, 576.
41. Berger, R., Protsch, R., Reynolds, R., Rozaire, C., Sackett, J. R., *Contributions of the University of California Archaeological Research Facility,* University of California, Berkeley (1971) **12**, 43.
42. Dungworth, G., Kessels, H., University of Nijmegen, unpublished results.
43. Romer, A. S., "Vertebrate Paleontology," 3rd ed., p. 267, University of Chicago, Chicago–London, 1966.
44. Carter, G. F., "Pleistocene Man at San Diego," p. 51, Johns Hopkins, Baltimore, 1957.
45. Karrow, P. F., Emerson, W. F., *The Veliger* (1974) **16**, 374.
46. Hare, P. E., unpublished results, cited in Ref. *45.*
47. Karrow, P., Bada, J. L., unpublished results.
48. Rubin, M., Suess, H. E., *Science* (1955) **121**, 481.
49. Hubbs, C. L., Bien, G. S., Suess, H. E., *Radiocarbon* (1962) **4**, 204.
50. Ku, T.-L., Kern, J. P., *Geol. Soc. Am. Bull.* (1974) **85**, 1713.
51. Kern, J. P., *J. Paleontol.* (1971) **45**, 810.
52. Valentine, J. W., *Ecology* (1969) **41**, 161.
53. Shackleton, N. J., Opdyke, N. D., *Quat. Res.* (N.Y.) (1973) **3**, 39.
54. Felton, E. L., "California's Many Climates," p. 74, Pacific Books, Palo Alto, 1965.

RECEIVED September 19, 1977.

Chemical Investigations on Ancient Near Eastern Archaeological Ivory Artifacts. Fluorine and Nitrogen Composition

N. S. BAER—Conservation Center, Institute of Fine Arts, New York University, 1 East 78th St., New York, NY 10021

T. JOCHSBERGER—Arnold & Marie Schwartz College of Pharmacy and Health Sciences of Long Island University, Brooklyn, NY 11216

N. INDICTOR—Chemistry Department, Brooklyn College, CUNY, Brooklyn, NY 11210

Specimens of ivory excavated at Ancient Near Eastern sites (Acem Hüyük, Anatolia; Hasanlu, Northwest Iran; Khorsabad, Iraq, Megiddo, North Palestine; and Nimrud, Iraq) have been analyzed for carbon, hydrogen, fluorine, nitrogen, and ash. The decrease in nitrogen content as collagenous material was removed and replaced, in part, by inorganic salts was general to all sites. Similarly, hydrogen and carbon contents decreased, and ash increased. The results of fluorine analyses were less general with some degree of site specificity. These data are applied to the development of criteria for artifacts of uncertain provenance. The possibility of chemical alteration of the composition of modern ivory to obtain compositions similar to those of ancient ivory is considered.

In a recent study we examined samples taken from ivory artifacts excavated at Hasanlu, Acem Hüyük, and Nimrud. Elemental analyses were obtained for carbon, hydrogen, and nitrogen (1). The results for specimens of known provenance were used to develop preliminary criteria for Ancient Near Eastern archaeological ivory artifacts of uncertain provenance. The useful results obtained in that work suggested an extension of similar observations to other excavation sites in the Ancient Near East (Megiddo and Khorsabad). The range of analyses was extended to include fluorine.

0-8412-0397-0/78/33-171-**139**$05.00/1

Table I.   Ancient Near Eastern Archaeological Sites and
Period Attributions of Specimens Examined

Site	Period (Century B.C.)	Description
Hasanlu	9th	Northwest Iran, Period IV
Acem Hüyük	19th–18th	Anatolia
Nimrud	9th–8th	Iraq
Khorsabad	8th	Ancient Dûr Sarrukin, Iraq
Megiddo	13th–12th	North Palestine

A variety of analytical methods have been applied to dating prehistoric bone (2). Included among them are changes in elemental composition accompanying systematic changes in buried bone. The slow exchange of ionic species from groundwaters with hydroxyapatite, the major inorganic component of bone, is accompanied by an increase in fluoride ion concentration with the partial conversion to fluorapatite. The nitrogen and carbon content decrease as the collagenous material is removed by several mechanisms including hydrolysis and biological attack (3, 4, 5). The many environmental factors—including pH, temperature, moisture, and groundwater ion content—which affect the rate of protein decay suggest that systematic behavior may not be assumed for new sites whose environment may only superficially parallel that of sites previously investigated.

The results of carbon, hydrogen, nitrogen, fluorine, and ash analyses are reported here for five archaeological sites (Table I). These data are compared with those for ivories of unknown provenance.

Experimental

Sample Preparation. Samples from artifacts were obtained by drilling deeply (ca. 2 cm) into the artifact with a carbon steel spade drill, discarding the initial superficial material. About 2–3 mg per drilling was available for analysis. For expendable fragments, the surface was cleaned by mechanical scraping with a scalpel.

Analyses. Analyses of ash, carbon, and hydrogen by combustion; nitrogen by Dumas or micro-Kjeldahl; and fluorine by spectrometry were obtained from the Schwarzkopf Microanalytical Laboratory, Woodside, NY 11377. The radiocarbon date was obtained from Teledyne Isotopes, Westwood, NJ 07675.

Analytical Errors. The uncertainties for the analytical methods are: ash by combustion residue, $\pm$ 0.3% absolute; carbon, $\pm$ 0.05 mg; hydrogen, $\pm$ 0.3% absolute; nitrogen, by Dumas, $\pm$ 5 μL or by micro-Kjeldahl, $\pm$ 1 μg; and fluorine, by spectrometry, $\pm$ 1 μg (6).

Errors associated with impurities introduced into or present in the ivory matrix, e.g., inorganic salts from groundwater replacing collagenous material during burial or organic adhesive materials used in the field or conservation laboratory to strengthen friable specimens, have been discussed previously (1).

Results and Discussion

Related Materials. Table II gives results of elemental analyses for ivory, bone, and related materials. The modern specimens all demonstrated high levels of nitrogen and carbon similar to those of modern ivory. Severe aqueous extraction of beefbone with boiling water removed less than half of the organic matter originally present. This suggests that artificial removal of the proteinaceous material from ivory artifacts to simulate the effects of archaeological burial is difficult without fragmentation of the artifact. The mastodon, mammoth, and ancient walrus samples retained their proteinaceous material although they represented fossil specimens.

The fluorine levels observed for modern materials were, in general, comparable with those observed in other work (7, 8). Included are the data of Jaffee and Sherwood (9) for modern and fossil manatee rib specimens from Florida waters and land-pebble phosphate deposits, respectively. The modern specimen shows a somewhat higher fluoride level than generally observed for modern bone. Since this sample was taken from an articulated skeleton (USNM 228478), only a small sample was taken. It is possible that preparation of the skeleton for display or the small sample size reduced the representativeness of the measurement.

Table II. Carbon, Hydrogen, Nitrogen, Fluorine, and Ash Analyses for Specimens of Ivory and Related Materials[a]

Specimens	Ash (%)	C (%)	H (%)	N[a] (%)	F (%)
Elephant ivory (Africa)	53.32	16.25	3.51	5.52	0.04
Elephant ivory (Africa)—treated[b]	56.50	15.18	3.33	5.30	1.34
Mammoth tusk (N.E. Siberia) [c]	54.24	15.64	3.78	5.37	0.06
Mastodon[c]	56.04	16.85	3.24	5.78	0.02
Sperm whale tooth	66.02	12.28	2.49	4.23	0.07
Whale bone (rib)	71.62	11.85	2.11	3.03	0.15
Hippopotamus tooth	60.36	12.64	3.04	4.25	0.04
Boar tooth (Aitape, New Guinia)	65.57	12.25	2.49	3.86	0.02
Walrus tusk (Alaska) [d]—fossil	60.46	15.16	2.89	4.75	0.01
Beefbone[e]	68.68	10.94	2.19	3.52	0.02
Manatee rib (Florida)—modern[f]	59.98	—	—	—	0.21
Manatee rib (Florida)—fossil[f]	93.25	—	—	—	3.57

[a] All analyses by Dumas method.
[b] Treated in boiling aqueous 1.0F NaF 1 hr.
[c] Pleistocene Epoch, courtesy The American Museum of Natural History.
[d] Courtesy The American Museum of Natural History.
[e] Collagenous material extracted with boiling water for 2 hr.
[f] Miocene or Pliocene; after Jaffee and Sherwood (9).

Table III. Designations of Ancient Ivory Specimens from Khorsabad, Ancient Dûr Sarrukin, Iraq (8th Century B.C.) and Megiddo, North Palestine (13th–12th Century B.C.)

Specimen	Accession Number[a]	Description
Khorsabad VI	A 17611	white fragment
Khorsabad VII	A 22166	blue–black fragment
Khorsabad VIII	A 17610	white (encrusted) fragment
Megiddo I	A 22267	white–yellow fragment
Megiddo II	A 15601	white–yellow (encrusted) fragment

[a] Oriental Institute, University of Chicago.

Table IV. Analyses of Ancient Ivory Specimens from Khorsabad, Ancient Dûr Sarrukin, Iraq (8th Century B.C.) and Megiddo, North Palestine (13th–12th Century B.C.)

Specimen	Color	Ash (%)	C (%)	H (%)	N (%)	F (%)
Khorsabad VI	white	87.7	3.3	0.8	< 0.04	0.06
Khorsabad VII	white–black	93.3	2.6	0.4	0.2	0.08
Khorsabad VIII	white	85.3	3.9	1.0	< 0.06	0.03
Megiddo I	white–yellow	78.9	7.0	1.5	1.20	< 0.01
Megiddo II	white–yellow	81.3	6.0	1.4	0.57	< 0.01

Table V. Comparison of Composite Values for Specimens with Specimens

Specimen	Date	Color
Elephant ivory	modern	white
Elephant ivory (treated)[b]	modern	white
Mastodon	Pleistocene	white
Hasanlu	9th century B.C.	white–grey
Hasanlu	9th century B.C.	black
Nimrud	9th–8th century B.C.	white
Megiddo	13th–12th century B.C.	white–yellow
Acem Hüyük	19th–18th century B.C.	white
Khorsabad	8th century B.C.	white; grey
"Khorsabad"[c]	unknown	white
"Assyrian"	unknown[d]	white
"Gupta"	unknown	brown

[a] Composite value refers to the average of all sample analyses. Number of replicate analyses appears in parentheses. A complete listing of the data for the individual analyses is given in Table VI.
[b] Treated in boiling 1.0F NaF for 1 hr.

Of particular interest is the fluorine data for modern African elephant ivory treated in boiling aqueous $1.0F$ NaF for 1 hr. Although the specimen sustained obvious structural damage (spalling), considerable insoluble fluorine (1.34%) was incorporated in the specimen. This suggests that it is possible, at least on a superficial level, to introduce sufficient fluorine into a bone moiety to simulate the fluorine exchange between groundwater and the hydroxyapatite in bone materials on long term burial. Of the bone and tooth materials, ivory is least susceptible to such treatment because of its layered structure.

Khorsabad. Table III presents the designations for excavated samples from Khorsabad. The analytical data are given in Table IV. In a previous study (1) data for a series of artifacts said to be from Khorsabad, but of unknown provenance, were presented. Those data were consistent with those for archaeological ivories and were most similar to those for Hasanlu and Nimrud (*see* Table V for composite data for all analyses). The general disagreement of archaeologists with the dealer's Khorsabad assignment to these artifacts made the availability of excavated specimens from Khorsabad especially interesting. The white samples contained less than 0.05% nitrogen, but the white–black specimen (Khorsabad VII) contained 0.2% nitrogen, again demonstrating the tendency for gray/black ivories to retain more of their proteinaceous material and to be found in a better physical state of preservation (10). The order of magnitude difference in nitrogen concentrations for the white specimens (Khorsabad-excavated (N $<$ 0.05%), "Khorsabad"-purchased (N $\cong$

Analyses of Excavated Ancient Near Eastern Ivory of Unknown Provenance [a]

Ash (%)	C (%)	H (%)	N (%)	F (%)
53.32 (3)	16.25 (3)	3.52 (3)	5.52 (3)	0.04
56.50	15.18	3.33	5.30	1.34
56.04	16.85	3.24	5.78	0.02
86.4 (3)	2.78 (3)	0.73 (3)	0.53 (3)	0.09
82.6 (2)	6.46 (2)	0.83 (2)	1.23 (2)	1.45
84.2 (5)	4.36 (5)	1.03 (5)	0.19 (7)	0.87
80.1 (2)	6.5 (2)	1.5 (2)	0.89 (2)	$<$ 0.01
93.07 (8)	0.87 (8)	0.25 (7)	0 (6)	0.3 (2)
88.8 (3)	3.3 (3)	0.7 (3)	$<$ 0.2 (3)	0.06
84.9 (4)	4.4 (4)	1.0 (4)	0.4 (5)	0.06
60.7 (2)	10.83 (2)	2.77 (2)	3.65 (2)	0.02
65.89	8.14	0.7	1.25	0.04

[e] Quotation marks indicate designations for specimens of unknown provenance, i.e., not obtained from controlled excavations.

[d] Radiocarbon date 1370 $\pm$ 120 A.D.

0.5%)) suggests a place of origin other than Khorsabad for the latter. The lack of significant fluorine increase in both cases is inconclusive.

Megiddo. The designations for the ivory specimens excavated at Megiddo are given in Table III. The analytical data (Table IV) show

Table VI.   Analytical Data for Individual Analyses

Specimen	Date	Sample No.
Elephant ivory	modern	I-1973-7
		I-1970-12
		I-1971-9
(treated) [b]		I-1974-9
Mastodon	Pleistocene	I-1973-14
Hasanlu	9th century B.C.	I-1973-2
		I-1973-3
		I-1973-4
		I-1973-5
		I-1973-12
Nimrud	9th–8th century B.C.	I-1970-17
		I-1971-1
		I-1971-7
		I-1971-7
		I-1973-13
		I-1973-1
		I-1973-6
Megiddo	13th–12th century B.C.	I-1977-4
		I-1977-5
Acem Hüyük	19th–18th century B.C.	I-1970-13
		I-1970-14
		I-1970-15
		I-1970-16 [a]
		I-1970-16 [e]
		I-1970-16 [f]
		I-1977-6
		I-1977-7
Khorsabad	8th century B.C.	I-1977-1
		I-1977-2
		I-1977-3
"Khorsabad"	unknown	I-1971-10
		I-1973-8
		I-1973-9
		I-1973-10
		I-1973-11
"Assyrian"	unknown	I-1972-1
		I-1972-2
"Gupta"	unknown	I-1974-1

[a] nd denotes no data.
[b] Treated with NaF.
[e] (K) denotes micro-Kjeldahl; otherwise nitrogen by Dumas.

a generally high, although variable, level of retention of proteinaceous matter. These specimens, falling in age between those of Acem Hüyük and Nimrud, demonstrate the nonuniformity of environmental conditions throughout the Ancient Near East, obviating any absolute dating method

Reported as Composite Values in Table V

Ash (%)	C (%)	H (%)	N (%)	F (%)
54.86	15.06	3.43	5.44	0.04
55.13	16.29	3.56	5.96	nd[a]
53.50	16.20	3.48	5.17	nd
56.50	15.18	3.33	5.30	1.34
56.04	16.85	3.24	5.78	0.02
87.15	2.36	0.65	0.53	0.10
84.35	3.11	0.92	0.52	0.07
87.55	2.90	0.63	0.54	< 0.08
87.48	4.40	0.47	0.95	1.45
77.69	8.52	1.18	1.50 (K)[e]	nd
85.6	4.33	0.86	0	nd
82.2	4.39	1.21	0	nd
82.3	5.45	1.18	0.39	nd
nd	nd	nd	0.19 (K)	nd
83.62	5.12	1.04	0.25	0.87
87.16	2.52	0.87	0.36	nd
nd	nd	nd	0.13	0.08
78.9	7.0	1.5	1.20	< 0.01
81.3	6.0	1.4	0.57	< 0.01
78.43	2.65	0.76	0	nd
92.60	2.12	0.41	0	nd
97.34	0.41	0.23	0	nd
98.86	0.38	0.09	0	nd
98.81	0.33	0.05	nd	nd
nd	0.22	nd	nd	nd
97.31	0.52	0.14	0.04	0.13
98.68	0.32	0.12	0.02	0.50
87.7	3.3	0.8	< 0.04	0.06
93.3	2.6	0.4	0.2	0.08
85.3	3.9	1.0	< 0.06	0.03
90.4	4.1	1.5	0.5 (K)	nd
nd	nd	nd	0.4 (K)	0.06
84.21	5.21	1.1	0.19	nd
83.3	5.6	0 (?)	0.5 (K)	nd
81.8	2.9	1.4	0.4 (K)	nd
56.45	10.54	2.71	3.42	0.02
65.09	11.11	2.80	3.87	nd
65.89	8.14	0.7	1.25	0.04

[d] 20-mg sample.
[e] 70-mg sample.
[f] HCl treated.

based on total proteinaceous material retained. They also suggest that some difficulty will be encountered in attempts to apply racemization methods (11, 12, 13) without specific rate studies for each site. Even within a site, significant variation in decay rate is observed for ivories of different color (e.g., black vs. white).

Other Sites. Composite data for specimens from Hasanlu, Acem Hüyük, and Nimrud are presented in Table V. The analytical data for the individual analyses are presented in Table VI. The fluorine data indicate some interesting differences among them. The black Hasanlu ivory contains 1.45% fluorine, but the white specimens contained only 0.09% fluorine. Comparison with the only other black/white pair, that from Khorsabad, indicates a generally higher rate of fluorine retention for the black ivories. This observation will have to be tested for a larger number of sites and specimens before its generality may be asserted. Of all sites, only Nimrud gave high (0.87%) fluorine analyses for white ivories. Again, a larger number of specimen analyses will be required to assess the generality of this result.

Other Ivories of Unknown Provenance. Figure 1 depicts an "Assyrian" ivory of unknown provenance brought to our laboratory for examination. The data presented previously (1) and summarized in Table V were not consistent with those for any of the excavated specimens. The owner submitted 10 g of the sample for radiocarbon dating and obtained the date 1370 ± 120 A.D. The fluorine analyses indicated no significant difference from modern ivory. The radiocarbon data in combination with the elemental analyses and surface texture suggest that the artifact is a relatively recent carving of a large ivory fragment approximately 600 years old.

A specimen designated as "Gupta" is included since it was sold to the present owner in Cairo with an Ancient Egyptian provenance. The figure was identified art historically (14) as Indian of the Gupta period. The brown color suggests the action of heat (15, 16), leading to partial decomposition of the proteinaceous material. In the absence of comparative materials, the elemental analyses do not permit distinction among high temperature, archaeological burial, or a combination of the two as the mechanism for the loss of proteinaceous material.

Conclusions

Elemental Analyses. The data for sample ivories excavated at five Ancient Near Eastern sites (Acem Hüyük, Hasanlu, Khorsabad, Megiddo, Nimrud) demonstrate a uniform behavior similar to that of buried bone. The most general change is decrease in nitrogen content to ≤ 0.5%, except for black ivories from Hasanlu and one yellow–white sample from

Megiddo which had approximately twice as much nitrogen. The percent ash as combustion residue was > 80% for excavated ivory vs. 55% for modern elephant ivory. The fluorine concentrations were less general in their behavior, demonstrating some degree of site specificity. Unfortunately, it appears possible to introduce fluorine into the ivory matrix in the laboratory. Further study of the fluoride bearing components will be required to demonstrate the validity of fluorine-based dating methods. The many difficulties associated with individual sites noted since the inception of this method (*17, 18*) suggest that fluorine analyses will only be of limited value in determining provenance.

Figure 1. "Assyrian" ivory, provenance unknown

Additional Methods of Analysis. Racemization studies (*11, 12, 13*) hold considerable promise for the dating of ancient bone samples. If the difficulties of shorter time span compared with those already studied, limited sample universe, and nonuniform burial conditions can be overcome, the application of this procedure to Ancient Near Eastern ivory specimens would be useful. Also of interest is the possibility (*19, 20, 21,*

22) of reducing considerably the sample size required for radiocarbon dating. However, a decrease from the 50-g samples currently required (23) to the several-milligram range will be necessary before the radiocarbon dating of ivory artifacts will become practical.

Acknowledgment

We thank Vaughn E. Crawford, Oscar White Muscarella, and Prudence Oliver Harper of the Metropolitan Museum of Art for samples of ancient ivory, for permission to sample the "Khorsabad" ivories, and for stimulating discussions. We thank Barbara Hall of the Oriental Institute, University of Chicago for samples of excavated Khorsabad and Meggido ivories.

Literature Cited

1. Baer, N. S., Indictor, N., "Chemical Investigations of Ancient Near Eastern Archaeological Ivory Artifacts," ADV. CHEM. SER. (1975) 138, 236–245.
2. Baer, N. S., Majewski, L. J., "Ivory and Related Materials: An Annotated Bibliography, Section A: Conservation and Scientific Examination," Art Archaeol. Tech. Abstr. (1970) 8(2), 229–275.
3. Hare, P. E., "Organic Geochemistry of Bone and Its Relation to the Survival of Bone in the Natural Environment," in press.
4. Ortner, D. J., Von Endt, D. W., Robinson, M. S., "The Effect of Temperature on Protein Decay in Bone: Its Significance in Nitrogen Dating of Archaeological Specimens," Am. Antiq. (1972) 37(4), 514–520.
5. Von Endt, D. W., Ortner, D. J., "Chemical Alterations in Buried Bone: Their Effect on Paleoecology and Related Archaeological Problems," in press.
6. Schwartzkopf, Francine, private communication.
7. Olsen, R., "The Fluorine Content of Some Miocene Horse Bones," Science (1950) 112, 620–621.
8. Oakley, K. P., "Analytical Methods of Dating Bones," in "Science in Archaeology," p. 41, Prager, New York, 1969.
9. Jaffee, E. B., Sherwood, A. M., "Physical and Chemical Composition of Modern and Fossil Tooth and Bone Material," U.S. Geol. Surv. T.E.M.-149 (1951).
10. Crawford, V. E., "Ivories from the Earth," Metropolitan Museum of Art, Bulletin (1962) 21(4), 141–148; private communication.
11. Bada, J. L., "The Dating of Fossil Bones Using the Racemization of Isoleucine," Earth Planet. Sci. Lett. (1972) 15, 223–231.
12. Bada, J. L., "Amino Acid Racemization Dating of Bone and Shell," ADV. CHEM. SER. (1978) 171, 117.
13. Hare, P. E., "Amino Acid Dating—A History and An Evaluation," MASCA Newsletter (1974) 10(1).
14. Rosen, E., private communication.
15. Baer, N. S., Indictor, N., Frantz, J. H., Appelbaum, B., "The Effect of High Temperature on Ivory," Stud. Conserv. (1971) 16, 1–8.
16. Low, M. J. D., Baer, N. S., Chan, J., "The Microstructure of Calcined Ivories," in press.
17. Brinton, D. G., "Current Notes on Anthropology XXVII," Science (1893) 21, 262.

18. Carnot, A., "Researches on the General Composition and Fluorine Concentration in Modern Bone and in Fossil Bone of Different Ages," *Ann. Mines, Mem.* (1893) **3**, 155–195.
19. Bennett, C. L., et al., "Radiocarbon Dating Using Electrostatic Accelerators: Negative Ions Provide the Key," *Science* (1977) **198**, 508–510.
20. Bennett, C. L., et al., "Radiocarbon Dating Using Electrostatic Accelerators: Dating of Milligram Samples," submitted to *Science.*
21. Muller, R. A., "Radioisotopic Dating with a Cyclotron," *Science* (1977) **196**, 489–494.
22. Nelson, D. E., et al., "Carbon-14: Direct Detection at Natural Concentrations," *Science* (1977) **198**, 507–508.
23. Ralph, E. K., "Carbon-14 Dating," *in* "Dating Techniques for the Archaeologist," p. 5, MIT Press, Cambridge, 1971.

RECEIVED February 27, 1978.

10

Asphalts from Middle Eastern Archaeological Sites

ROBERT F. MARSCHNER

Standard Oil Co. (Indiana), Amoco Research Center, Naperville, IL 60540

HENRY T. WRIGHT

Museum of Anthropology, University of Michigan, Ann Arbor, MI 48109

Among the few materials available to early man was bitumen which he collected from seepages that are especially abundant in the Middle East. When mixed with two or three parts of mineral fines, bitumen makes asphalt, a crude but versatile adhesive. In the large riverine cities of ancient Mesopotamia, asphalt was used as mortar by the ton, but the technology was known thousands of years earlier in settlements towards the northeast in the foothills and valleys of the Zagros Mountains. Enough of the asphalts remain for thorough examination, and 70 samples from 14 sites now have been investigated. Various sites and periods can be compared by improved and standardized methods of analysis, but interpretation is hampered by lack of samples of the bitumen seepages.

Few materials were available to early man for the development of primitive technology; even fewer have survived the ravages of centuries to show when he first used them. Such inorganic materials as rock from which he fashioned crude tools and clay from which he molded fragile vessels remain abundantly as chips and shards. But such useful organic materials as wood and those parts of animals he could not eat have disappeared for the most part. A few other materials occupy an intermediate position. Familiar examples include shell and bone, inorganic substances derived from organic sources; another less familiar one is asphalt, a mixture of organic bitumen with inorganic mineral matter.

Bitumen seeps to the surface along with water at hundreds of places on the face of the earth. Presumably the black sticky tar was either

0-8412-0397-0/78/33-171-**150**$05.50/1

petroleum in migration or something that would have become petroleum if it had not escaped. Most bitumen seepage washes downstream and out into the ocean, but some may collect to thicken to pitch or to soak into surrounding silt or sand and set to a natural asphalt. Bitumen lakes and asphalt deposits are hazardous to animals in summer; attracted to the water, they become caught in the sticky mass and are attacked by predators, which are entrapped in turn. Early man witnessed the struggle, became intrigued by bitumen, and gradually learned how to put it to use.

Because of the abundance of bitumen seepages in the Middle East, no people made such extensive use of it as did those of Southeast Asia. An early use may well have been as a hafting cement—first to affix stone weapon points to wooden shafts and later to secure tool heads onto wooden handles. Traces of bitumen have been found indeed on a sickle tooth unearthed at the eighth-millennium village of Jarmo (1) and on stone tools from the fifth-millennium town of Tell Hassuna (2), both in Northern Iraq.

Geography

No one of the familiar geographical terms adequately defines the area in which man developed asphalt technology. Middle East is too inexact, Southwest Asia is too broad, and Mesopotamia is too narrow. Most of Iraq and much of adjacent Iran were involved. Figure 1 defines the area encompassed in the present work; it ranges south to north from the Persian Gulf to the Caspian Sea and includes (southwest to northeast) Mesopotamia proper, the central Zagros Mountains, and a small part of the Great Desert of Iran.

Within the area of Figure 1 are several hundred seepages of gas, oil, and bitumen or deposits of asphalt—large or small. Most of the oil fields of the Middle East were discovered by drilling at such locations, notably Masjid-i-Sulaiman (M-i-S) which opened up development of the area. A few bitumen seepages that are well known and have been described adequately are indicated by dotted circles. Others that have been destroyed or lost or cannot be located precisely are indicated by open circles.

Also within the area shown in Figure 1 are several thousand mounds or tells, each marking the site of an ancient settlement such as a town or village. The most significant are indicated by triangles, and those pertinent to the present study are named as well. In general, the Mesopotamian sites follow the present or former courses of the Tigris and Euphrates Rivers, and the Zagros sites follow the valleys of the major Tigris tributaries. The ancient settlements tend to be located near sources of bitumen, but the proximity may be an artifact of inadequate reconnaissance.

Figure 1.   Sources of bitumen and asphalt samples

No complete surveys of either bitumen sources or ancient townsites exist. Geographical coordinates of the sources of all samples described in the present work are given in Table I along with the names of the investigators who recovered them and from whom they were obtained. Of the 14 archaeological sites and 8 locations for possible source materials listed, asphalts from 7 had been studied previously, and 17 were included in the present investigation.

Previous Analyses of Asphalts

The conventional approach to asphalt analysis has been to extract the bitumen with an organic solvent and then to characterize bitumen and minerals separately. Sulfur content and melting point are commonly measured properties of bitumen. Minerals are generally screened and then subjected to simple quantitative tests. Measurement of weight loss on ignition helps to distinguish between calcium carbonate, which loses

Table I.   Identification of Samples

Source or Site	Coordinates E long	N lat	Samples	Collector	Analyst
Lower Mesopotamian Sites					
Sakheri Sughir	46 03	31 00	8	[a]	[a]
Ur (Tell al Mughair)	46 05	30 58	5	various	various
Middle Mesopotamian Sites					
Abu Salabikh	44 50	32 16	2	Biggs	[a]
Babylon	44 26	32 30	3	various	various
Eshnunna (Tell Asmar)	44 41	33 30	9	Frankfort	Forbes, Nellensteyn
Khafage	44 32	33 17	1	Delougaz	[a]
Upper Mesopotamian Site					
Jarmo	44 58	35 33	4	Braidwood	[b]
Zagros Valley Sites					
Chogha Mish	48 18	32 31	5	Kantor	[a]
Djaffarabad	48 15	32 17	3	Dollfus	[a]
Farukhabad	47 14	32 35	28	[a]	[a]
Sharafabad	48 18	32 21	10	[a]	[a]
Susa	48 16	32 12	2	deMorgan	Le Chatelier
			1	Perrot	[a]
Zagros Mountain Sites					
Anshan (Tall-i-Malyan)	52 27	29 57	1	Sumner & Carter	[a]
Tappeh Zabarjad	49 43	35 31	1	[a]	[a]
Native Bitumens and Asphalts					
Abu Gir	42 55	33 15	1	—	Forbes
Hit	42 51	33 37	1	—	Forbes
Ain Gir	47 21	32 41	4	[a]	[a]
Masjid-i-Sulaiman (M-i-S)	49 18	31 57	1	[a]	[a]
Mordeh Fel	49[c] 46	31[c] 45	4	Lay	[a]
Rijab	46 00	34 27	1	[a]	[a]
Bebehan	50[c] 09	30[c] 48	3	Lay	[a]
Mamatain	49[c] 40	31[c] 20	1	Lay	[a]

[a] Collected or analyzed by the authors.
[b] Analyzed at Iraq Petroleum Co. laboratory in Kirkuk in 1951 and 1955.
[c] Location approximate.

44%, and silica, which loses none. A sample that loses less than 44% must contain silica or some other inert mineral, and a sample that loses more than 44% must contain combustible material such as kerogen, vegetal matter, or unextracted bitumen. Accordingly, when both silica and combustible matter are present, the amounts are indeterminate.

This dilemma can be resolved partly by treating the minerals with hydrochloric acid. Carbonates dissolve, whereas silica and combustible matter do not. But the presence of bitumen and even vegetal matter interferes more or less with the reaction with hydrochloric acid, so the cleanest way to carry out acid treatment is on ignition residues. The ratio of acid solubles to ignition loss will then be below 56:44 in proportion to the amount of combustibles present. But because some common minerals dissolve in hydrochloric acid whereas other do not, the conventional method of mineral analysis is approximate at best.

Early analyses for 14 archaeological asphalts from various sources are collected in Table II. The late R. J. Forbes of the Shell Laboratories in Amsterdam contributed most, with samples from Ur and Babylon as well as from Mohenjo Daro in the Indus Valley far to the east (3). Analytical methods were not described clearly, yet several similarities in the results of different analysts are evident. The bitumen content of nine

Table II. Analyses of Asphalts

Name of site	Susa[a]	Ur (Tell al Mughair)[b]				
Nature of sample	Sculpture	Ring	Mortar	Lump		
Approximate age, B.C.	2500	—	3500	3500		
Analysis of asphalt solvent	—[f]	[g]	CS$_2$	CHCl$_3$ + pyridine		
bitumen (soluble organic) (%)	26.8	67.5	11.2	39.5		
insoluble organic (%)	—	3.2	31.9	37.0		
mineral matter (%)	73.2	29.3	56.9	23.5		
Analysis of bitumen (%)						
sulfur	6.3[d]	7.9	—	6.8		
ash	—	5.6	—	1.0		
Analysis of mineral matter (%)						
smaller than 74 μm (200 mesh)	—	—	—	68.6	39.4	40.1
larger than 250 μm (60 mesh)	—	—	—	13.3	27.8	5.2
CaCO$_3$ (+ MgCO$_3$)	63.5	(41.4)	(99)	33.6	39.5	39.7
CaSO$_4$	4.9	—	—	—	—	—
other	1.1	—	—	—	—	—
SiO$_2$ (+ Fe$_2$O$_3$ + Al$_2$O$_3$)	30.5	58.6	(1)	66.4	60.5	60.3
Total	100.0	(100.0)	(100)	100.0	100.0	100.0

[a] Refs. 8, 9.
[b] Refs. 10, 11, 12.
[c] Refs. 3, 11.
[d] Refs. 3, 4.
[e] Ref. 1.
[f] Moisture-free.

samples fell in the range of 25–40%, the sulfur content of the bitumen of most samples analyzed was 5–9%, and the most common calcium carbonate content was 34–40%.

There are also evident differences among the results of analyses given in Table II. Least bitumen appears to be removed by ethyl ether and carbon bisulfide and most by chloroform and pyridine. The contents of sulfur and especially minerals in the bitumens seem erratic. The ratio of carbonate to silica is about 2:1 at Susa but is much higher at Babylon and much lower at Jarmo, Ur, and Mohenjo Daro. These differences and the high gypsum content at Mohenjo Daro suggest that each site had separate sources of asphalt.

Analyses of a series of asphalts from layers spanning more than a millennium at the single site of Eshunna are presented in Table III. All nine samples were analyzed by Forbes and the seven largest by Nellensteyn (4) as well. The asphalts averaged 19–36% bitumen (a little less than in Table II) and about 10% vegetal matter. Sulfur contents of the bitumens were more consistent than in Table II; with only two exceptions for Forbes (D and H) and one for Nellensteyn (5), all were 7.3 ± 1.0%. Minerals content of the bitumens was 5.5 ± 2.5%. The ratio of carbonate to silica was 2:1 to 4:1 except for samples A (and 1) and C (and 3) in

from Various Middle Eastern Sites

Babylon[a]			Mohenjo Daro[d]			Jarmo[e]			
Mortar		Pavement Joint	Sealant			Scythe Blade	Mat Imprint	Shiny Black	Earthy Brown
—	600	—	—	3000	—	7000	7000	7000	7000
CS_2	—	—	$CHCl_3$[h]	pyridine	—	CCl_4	benzene	CS_2	CS_2
17.0	35.6	37.0	23.8	24.2	30.0	9.8	30.0	1.3	31.2
26.4	—	—	—	4.1	4.3	—	—	—	—
56.6	64.4	63.0	—	71.7	70.0	—	—	—	68.8
—	7.4	9.0	7.4	7.1	2.2	5.2	4.5	9.6	9.3
—	—	3.0	—	—	0.0	—	—	9.5	—
—	29.0	—	[j]	35.4	65.1	—	—	—	—
—	30.5	—	(36)	[j]	17.0	—	—	—	—
(99)	82.6	90.3	17.5	8.6	34.1	—	—	—	37.1
—	3.6	—	14.3	13.0	—	—	—	—	1.6
—	—	(4.5)	(5.4)	(8.6)	(6.0)	—	—	—	—
(1)	13.8	5.2	62.8	69.8	55.6	—	—	—	61.3
(100)	100.0	(100.0)	(100.0)	(100.0)	(95.7)	—	—	—	100.0

[a] Ethyl ether (27.1% soluble), followed by benzene (10.4%), followed by pyridine (30.0%).
[h] Carbon bisulfide gave 8.3% soluble.
[i] Percent wax.
[j] After extraction of bitumen with carbon bisulfide.

Table III.   Analyses of Asphalts

Nature of sample	Mortar		Pavement		Mortar	
Period	Jemdet-Nasr		Protodynastic		Early Dynastic	
Approximate age, B.C.	3200–3000		3000–2900		2750	
Sample number	A	1	B	2	C	3
Analysis of asphalt (%)						
soluble in CS_2	—	33.1	—	14.5	—	22.9
soluble in pyridine	—	5.8	—	7.3	—	10.0
bitumen (total)	29.2	(38.9)	24.5	(21.8)	39.0	(32.9)
insoluble organic (vegetal)	4.5	[b]	2.7	17.5	3.4	[b]
inorganic or mineral matter	66.3	61.1	72.8	60.7	57.6	61.1
Analysis of bitumen (%)						
sulfur in CS_2 extract	[a]	6.6	[a]	8.1	[a]	6.8
sulfur in pyridine extract	8.1	7.2	8.2	7.8	7.7	8.0
ash (as $CaCO_3$)	3.0	—	5.4	—	5.9	—
Analysis of mineral matter (%)						
smaller than 74 μ m (200 mesh)	38.8	84.6	50.6	52.0	58.3	83.7
larger than 175 μ m (80 mesh)	43.4	4.3	23.7	10.9	21.2	3.0
CaO (+ MgO)	24.1	(20.4)	44.3	(39.5)	21.0	(19.7)
CO_2	19.2	9.6	33.7	26.6	26.7	9.0
$CaCO_3$ (+ $MgCO_3$)	43.3	(30.0)	78.0	(66.1)	47.7	(28.7)
SO_2		5.5		5.2		6.9
silica (+ insolubles)	56.7	(52.3)	22.0	(22.9)	52.3	(52.1)
Fe_2O_3 + Al_2O_3		9.1		4.7		8.7
noncarbonates		66.9		32.8		67.7
accounted for (or assumed)	(100.0)	96.9	(100.0)	98.9	(100.0)	96.4

[a] Collected by Henri Frankfort and analyzed by Forbes (letters (3)) or Nellensteyn (numerals (4)).

which silica predominated. The difference in mineral composition is pronounced enough to infer a deliberate modification in either source or preparation.

Why Forbes and Nellensteyn did not agree more closely is puzzling. Especially pronounced is the discrepancy in particle size of the minerals. Unquestionably Nellensteyn used material of smaller screen size than Forbes, and presumably his samples had been crushed more thoroughly before extraction. His much higher yields of bitumen on samples 1 (vs. A) and 5 (vs. E) might thus be accounted for. On the other hand, the fair agreement in mineral analyses despite the differences in procedure gives reassurance that such errors as switched samples were not involved.

Work of early analysts in Tables II and III demonstrates that useful information can be obtained by the conventional procedures. Even better results should be possible with refinements at three points:

(1) Use of various solvents for separating bitumen and minerals leads to confusion of both yields and compositions.

(2) Presence of unfilterable suspended minerals in the bitumen gives fictitiously high yields and affects the properties of both bitumen and minerals.

from Eshnunna (Tell Asmar)[a]

Sealant Early Dynastic 2750		Floor Akkadian 2700–2500		Pavement 3rd Dynasty 2300–2200		Mortar Larsa 2225		Threshold Larsa-isin 2200		Steps Ibiq-Adad 2000	
D	4	E	5	F	—	G	—	H	6	K	7
—	6.9	—	12.7	—	—	—	—	—	10.6	—	8.8
—	6.1	—	37.4	—	—	—	—	—	11.2	—	4.5
11.6	(13.0)	22.4	(50.1)	32.6	—	19.3	—	16.6	(21.8)	25.8	(13.3)
0.3	10.5	15.2	[b]	13.1	—	7.7	—	1.6	25.2	4.5	9.0
88.1	76.5	62.4	49.9	64.3	—	73.0	—	81.8	53.0	69.7	77.7
[c]	8.2	•	5.9	[c]	—	[c]	—	•	7.1	•	7.6
9.9	7.2	7.8	4.6	7.0	—	6.3	—	9.7	7.0	6.5	7.6
4.6	—	3.2	—	8.0	—	7.0	—	4.3	—	4.1	—
22.5	65.9	7.2	80.4	30.6	—	15.6	—	29.4	51.5	40.5	37.7
50.4	8.2	80.8	3.7	49.9	—	69.9	—	36.5	18.5	42.7	23.2
47.2	(38.3)	41.7	(31.8)	45.5	—	37.8	—	27.3	(32.7)	43.0	(29.9)
30.1	24.3	37.3	18.3	29.4	—	25.0	—	29.7	19.7	33.8	18.7
77.3	(62.6)	79.0	(50.1)	74.9	—	62.8	—	57.0	(52.4)	76.8	(48.6)
	4.5		10.6						8.6		3.4
22.7	(24.4)	21.0	(25.7)	25.1		37.1		43.0	(30.0)	23.2	(39.4)
	5.0		6.3		—		—		6.2		5.1
	33.9		42.6		—		—		44.8		47.9
(100.0)	96.5	(100.0)	92.7	(100.0)	—	(100.0)	—	(100.0)	97.2	(100.0)	96.5

[b] Presumably included in minerals and accounted for as insoluble minerals.
[c] Number beneath is % sulfur in entire extract.

(3) Different conventional procedures for determining the relative amounts of carbonates and silicates do not agree, and the best choice is not clear.

In the further work, attempts were made to improve the analysis at these critical points.

Extraction Solvents

Selection of a standard solvent for extracting bitumens from asphalts was made on the basis of experience with commercial asphalts and comparative tests with archaeological samples. Four of the solvents commonly used in the past were compared by extracting aliquots of a crushed asphalt from Tepe Farukhabad. Toluene was substituted for benzene because of its higher boiling point and lower toxicity. Ten milliliters of solvent was used per gram of asphalt, and a single extraction was carried out for 20 hr at room temperature. The results are given in Table IV.

No two solvents were identical but, under these conditions, chloroform and pyridine removed twice as much material as carbon disulfide

Table IV. Comparison of Extraction Solvents

Solvent	Filtration Rate	Wt %			Mineral Color
		Gross Extract	"Asphaltenes"	Net Extract	
Carbon bisulfide	fast	8.5	2.0	6.5	medium dark
Toluene	moderate	7.5	3.5	4.0	medium dark
Chloroform	slow	23.5	20.0	3.5	very light
Pyridine	very slow	16.5	14.0	2.5	very light

and toluene. However, when the gross extracts were dissolved in toluene and thrown into 10 volumes of isooctane, the additional extracts, instead of consisting of total bitumen, resembled asphaltenes contaminated with mineral matter. Because toluene and chloroform were the less objectionable pair, they were compared further with natural asphalts. At room temperature chloroform always extracted more material faster, but at the reflux temperatures toluene did nearly as well.

Toluene appeared to be the best compromise from the standpoints of volatility, ease of handling, toxicity, and halogen contamination. In all subsequent work, bitumen was extracted with refluxing toluene. Typically, three successive extractions with 100 mL toluene each were used for 10 g

Table V. Analyses of Asphalts

Name of site	Abu Salabikh[a]		Khafage[b]
Period	—		Dynastic III
Approximate age, B.C.	2750		2100
Sample number	1966	1963	000
Nature of sample	Imprint of Mat	On Clay Tablet	Pit Lining
Analysis of asphalt (%)			
soluble in hot toluene	14.7	14.4	21.2
coarse minerals	75.6	—	—
suspended minerals	9.7	—	—
Analysis of bitumen (%)			
sulfur: by high-temp combustion	6.8	7.5	8.2
minerals by adsorption	17.0	0.0[f]	2.3[f]
Analysis of minerals			
smaller than 74 μ m (200 mesh) (%)	29.0	38.7	44.0
larger than 250 μ m (60 mesh) (%)	36.5	30.7	25.6[g]
color[h]	D	D	D
loss on ignition (%)	42.2	49.0	44.8
maximum CaCO₃ (%)	96.0	11.2[i]	1.9[i]
insoluble in 10% HCl (%)	55.1	—	—

[a] Samples excavated and contributed by Ref. 13.
[b] Ref. 14.
[c] Ref. 15.
[d] Sample excavated by Ref. 16 and contributed by the University of Chicago Oriental Institute.
[e] 8.8% by bomb method; 7.5% by x-ray.

of asphalt. These numbers were often modified to suit the natures of particular samples. Chloroform might have been as satisfactory as toluene, but a more definitive choice would probably require an extensive program with additional solvents under a wider variety of conditions.

New Analyses of Asphalts

Most asphalts when dissolved in such concentrations in organic solvents can hardly be filtered. Even centrifuging some fails to separate minerals from the bitumen solution cleanly. Application of the Duffy standardized procedure for analyzing asphalts (5) by adsorbing on suitable substrates and eluting successively with appropriate solvents proved to be a more satisfactory way of attacking the problem. Bitumens are thereby resolved into three fractions: hydrocarbons, resins, and asphaltenes; any insoluble components remain behind on the adsorbent and are determined by difference. The procedure can be used for entire asphalts as well as bitumens, and the difficulty is thus circumvented.

Failure of ignition and acid treatment to cope satisfactorily with asphalt minerals was awkward, and an alternative approach was indicated. The main need in archaeology is evidence that samples are either alike or different, so that repeatability is more important than accuracy. The

from Four Sites in Mesopotamia

| | | | *Sakheri Sughir* [g] | | | | | *Ur* [d] |
| | | | *2800* | | | | | *Dynastic III* *2100* |
107 *Lump*	*072* *Shaped Lump*	*051* *Imprint of Sherd*	*044*	*107* *Imprint of Scerifa Mat*	*005*	*060* *Lashing Imprint*	*050* *Reed Imprint*	*000* *Mortar*
27.5	15.5	20.7	12.6	22.4	17.5	16.3	17.3	18.0
55.3	77.5	61.7	77.6	55.2	66.5	71.5	78.4	78.0
17.2	7.0	17.6	9.8	22.4	16.0	12.2	4.3	22.0
9.6 [e]	7.4	7.9	9.0	8.0	7.6	7.9	7.9	8.5
—	—	12.3	—	17.3	—	13.4	—	0.0 [f]
38.1	17.8	33.6	26.2	24.7	37.2	40.0	10.9	24.0
38.2	55.6	42.3	51.8	41.7	34.3	30.0	61.2	48.6
D	MD	D	D	MD	MD	D	MD	D
29.9	38.4	—	38.4	38.5	—	—	—	43.6
67.9	87.3	—	87.3	87.5	—	—	—	99.2
—	46.4	—	—	—	—	—	—	32.2 [j]

[f] Percent ash, calculated as $CaCO_3$.
[g] Exclusive of discrete pieces of charcoal, wood and pottery picked out before screening.
[h] D = dark, MD = medium dark; compare with subsequent tables.
[i] Minimum % insoluble organic matter.
[j] Contained black organic material.

unique ability of x-ray diffraction to identify minerals qualitatively offered the hope of fingerprinting precisely the mixture of extracted minerals. Until a routine procedure could be developed, however, use of the old methods was continued.

Lowland Asphalts. New analyses from four sites in the Mesopotamian lowland of Iraq are presented in Table V. Asphalts from Ur had been examined previously by three investigators, but none from the other three sites had been studied before. There was usually evidence of use. For example, three of the eight samples from Sakheri Sughir bore impressions of woven articles. All samples came from the third millennium B.C. and are therefore roughly contemporaneous with those from Eshnunna in Table III.

These lowland asphalts were alike in many respects. With one exception, all contained $17.5 \pm 5.0\%$ bitumen, and the bitumens contained $7.5 \pm 1.0\%$ sulfur; both of these values closely resemble those from Eshnunna in Table III. Furthermore, the minerals were dark in

Table VI.   Analyses of Asphalts from

Name of site	Chogha Mish[a]			Farukhabad[b]	
Sample number	508	502c	513	635	479
Period	Archaic	Uruk	Uruk	'Ubaid	Uruk
Approximate date, B.C.	6000	3300	3300	4500	3400
Nature of sample	Plant Imprints	Melted	from Jar	Sphere	Cigar
Analysis of asphalt					
loss on ignition (%)	44.8	52.9	49.3	—	—
soluble in hot toluene (%)	21.1	26.2	14.5	20.2	23.0
CO_2 from minerals	23.7	26.5	29.8	—	—
Analysis of bitumen (%)					
sulfur	—	5.4	5.8	5.7	5.3
minerals by adsorption	—	1.2	0.5	5.0	—
ash on ignition (as $CaCO_3$)	—	—	—	2.7	4.5
Analysis of minerals					
smaller than 14 μ m (200 mesh) (%)	53.7	46.2	44.4	49.4	28.6
larger than 250 μ m (60 mesh) (%)	32.8	34.9	38.9	25.2	41.8
color[g]	M	M	M	ML	ML
loss on ignition (%)	45.2	43.4	44.7	—	—
CO_2 from minerals (%)	35.7	32.0	36.0	—	—
calculated $CaCO_3$ (%)	81.1	72.7	81.8	—	—
Corrected analysis (%)					
bitumen	18.4	20.8	18.4	—	—
limestone	83.8	78.1	82.9	—	—
sand	−2.2	1.1	−1.3	—	—
total	100.0	100.0	100.0	—	—

[a] Ref. 17.
[b] Ref. 18.
[c] Sample contributed by Ref. 19.
[d] Ref. 20.

color. Because all ignitions gave light residues, however, the dark color must have been associated with organic rather than with inorganic material. Also, after adsorption of the bitumen, about 15% of residues remained, whereas after ignition little or no residue was left. The presence of combustible vegetal material would readily explain this difference.

Highland Asphalts. New analyses from seven sites in the foothills and valleys of the Zagros Mountains in southwest Iran are presented in Table VI. Susa is the only site from which an asphalt had been examined previously, and our analysis agrees well with that of LeChatelier in Table II. These samples date from the Archaic to the Elamite period, a span of nearly 5000 years, although none are quite as old as those from Jarmo or as young as some from Babylon in Table II.

Like the lowland asphalts, the highland asphalts tend to be alike. With two exceptions all contained 27 ± 9% bitumen, the bitumens contained 5.8 ± 0.5% sulfur and typically 5% or less minerals, and the mineral color was medium in shade. Thus, they differ from the lowland

Seven Sites in Southwest Iran

Susa [e]	Djaffarabad [d]			Sharafabad			Tappeh Zabarjad	Tall-i-Malyan [*] (= Anshan)	
000	1139	1151	1277	005A	151	102	—	B	A
'Ubaid	'Ubaid	'Ubaid	'Ubaid	'Ubaid	Uruk	Elamite	Uruk	Elamite	Elamite
4200	4200	4200	4200	4200	3300	1600	3500	1200	1200
Mortar	With Pebbles	Bar or Ring	Plant Remains	from Pit	Rock Asphalt	Rock Asphalt	Small Sample	Rock Asphalt	Flat Melted
—	51.5	—	—	—	41.1	—	—	—	—
27.7	34.1	32.2	62.1	25.9	18.0	32.6	35.0	30.1	26.1
—	17.4	—	—	—	23.1	—	—	—	—
5.3	5.8	6.0	6.2	5.2	5.8	5.7	—	—	—
5.4	—	—	14.0	5.2	3.0	5.7	—	—	—
5.7	0.5	4.1	—	—	—	—	—	—	—
27.1	36.2	56.7	47.0	66.4	49.3	68.5	10.0[f]	70.0[f]	70.0[f]
47.0	45.5	25.2	39.3	18.1	30.4	15.9	90.0[f]	30.0[f]	30.0[f]
MD	ML	M	M	L	ML	L	D	L	M
41.5	36.5	36.0	36.2	39.5	34.0	28.9	—	—	—
30.0	24.0	24.4	13.7	29.2	27.9	19.5	—	—	—
68.2	54.5	55.5	31.1	66.4	63.4	44.3	—	—	—
22.2	33.6	28.1	48.1	20.7	15.0	26.9	—	—	—
73.7	55.0	49.6	45.1	71.6	66.4	50.0	—	—	—
4.1	11.4	12.3	6.8	7.7	18.6	23.1	—	—	—
100.0	100.0	100.0	100.0	100.0	100.0	100.0	—	—	—

[e] Sample excavated and contributed by Ref. *21*.
[f] Approximate.
[g] M = medium, ML = medium light; compare with Tables V and VII.

asphalts in containing half again as much bitumen, about 1.7% less sulfur in the bitumen, and minerals that were usually lighter in color.

In some respects, differences can be seen among the asphalts from the several highland sites. Chogha Mish and Farukhabad are lowest in bitumen, Djaffarabad is highest in sulfur, and Sharafabad minerals are lightest in color. Chogha Mish and Susa contain little or no sand, Djaffarabad 7–12%, and Sharafabad 8–23%. In general, however, these differences are considerably smaller than those between the lowland asphalts of Table III and the highland asphalts of Table V.

These analyses support a significant series of inferences with respect to the sources of asphalt. Differences in asphalt composition, mineral color, and especially sulfur content of the bitumen indicate that lowland and highland asphalts were not derived from the same raw materials. Mesopotamian asphalts simply do not match those found in the usually earlier settlements along the Tigris tributaries. Presumably new sources upstream along the Euphrates were used. They would certainly be more convenient than overland shipment from the east. Lesser differences

Table VII. Analyses of Asphalts

Period	'Ubaid				
Excavation[a]	A			B	
Sample Number	668	585	451	708	690
Approximate date, B.C.	4800	4500	4500	4500	4500
Nature of sample	Rock Asphalt	Flat Melted	Angular Melted	Rock Asphalt	Flat Melted
Analysis of asphalt					
loss on ignition (%)	—	52.1	—	41.6	46.4
soluble in hot toluene (%)	32.6	28.4	24.8	19.5	21.9
CO_2 from minerals	—	23.7	—	22.1	24.5
Analysis of bitumen					
sulfur (%)	—	—	—	5.3	5.6
ash on ignition (%)	3.1	2.0	—	1.4	—
calculated $CaCO_3$	7.0	4.5	—	3.2	3.7[c]
Analysis of minerals					
smaller than 74 μ m (200 mesh) (%)	39.5	44.9	45.1	63.4	33.2
larger than 250 μ m (60 mesh) (%)	38.9	35.4	39.6	13.6	54.7
color[b]	VL	VL	ML	VL	L
loss on ignition (%)	—	39.3	—	—	37.3
CO_2 from minerals (%)	—	28.1	—	—	29.2
calculated $CaCO_3$	—	63.9	—	—	66.4
Corrected analysis (%)					
bitumen	25.6	23.9	—	16.3	18.2
limestone	—	68.4	—	60.7	70.1
sand	—	7.7	—	23.0	11.7
total	—	100.0	—	100.0	100.0

[a] Collected by Wright (18), from Excavation A among large structures and Excavation B among small structures.

among the asphalts of the Zagros settlements further emphasize multiple sources. Regional distribution of lowland bitumen and even local distribution of highland bitumen is the pattern suggested rather than widespread distribution of asphalt.

Asphalts of Farukhabad and the bitumen of Ain Gir assume special importance in view of these inferences. Farukhabad and neighboring settlements were so close to Ain Gir and other nearby sources that they could hardly fail to interact. Evidence for development of asphalt technology would more likely be found in this area than in any other studied.

Asphalts from Tepe Farukhabad. Analyses for 18 asphalts from Tepe Farukhabad are presented in Table VII. These samples had more the appearance of asphalts in process of preparation than of asphalts in use as has been the case in the previous tables. They came from two distinct areas: Excavation A among large buildings that might represent a center for some community activity and Excavation B among small buildings that might instead have been private habitations. In appearance, the samples fell into three categories: rock asphalts that were relatively

from Two Excavations at Farukhabad

Uruk					Jemdet-Nasr				
A			B		A			B	
305	298	329	423	404	127	800	086	176	352
3200	3200	3200	3200	3200	3000	3000	3000	3000	3100
Rock	Flat	Angular	Rock	Angular	Rock	Flat	Angular	Rock	Flat
Asphalt	Melted	Melted	Asphalt	Melted	Asphalt	Melted	Melted	Asphalt	Melted
51.9	—	—	—	45.3	—	43.1	—	51.9	44.5
27.0	7.6	33.4	18.4	23.4	69.6	25.1	32.0	31.3	24.6
24.9	—	—	—	21.9	—	18.0	—	20.6	19.9
—	6.0	6.1	5.8	—	6.1	—	5.7	5.9	—
—	—	—	—	—	3.9	1.4	—	2.8	—
—	—	—	—	3.9 ᶜ	8.9	3.2	—	0.4	—
46.0	17.4	41.8	41.2	49.7	38.6	33.2	57.0	66.5	44.5
36.0	71.6	49.9	46.3	25.3	41.4	44.5	16.3	16.2	39.8
VL	L	L	L	L	M	MD	VL	VL	M
—	28.1	—	—	30.4	38.3	30.2	40.2	—	—
—	26.0	—	—	23.3	11.6	22.6	27.3	—	—
—	59.1	—	—	53.0	26.4	51.4	62.0	—	—
—	7.6	—	—	19.5	60.7	21.9	32.0	24.9	20.7
—	59.1	—	—	56.9	35.3	54.6	62.0	67.8	58.0
—	33.3	—	—	23.6	4.0	23.5	6.0	7.3	21.3
—	100.0	—	—	100.0	100.0	100.0	100.0	100.0	100.0

ᵇ L = light, VL = very light; compare with previous tables.
ᶜ % minerals by adsorption.

homogeneous, flat-melted pieces that resembled Forbes' descriptions of Eshnunna asphalts, and angular-melted pieces that seem to have been distorted further. Most of these asphalts immediately predate those from Eshnunna in Table III, although some are contemporaneous, and one from Excavation B antedates them.

With the exception of three samples, one much leaner and two much richer, all samples in Table VII contained 21 ± 5% bitumen. All bitumens contained 5.7 ± 0.4% sulfur, and most minerals were lighter in color than those in the previous tables. The one lean sample (298) contained both the largest amount of coarse material and the largest amount of sand—anomalies that the presence of coarse vegetal material would account for. Otherwise, all samples fall into only two mineral groups: seven with 8 ± 5% sand and five with 20 ± 4%.

Four asphalts from highland sites were atypically high in bitumen: the two from Farukhabad mentioned above and two others from Djaffarabad in Table VI. One of each pair had a bitumen content near 33%, only moderately higher than typical asphalts, but the other two had 48 and 61%, two or three times the typical percentages. They were probably raw or hardened seepages, an indication supported by the presence of vegetal matter. They provide direct evidence that at least some asphalts

Table VIII.　Analyses of

	Original % Bitumen	Bitumen	
		% S	% Ash
Hit seepage[a]	100.0[b]	8.3	0.5
Abu Gir asphalt (3)	89.7	7.3	—
Abu Gir asphalt[a]			
raw bitumen	91.0[b]	6.9	7.3
hardened bitumen	37.3	9.0	2.3[c]
rock asphalt	42.1	7.3	6.5
Masjid-i-Suleiman (M-i-S)	4.5	3.6	0.7
Mordeh Fel[f]			
unweathered	3.2	—	0.3[c]
weathered	5.2	—	2.8[c]
Rijab rock asphalt	2.0	—	—
Bebehan[f]	72.0	—	—
main exposure surface	36.0	—	—
Mamatain[f]	23.0	—	—

[a] Ref. 3.
[b] After separating all water.
[c] Ash calculated as $CaCO_3$.

were made by mixing bitumen with mineral matter rather than being derived ready-made from natural sources.

Bitumen Sources. Analyses of possible source materials are given in Table VIII. The seepages of Hit and such bitumen lakes as Abu Gir are the traditional sources for asphalts of the cities of lower Mesopotamia. Despite repeated attempts, samples could not be obtained for analysis.

All other sources in Table VIII are from Zagros locations, especially the foothills and valleys from Ain Gir to Bebehan from which multiple samples were obtained. The seepages of Ain Gir contained more bitumen and less sand than the dried and rock asphalts, which are much alike. Weathered and unweathered rock asphalts from Mordeh Fel were hardly distinguishable, as were the Bebehan samples in other respects than bitumen content. The samples from M-i-S, Mordeh Fel, and Rijab are probably too lean to have been used as source materials. But the similarities at Ain Gir and Mordeh Fel suggest that the bitumen in any seepages at other sites would have resembled that in the asphalts there.

Settlements between Ain Gir and Bebehan would have a choice of bitumens from which to produce asphalts for various purposes. A resemblance to Ain Gir and such intermediate sources as M-i-S is detectable in many asphalts, especially those that contain more fines than Bebehan.

Possible Source Materials

Minerals		
% Fines[d]	Color[e]	*Additional Comments*
—	—	softening point of bitumen 47.5°, minerals rich in gypsum
—	—	softening point of bitumen 127°, minerals rich in dolomite
76.4	—	mineral fraction included some vegetal matter
60.1	VL	mineral fraction included considerable vegetal matter
53.9	L	
54.3	MD	insoluble organic matter present
—	M	crushed before extraction
—	M	crushed before extraction
27.8	D	insoluble organic matter probably present
36.0	L	
38.5	L	
26.0	VL	

[d] Smaller than 74 μ (200 mesh).
[e] Compare with previous tables.
[f] Samples recovered and contributed by Ref. 22.

Other samples are difficult to correlate with the sources examined in Table VIII because many sources of minerals as well as bitumens that could have been used are as yet unknown or unstudied.

Characterization of Bitumens

Bitumens from archaeological asphalts were glassy black pitches that acquired electrical charges when broken into small pieces. They were nearly insoluble in petroleum ether or *n*-pentane—the classical criterion for the presence of asphaltenes. Because asphaltenes can be formed by oxidizing hot petroleum residues, indications that they were also the end-products of weathering seemed reasonable.

Liquid Chromatographic Analyses. Analyses of about half the samples by the standardized method of liquid chromatographic separation are presented in Table IX. In addition to asphaltenes and minerals, all contained large amounts of resins and small amounts of oils that are mostly hydrocarbons. Similar results were obtained whether the samples were analyzed as bitumens or as asphalts. In general, however, the bitumens gave somewhat more resins and hydrocarbons, whereas asphalts gave more asphaltenes. With Farukhabad 404, which was analyzed both as bitumen and as asphalt, the difference in asphaltenes was 8%.

Source bitumens also were analyzed satisfactorily by liquid chromatography. Ain Gir seepage averaged 35% hydrocarbons and 10% asphaltenes, whereas its hardened seepage and asphalt both contained about 25% of each. M-i-S and Mordeh Fel bitumens resembled instead those extracted from archaeological asphalts in having 40% and 60% asphaltenes and less than 10% hydrocarbons. This similarity suggests that the terminal product of exposure to the elements is much the same whether it occurs over geological ages deep underground or over archaeological millennia near the surface.

Vanadium-to-Nickel Ratios. An item of bitumen composition that should not be affected by weathering is the amount of metals present. Nickel and vanadium, the metals most characteristic of petroleum, can

Table IX. Liquid Chromatographic Analyses of Asphalts and Bitumens[a]

Sample No.	Hydrocarbons		Resins		Asphaltenes		Total[c]
Asphalts from Farukhabad							
086 Area A	1.2	(5.7)	7.5	(35.5)	12.4	(58.8)	21.1
404 Area B[b]	0.6	(3.3)	7.7	(41.8)	10.1	(54.9)	18.4
018 Area B	1.3	(5.1)	11.2	(44.3)	12.8	(50.6)	25.3
176 Area B	1.2	(4.7)	9.6	(38.0)	14.5	(57.3)	25.3
708 Area B	0.8	(4.5)	6.8	(38.2)	10.2	(57.3)	17.8

Table IX. Continued

Sample No.	Hydrocarbons	Resins	Asphaltenes	Total[c]
Asphalts from Other Sites				
Djaffarabad 1139	1.5 (5.5)	10.9 (39.9)	14.9 (54.6)	27.3
Sharafabad 292	1.6 (5.4)	8.8 (29.8)	19.1 (64.8)	29.5
Bitumen Fractions from Farukhabad				
127 Area A	3.1 (4.5)	33.1 (48.6)	32.0 (46.9)	68.2
404 Area B	4.0 (4.8)	40.1 (48.0)	39.4 (47.2)	38.5
479 Area B	4.8 (6.2)	35.9 (46.2)	36.9 (47.6)	77.6
635 Area B	6.2 (8.4)	28.4 (38.3)	39.4 (53.3)	74.0
same, rerun	6.5 (8.5)	27.3 (35.9)	42.3 (55.6)	76.1
690 Area B	4.3 (5.2)	40.7 (48.8)	38.3 (46.0)	83.3
Bitumen Fractions from Other Sites				
Abu-Salabikh 1963	5.9 (6.6)	45.1 (50.6)	38.1 (42.8)	89.1
Abu Salabikh 1966	8.8 (10.6)	51.8 (62.4)	22.4 (27.0)	83.0
Khafage	5.3 (6.1)	39.9 (45.8)	42.0 (48.1)	87.2
Sakheri Sughir 051	4.0 (4.6)	44.9 (51.2)	38.8 (44.2)	87.7
Sakheri Sughir 060	2.4 (2.8)	41.8 (48.3)	42.4 (48.9)	86.6
Sakheri Sughir 107	3.8 (4.6)	42.0 (50.8)	36.9 (44.6)	82.7
Ur	3.6 (4.0)	50.7 (56.8)	35.0 (39.2)	89.3
Choga Mish 502	2.2 (2.8)	27.7 (34.9)	49.4 (62.3)	79.3
Djaffarabad 1277	3.1 (4.0)	31.3 (40.4)	43.0 (55.6)	77.4
Sharafabad 102	4.0 (4.9)	37.9 (46.0)	40.5 (49.1)	82.4
Sharafabad 151	6.0 (7.2)	27.4 (33.0)	49.7 (59.8)	83.1
Sharafabad 005	4.9 (6.1)	29.5 (36.9)	45.6 (57.0)	80.0
Susa Platform	5.1 (6.3)	34.2 (42.5)	41.2 (51.2)	80.5
Native Bitumens and Asphalts				
Ain Gir Seepage 1	35.0 (35.4)	56.0 (56.5)	8.0 (8.1)	99.0
same, rerun[d]	27.6 (30.2)	55.5 (60.8)	8.2 (9.0)	91.3
Ain Gir Seepage 2	32.0 (34.6)	51.3 (55.4)	9.3 (10.0)	92.6
Ain Gir Hard Seepage	20.2 (27.3)	39.4 (46.5)	19.4 (26.4)	74.0
Ain Gir Asphalt 1	20.2 (23.3)	45.9 (52.8)	20.8 (23.9)	86.9
Ain Gir Asphalt 2	18.3 (22.3)	46.2 (56.3)	17.6 (21.4)	82.1
Masjid-i-Suleiman 1	5.6 (6.6)	46.8 (54.8)	33.0 (38.6)	85.4
Masjid-i-Suleiman 2	8.2 (8.5)	50.6 (52.1)	38.3 (39.4)	97.1
Mordeh Fel Asphalt				
unweathered	3.9 (4.9)	32.9 (41.7)	42.1 (53.4)	78.9
weathered	2.1 (2.4)	32.5 (37.4)	52.4 (60.2)	87.0
Bebehan (surface)	19.5 (21.4)	36.3 (39.8)	35.5 (38.8)	91.3
Mamatain	12.3 (14.9)	47.7 (58.0)	22.3 (27.1)	82.3

[a] Wt % on sample (on bitumen in parentheses).
[b] Average of seven analyses in close agreement.
[c] Parenthetical percentages all add to 100.0.
[d] Rerun at higher temperature to demonstrate volatility of hydrocarbon fraction.

be neither volatilized nor oxidized away as organic material can. The few parts per million in which they are present, however, makes the determination tricky. Contamination from impurities, handling, and reagents is probably the worst problem.

Metals in several of the asphalts were solubilized by incinerating with 10% sulfur, coking, ashing, and dissolving in hydrochloric acid (7). Nickel was determined colorimetrically on one aliquot with dimethylglyoxime, and vanadium was determined on another by amperometric titration with ferrous ammonium sulfate. Ratios of one metal to the other should remain constant regardless of concentration, so that plotting one against the other reveals changes and differences clearly. The amounts may be plotted logarithmically for convenience without losing the linear nature of lines of constant ratio.

Figure 2 presents the metals data for archaeological asphalts as solid points against a petroleum background of open points. The dotted point for Ain Gir falls at a V:Ni ratio near 3:1, and Sharafabad and Djaffarabad fall on either side in fair agreement at 2:1 and 4:1. Two points for

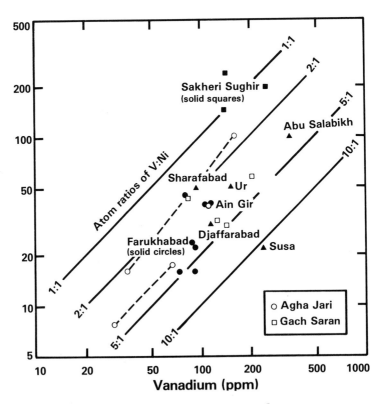

Figure 2.   Metals concentrations in bitumens

Farukhabad agree excellently with Ain Gir, but five others range from below 2:1 to above 5:1. Points for Sakheri Sughir straddle 1:1, Abu Salabikh and Ur lie near 3:1, and Susa exceeds 10:1. Multiple sources of asphalts are indicated.

Data on the metals content of petroleum residues suggest that conclusions be drawn from Figure 2 with caution. The four open squares for Gach Saran petroleum, obtained on the same sample in different laboratories, spread almost as much as the Farukhabad points. Both pairs of open circles for Agha Jari petroleum and its distillation residue agree well, but the two pairs are as far apart as Sharafabad and Djaffarabad. The ratio of the metals vandium and nickel can become a valid test for distinguishing asphalts and bitumens only when duplicate analyses no longer disagree.

Characterization of Minerals

Mineral fractions from all sources have been examined in a preliminary way by x-ray diffraction. Most helpful has been x-ray spectrometry before and after treatment with acid. Beforehand, the main constituents were calcite and quartz with some dolomite; afterwards, they were largely quartz with some feldspar and occasional traces of gypsum and clay. Because of its high crystallinity, quartz tends to dominate all spectra, and methods of subtracting out its contributions are being studied. The remaining difference spectrum would be an adequate fingerprint of the minerals for comparison purposes. Mineral analyses that are more accurate than those in the tables as well as far more repeatable can be expected eventually.

Insoluble organic matter that may accompany the minerals poses the most serious remaining problem. There is no quantitative way to distinguish among three kinds of such materials: whatever kerogen may have been present in the original asphalt or bitumen; any deliberately added or adventitious vegetal material; and possible traces of bitumen—presumably largely asphaltenes—that were not completely extracted. However the individual minerals are determined, this group of substances can be determined only by difference. The greater the amount present, the larger the errors that can be expected; fortunately, only the vegetal matter is likely to ever reach as much as 10%.

Conclusion

Use of asphalt in the Zagros settlements dates at least as far back as 7000 B.C., four millennia before it became an essential construction material in the Mesopotamian cities. By 5000 B.C., seepage bitumen was converted to asphalt with considerable sophistication, the only subsequent innovation being the incorporation of vegetal matter together with reduction of the bitumen content. Preparation of asphalt was centralized in locations convenient to sources of bitumen and can be considered an early example of commercial enterprise.

The versatility of asphalt as an adhesive, sealant, and protectant provided early innovators with a unique material with which to advance their primitive technology. Despite the ravages of thousands of years, the main components are still recognizable. Minerals of course remain largely unchanged. Although bitumen is converted extensively to asphaltenes, many resins and even some hydrocarbons are still extant and available for examination. Often the use to which the asphalt was put is still obvious—but not always.

Standardization of analytical techniques has helped greatly to compare asphalts of different areas and different periods. Repeated extractions with boiling toluene has improved consistency of data, whether or not it is the best way of separating bitumen and minerals. Liquid chromatographic analysis not only determines the amount of minerals suspended in the bitumen but also provides fractions suitable for further characterization. X-ray diffraction promises to improve vastly analysis of the minerals. No good way of resolving whatever usually small amounts of insoluble organic matter may be present has been devised.

Asphalts that are found in many ancient sites of the Middle East deserve continuing study. Closer examination by sharper techniques promises to better define where the bitumens were found, with what they were mixed, and how they were used. That information on various sites would help to outline areas of influence, to trace favored trade routes, and to suggest patterns of exchange. The greatest deterrent to such research is a lack of samples of the bitumen seepages that were available to ancient technologists.

Acknowledgments

This work depended heavily on the generosity of a host of experts: archaeologists for samples, chemists for analyses, and both—as well as others—for advice, data, and assistance.

Literature Cited

1. Braidwood, Robert, private communication (1966).
2. Lloyd, Seton, Safar, M., "Tell Hassuna," *J. Near Eastern Stud.* (1945) **4**, 255.
3. Forbes, R. J., "Oldest Uses of Bitumen in Mesopotamia," *Bitumen* (1935) **5**, 9.
4. Nellensteyn, F. J., Brand, J., "Asphalt Found in Mesopotamian Excavations," *Chem. Weekbl.* (1936) **33**, 261.

5. Duffy, L. J., "Liquid-Chromatographic Analysis of Asphalts," presented to Meeting of Am. Pet. Inst. Proj. 60, Laramie, Wyoming, 1971.
6. Marschner, Robert F., Duffy, L. J., Wright, H. T., "Asphalts from Ancient Townsites in Southwestern Iran," *Paleorient* (1978), in press.
7. Agazzi, E. F., et al., "Trace Metals in Oils by Sulfur Incineration," *Anal. Chem.* (1963) **35**, 332.
8. Le Chatelier, Henri, "Memoirs," Vol. VIII, p. 162, Colin, Paris, 1913.
9. Abraham, Herbert, "Asphalts and Allied Substances," Vol. I, p. 15, Van Nostrand, Princeton, 1960.
10. Hackford, J. E., Lawson, S., Spielman, P. E., "Asphalt Ring from Ur," *J. Inst. Pet. Technol.* (1913) **17**, 738.
11. Parkhurst, R. W., "Assyrian Engineering, Ancient and Modern," *Civ. Eng.* (1932) **2**, 345.
12. Forbes, R. J., "Lump of Asphalt from Ur," *J. Inst. Pet. Technol.* (1936) **22**, 180.
13. Biggs, Robert, sample excavation and contribution.
14. Delougaz, Pinhas P., "The Temple Oval at Khatajah," Oriental Institute Publication, Vol. LIII, Univ. Chicago (1940).
15. Wright, Henry T., "Administration of Rural Production in an Early Mesopotamian Town," Anthropological Paper No. 38, University of Michigan (1968).
16. Frankfort, Henri, sample excavation.
17. Delougaz, Pinhas P., Kantor, Helene, "Chogha Mish: First Five Seasons of Excavation 1967–1971," University of Chicago Oriental Institute Comm. XXIII, 1977.
18. Wright, Henry T., Johnson, Gregory A., "Population, Exchange and Early State Formation in Southwestern Iran," 1975.
19. Perrot, Jean, sample contribution.
20. Dollfus, Genevieve, "Djaffarab: Cahiers de la Delegation Archaeologique Francaise en Iran," V (1976).
21. Sumner, William, Carter, Elizabeth, sample excavation and contribution.
22. Lay, Douglas M., sample recovery and contribution.

RECEIVED September 19, 1977.

11

The Identification of Dyes in Archaeological and Ethnographic Textiles

MAX SALTZMAN

Institute of Geophysics and Planetary Physics, University of California, Los Angeles, CA 90024

Solution spectrophotometry in the visible and ultraviolet can be used to determine the nature of dyes used to color textiles of archaeological or historical importance. The curves obtained are compared with those from solutions of known materials. The method requires dyeings made with known natural dyes from the region of interest. Collections of such materials have been obtained from Peru. The principle red dyes of Peru can be identified by means of a single solution curve. Blue, from indigo, can easily be distinguished from the purple dibromoindigo obtained from marine molluscs. Yellow dyes are more difficult to identify and require, in most cases, curves in more than one solvent.

The value of identifying the materials used to produce artifacts of archaeological interest is widely accepted. Among these objects are the colorants, dyes, and pigments used to color ceramics, painted materials, and textiles. The inorganic colorants which are used in most ceramics and paints can be analyzed by well established methods which can be carried out in a great number of laboratories. With organic colorants, however, there are some complications. While suitable methods exist, the number of laboratories which use these methods and which are available to the archaeologist or museum conservator is much more limited. Laboratories connected with the dye and pigment industries perform such analysis on a routine basis, but there is little time available at such facilities for nonindustrial work. To our knowledge, the only facilities available to archaeologists for the analysis of organic dyes and pigments are in Europe. It has been possible, however, with the support of the Munsell Color Foundation, to set up such a laboratory at the University of California.

0-8412-0397-0/78/33-171-**172**$05.00/1

Analytical Techniques of Dye Identification

This laboratory has examined the many techniques which are available. These range from the classic organic analytical methods of the 19th century through chemical spot tests, solution spectrophotometry, infrared, and other optical spectroscopic techniques through mass spectrometry. Thin-layer chromatography is, by itself, a separation technique which allows identification of the separated components by some appropriate technique. In many cases the patterns obtained may be sufficient for identification, and in the hands of such workers as H. Schweppe (*1*), this technique has proven to be extremely useful. It has the distinct advantage of requiring very small samples and relatively simple equipment. IR spectra can give accurate and unambiguous results. It has been used with great success by Abrahams and Edelstein on Israeli finds (*2*), but it requires rather large samples and a great deal of careful separation.

Based on 30 years of industrial experience, we prefer to use solution spectrophotometry (*3, 4*). The techniques are simple and the equipment readily available. Sample requirements are relatively small; on the order of several milligrams. Provided that one has an adequate reference file of known materials, the method is rapid and reliable. The method we use is based on the work of Formanek (*5*) who published the first large systematic study of the spectrophotometric curves of solutions of dyestuffs. His work was modified by Stearns and his colleagues at the American Cyanamid Co. (*6*). In order to facilitate the comparison of curves it is necessary to obtain curves whose shapes are independent of concentration. This can be done by plotting the logarithm of the absorbance vs. wave length. Provided that there are no concentration-dependent anomalies, such a plot gives a curve shape which is constant over at least a 10-fold absorbance range. In our industrial work we used the Hardy–General Electric recording spectrophotometer (now made by the Diano Corp.) which could plot this function of $\log (\log 1/T)$ by an auxiliary cam. We now use a Cary model 11 to which we have added a transmitting potentiometer which feeds the absorbance signal to a logarithmic recorder. Figure 1 illustrates the advantages of plotting the function of $\log (\log 1/T)$ as compared with T. The use of a log density or log absorbance enables us, in this example, to detect one tenth the amount of material by examining the curve shape and the positions of maximum absorbance as compared with normal transmittance.

To our knowledge, the first use of solution spectrophotometry to study dyes of archaeological interest was made by Fester and his associates in their work in the colorants of Paracas (*7, 8*). The method requires the preparation of solutions—obtained by stripping the dye from the fiber

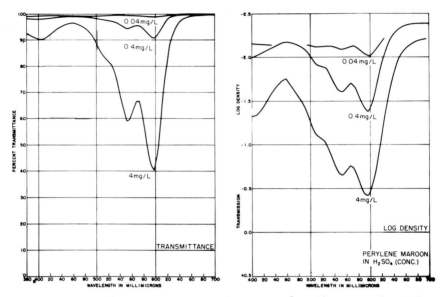

Figure 1. Normal transmission curves (T vs. wavelength) compared with log density or log absorbance curves [log(log 1/T) vs. wavelength]

—in one or more solvents, running the spectrophotometric curve of the solution over the range of interest, and comparing the curves with those of known materials run under the same conditions. With the Cary 11 we run from about 220 nm in the near-UV to 800 nm in the near-IR. Our curves are prepared in a standard format. One log cycle (absorbance from 0.1 to 1.0) covers 5 in. The wave length scale is plotted at 40 nm/in. Curves are run on transparent paper, and comparisons are made by simple overlay of the unknown on the knowns. Solutions of the knowns are classified first by solvent and then by the use of the curve-shape index method of Shurcliff (8, 9).

Reference File of Known Materials

The key to the use of the method lies in the preparation of a file of known materials. In the case of the coloring matters of earlier times we have both a simple and a complex problem: simple because the number of natural organic dyes and pigments which have survived to the present time is rather small; complex because these materials are of animal or botanical origin, and we may or may not have authentic samples of the colorants of interest. No dyer's sample swatches have come down to us from the preliterate societies of Precolumbian Peru or any other part of the Americas as did the pots of pigment from Pompei (10). If we do not have authentic samples of known material which might have been used, we can only supply a negative answer, since the method is one of comparison. That is, the sample is not one with which we are familiar. We are very fortunate in having obtained from Barbara Mullins of Sussex, England, samples of 80 dyeings which she made in Peru from native materials under village conditions. This collection is being augmented by Kay Ketchum Antunez de Mayolo, a trained botanist. She and her husband Erik are gathering materials not collected by Miss Mullins, paying special attention to the various species of *Relbunium*. In addition to plants, they are gathering samples of insect dyes from various regions of Peru and, through the help of the Instituto del Mar, are collecting molluscs of the *Muricidae* family from which come the dye known generically as Tyrian Purple.

Experimental Details and Discussion

We examined first the archaeological materials for dyestuff identification at the request of Junius Bird of the American Museum of Natural History on behalf of Peter Gerhard who was studying the distribution of the shellfish purple dye. This dye is believed to have been used in ancient times in Central and South America and in Mexico and is still in use on the west coast of Mexico. The results of his work have been published (*11*), and we will confine our comments to the analytical

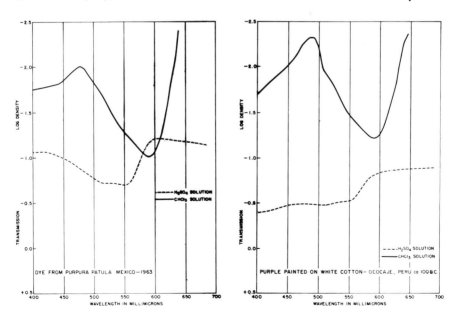

Figure 2.   Known shellfish purple dye solutions compared with those obtained from a Peruvian textile

results. Using a modern yarn dyed with the secretion from *Purpura patula pansa* Gould as a standard, we attempted to analyze about 15 samples.

The samples ranged in size from single threads about 3 in. long (about 30 mg in weight) to some very fine powder weighing less than 10 mg. One sample, a dot of color about 2 mm in diameter, proved to be dyed with shellfish purple. Figure 2 shows the spectrophotometric curves in chloroform and in concentrated H_2SO_4 of the dye from modern *Purpura patula* (*left*) and those from the sample from Ocucaje (south coast of Peru–Paracas region, ca. 100 B.C.) (*right*). The similarity is quite evident. Chemically, the dye is 6,6′-dibromoindigo which is the main component of shellfish purples from all sources. For comparison, Figure 3 shows the chloroform solutions of the shellfish purple dye with its absorption maximum at 585–590 nm compared with that of indigo with its absorption maximum at 600–605 nm. The curves of both dyes in concentrated H_2SO_4 are shown in Figure 4. These two sets of curves differentiate clearly the indigo dye obtained from plants from the brominated indigo dye obtained from the shellfish. As both these dyes were thought to have been used in ancient America, it is useful to have a simple test to distinguish them from each other. Purple colors made with indigo and a red dye can be identified clearly as compared with those made from secretion of the marine snail.

In examining these samples, we found evidence of indigo and a red dye in several samples, and the same red dye was found in other fibers. A search of our files of modern dyes showed this to be similar to the modern food colorant, Carmine CI Natural Red 4, which is obtained from the cochineal insect *Dactylopius cacti* (formerly called *Coccus cacti*). Since it is believed that cochineal was known to the Precolumbian Peruvians, we made the comparison, and the curves are shown in Figure 5. Later, we found that William J. Young of the Museum of Fine Arts in Boston had found cochineal in Nazca textiles (*12*).

The value of solution spectrophotometry is clearly shown in the examination of the two Peruvian fabrics shown in Figure 6. The light brown sample (*left*) gives a reflectance curve that is different from that of the bluish-red sample (*right*). The absorption maxima of the curves of the solutions of the extracted dyes in concentrated H_2SO_4 shows that both of them were dyed with cochineal. The light brown sample may also have another dye in it, but if so, it was not identified. The colors obtained with cochineal vary with the mordant used from a bright bluish-red to a chocolate brown (almost black at a high concentration). By dissolving the dye, the effect of the mordant is eliminated, and the solution of the organic dye is available for examination. Figure 7 shows a pair of samples which are not the same color but which are dyed with

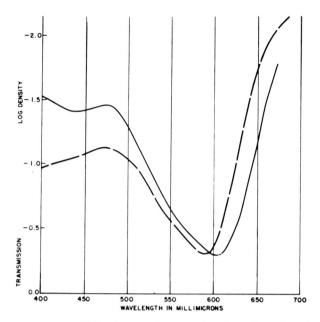

Figure 3. Chloroform solutions of indigo (———) and shellfish purple dye (dibromoindigo) (— —). The chloroform solutions are obtained by extraction from the diluted H_2SO_4 solutions ("drowns") (6).

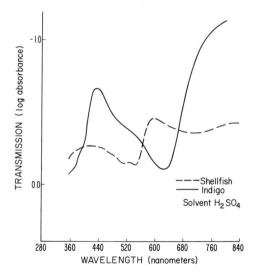

Figure 4. H_2SO_4 (conc.) solutions of indigo (———) and shellfish (— —) purple dye (dibromoindigo)

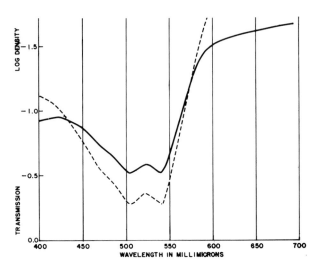

Figure 5. H_2SO_4 (conc.) solutions of cochineal (– – –) and red dye extracted from Nazca fiber (——)

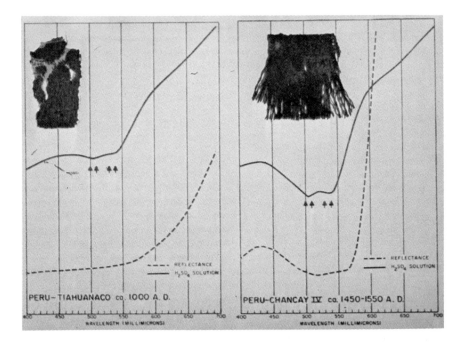

Figure 6. Two Peruvian fabrics showing both their reflectance curves (– – –) and the curves of the H_2SO_4 solutions of the extracted dye (——). Both solutions show the characteristic double absorption maximum of cochineal.

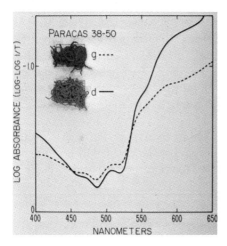

Figure 7. *Ethyl acetate solutions of dyes extracted from the fibers found in a Paracas mummy bundle. The fibers are first wet with a few drops of H_2SO_4 (conc), and then the acid solution is diluted with distilled water. The dyes are then extracted from the dilute H_2SO_4 solutions ("drowns") (6) with ethyl acetate.*

the same dyestuff as evidenced by the close agreement between the curves of ethyl acetate solutions of the extracted dyes. The curve shape and position of the absorption maxima correspond to the solution curve obtained from yarn dyed with a species of relbunium, a madder-like plant which is native to Peru.

Following this work, we obtained a set of samples from the Ocucaje (13) region of Peru from Mary Elizabeth King, then at the Textile Museum. Of the 22 samples examined, seven were found to be dyed with indigo, two with shellfish, 12 with a red dye similar to madder or relbunium, and one with both indigo and relbunium. We are confident that the red dye is not cochineal but are not certain whether it is indeed relbunium or is another plant which gives a similar dye. We have examined only two samples of relbunium. One, *R. hypocarpium,* was obtained from a reputable botanical house, and the other, *R. microphyllum,* was obtained by Kay Antunez de Mayolo in Peru. *R. hypocarpium* may or may not be native to Peru, but there are other varieties which are, as well as other madder-like plants which will be examined after the Antunez de Mayolos return. On the question of the use of relbunium, Fester, in several publications, reported finding the dye from this plant in many textiles from Paracas. However, a note by Yacovleff and Muelle at the end of their paper commented on Fester's work (14), stating that "they have been advised by Dr. Fester that his latest analysis shows wide use of cochineal in the Paracas textiles and increases the doubts with respect to relbunium."

It is our opinion that both of these dyes were used in Paracas but that one may have preceded the other. In our analyses, we have not found any cochineal in the early Paracas textiles from Ocucaje. We have found it in Nazca which is also on the south coast, but the culture is later than that of Ocucaje. It is necessary to obtain well documented samples from a long time sequence in the same area to determine whether the pattern of dye used changed with time and, if so, where the change came from. For this purpose we have obtained a set of samples from the Museo Nacional de Arqueología y Antropología in Lima, which we will analyze in the next few months.

The problem of the correct sequence of excavated Peruvian textiles is a complex one. Until we have an accepted seriation, it will be difficult to draw any conclusions regarding the temporal distribution of dyes in any area.

In addition to finding cochineal in Nazca material, we have also found it in samples of Moche material (AD 200–500) from controlled excavations.

Problems of Using Solution Spectrophotometry

There are two general problems associated with the use of solution spectrophotometry. The first, which at this time is relatively unimportant, is the fact that the technique requires the destruction of some of the sample. Where the colorant has a high tinctorial strength in concentrated H_2SO_4, the amount required can be extremely small. In one case, a radiocarbon-dated sample from Paracas Cavernas, the test used less than 5 mg of yarn from a damaged area. Should the time come where relatively perfect pieces must be examined, it will be possible to sample woven material so that the sampled area does not show. With micro cells it may be possible to reduce the sample size to one-tenth of that used now. Small samples, however, bring up the question of how representative the sample is. At any rate, with our present state of knowledge about Peruvian material, we can learn much by using samples from already damaged pieces.

The second problem associated with solution spectrophotometry is our need for samples of known origin. For dyes from plants this requires, in addition to complete botanical information, samples of yarn dyed under known conditions from these plants. While the dyed yarn cannot contain any material which was not in the plant (except for mordants which may be used), there is much in the plant which may alter a curve but which will not dye the plant. The same is true of insect or other animal material.

Our collection of knowns has been well started wtih the Mullins material and should be quite complete after the return of our collectors from Peru. We should have most, if not all, of the plants mentioned in the writings of the early chroniclers and the ethnobotanists.

After we found shellfish purple on the Paracas Cavernas piece, Bird found an ancient breech cloth from about the same period, from which we were able to obtain samples of purple stains which, when analyzed, proved to be 6,6'-dibromoindigo. This does not prove that the ancient Peruvians ate the molluscs which secrete the purple or even that they dyed cloth with them; it does show that the habit of wiping ones hands on ones clothes is an old one and while you may be able to hide it from your contemporaries, someone may come along 19 or 20 centuries later and find out that you did it! We do know that at least three species which secrete a purple dye (actually they secrete a precursor which, when exposed to sunlight and oxygen, converts to a purple dye) are widely available in Peru. One of them, *Concholepas concholepas* (formerly *C. peruviana*) is gathered in commercial quantities today off the coast of Paracas and is sold as "pata de burro" (donkey's hoof). We have examined a sample of the dye from this species, and it is the same as those extracted from the ancient Paracas textiles.

Other dyes may require larger samples to get satisfactory curves. The amount of sample to be sacrificed for the analysis is a function of the importance of the information to be obtained and possibly of the state of preservation of the object; no one likes to damage any ancient artifact.

Value of Dyestuff Identification to Museum Curators

As mentioned earlier, any information obtained from the artifacts of a preliterate society helps us to understand them. There is another very important reason to determine the nature of the dyes used in museum textiles. This applies particularly to textiles which are to be exhibited. Thanks to the work of Padfield and Landi (15), we have a reasonable amount of information about the lightfastness of natural dyes. With some exceptions, indigo, madder, and cochineal are the most important; they have relatively poor lightfastness. We shall have to determine the fastness of the yarns dyed from Peruvian plants. When we know these facts, we can then help the curator or conservator to make decisions regarding exhibition, lighting, conditions of storage, and safety of conservation treatments.

An excellent example of the value of this information comes from our recent examination of a collection of native Mexican materials which were made by the Mixtec Indians of southwest Mexico. While we had examined samples of purple yarns from this region (shellfish purple is still gathered and used in this area), we had never looked at any red yarns. Cochineal is native to Mexico and this area is "cochineal country." We examined 23 samples which were generally believed to be cochineal, but we determined that only three samples were dyed with cochineal! Seven were dyed with an early synthetic dye, fuchsine or magenta, similar to CI Basic Violet 14 and the rest with four or five different synthetic dyes. Three of these yarns are shown in Figure 8, and the corresponding curves of solutions in concentrated H_2SO_4 are shown in Figure 9. This clearly shows that colors that look alike can often be produced with many different dyes. We have not yet identified the synthetic dyes except for fuchsine. Fuchsine is a dye with a lightfastness of 1 which means that it will fade very quickly (within weeks or less, depending on the level of illumination) and therefore any textile dyed with this dye should not be exhibited but should be reserved for a study collection.

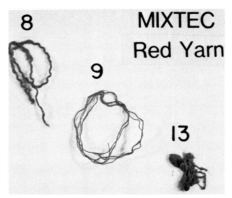

Figure 8.   Three Mixtec red yarns. The H_2SO_4 (conc.) solutions of the dyes extracted from these yarns are shown in Figure 9.

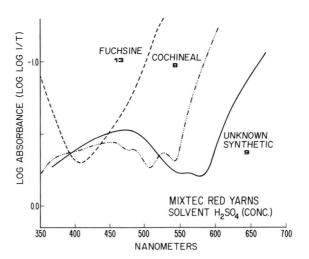

Figure 9. H_2SO_4 (conc.) solutions of dyes extracted from Mixtec red yarns in Figure 8

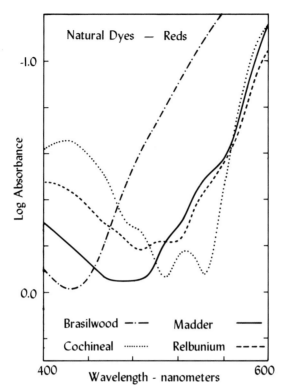

Figure 10. H₂SO₄ (conc.) solutions from red natural dyes which may have been used in the textiles of Precolumbian Peru

Determination of Dyes in Peruvian Textiles

To get back to our Peruvian dyes, from our work to date it seems that it should be possible to determine which of the four red dyes, believed to have been used in Peru in Precolumbian times, was actually used on a given textile. Figure 10 shows the curves of brasilwood, madder, relbunium, and cochineal in concentrated H_2SO_4. They are clearly different from each other even in a single solvent. Their curves in other solvents such at NaOH are also different and resolve any question of differentiating among them.

This may not be as easy in the case of the yellows and browns shown in Figure 11. With yellows, the sulfuric acid curves are of less value, and milder solvents such as methanol with a trace of acid give better results. To date, no one has identified the source of any of the yellow dyes in Precolumbian Peruvian textiles. A recent paper by Kashiwagi points to logwood, but this has not been confirmed (*16*).

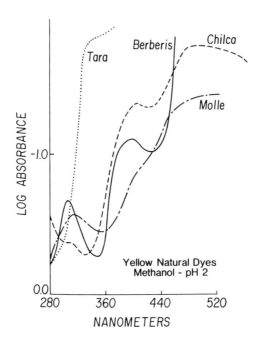

*Figure 11.   Methanol solutions of yellow natu-
ral dyes from Peruvian plants*

In summary, we believe, based on almost 30 years of experience with the technique of colorant identification using solution spectrophotometry, that it is a satisfactory, simple, and rapid technique for identifying the natural dyes used in early America. As we complete our file of reference material and as we obtain samples of known provenence, we can attempt to compile lists of dye usage as related to time and place in Precolumbian America.

Other workers in the field feel that is important to obtain positive identification with a minimum amount of material. In such cases the most desirable method would be that of spectroscopy. DePuy and Quinlivan have presented a paper on the use of this technique to analyze dyes in archaeological material (17). The combination of thin-layer chromatography followed by mass spectrometry (with or without the formation of a derivative) is, without doubt, the method of the future. Now, however, it is not possible to detect cochineal, one of the most important dyes of early America, by this method. We leave this to the modern chemists who are more familiar with the advances in this special field and wish them luck. Until then we feel that much valuable information can be obtained by the simple and readily available techniques which we use.

Acknowledgments

We wish to thank Nora Fisher of the Museum of International Folk Art, Sante Fe, New Mexico, for permission to report on the results obtained on the samples from her collection of Mixtec textiles. We thank Barbara Mullins for sharing with us her samples of dyeings from Peruvian plants.

Literature Cited

1. Schweppe, Helmut, "Nachweis von Farbstoffe auf alten Textilen," *Z. Anal. Chem.* (1975) **276**, 291–296.
2. Abrahams, D. H., Edelstein, S. M., "A New Method for the Analysis of Ancient Dyed Textiles," *Am. Dyest. Rep.* (1964) **53**, 19–25.
3. Saltzman, Max, Keay, A. M., Christensen, Jack, "The Identification of Colorants in Ancient Textiles," *Dyestuffs* (June 1963) **44**(8), 241–251.
4. Saltzman, Max, Keay, A. M., "Colorant Identification," *J. Paint Technol.* (1967) **39**, 360–367.
5. Formanek, Jaroslav, "Untersuching und Nachweis Organischer Farbstoffe auf Spektroskopischen Wege," 2 vols., Springer, Berlin, 1908 and 1911.
6. Stearns, E. I., "Spectrophotometry and the Colorist," *Am. Dyest. Rep.* (1944) **33**, 1–6.
7. Fester, G. A., "Los Colorantes del Antiguo Peru," *Archeion* (1940) **22**, 229–241.
8. Ibid. (1943) **25**, 195–196.
9. Shurcliff, Wm. A., 'Curve Shape Index for Identification by Means of Spectrophotometric Curves," *J. Opt. Soc. Am.* (1942) **32**, 160–163.
10. Augusti, Selim, "Analysis of the Material and Technology of Ancient Mural Paintings," in "Application of Science in the Examination of Works of Art," Procedings of the Seminar, Sept. 7–16, 1965, Research Laboratory, Museum of Fine Arts, Boston (1967), 67–70.
11. Gerhard, Peter, "Shellfish Dye in America," XXXV Congresso Internacional de Americanistas, Mexico, 1962, *Actas y Memorias* (1964) **3**, 177–191.
12. Young, William J., "Appendix III: Analysis of Textile Dyes "Late Nazca Burials in Chavina, Peru," Samuel K. Lothrop and Joy Mahler. Papers of the Peabody Museum of Archaeology and Ethnology, Harvard University (1957) **50**(2).
13. King, Mary Elizabeth, "Textiles and Basketry of the Paracas Period, Ica Valley, Peru," Ph.D. Thesis, University of Arizona (1965).
14. Yacovleff, Eugenio, Muelle, Jorge C., "Notes al Trabajo 'Colorantes de Paracas,'" *Revista del Museo Nacional* (Lima) (1934) **3**(1,2) 157–163.
15. Padfield, Tim, Landi, Sheila, "The Light-Fastness of the Natural Dyes," *Stud. Conserv.* (1966) **11**, 181–196.
16. Kashiwagi, K. Maresuke, "An Analytical Study of Pre-Inca Pigments, Dyes, and Fibers," *Bull. Chem. Soc. Jpn.* (1976) **49**, 1236–1239.
17. DePuy, C. H., Quinlivan, Sandra C. B., "The Analysis of Dyestuffs by Mass Spectrometry," presented at 39th Annual Meeting of the Society for American Archaeology, Washington, D.C., May 1974.

RECEIVED September 19, 1977. This work was supported in part by a grant from the Munsell Color Foundation, Inc.

Ceramics

Analysis of Early Egyptian Glass by Atomic Absorption and X-Ray Photoelectron Spectroscopy

JOSEPH B. LAMBERT and CHARLES D. McLAUGHLIN

Department of Chemistry, Northwestern University, Evanston, IL 60201

Analysis of nine XVIIIth Dynasty Egyptian glass fragments by atomic absorption shows that the silica matrix contains soda ($\approx 18\%$), lime (9%), magnesium oxide (4%), potassium oxide (2%), and alumina (1%). Colors result from the presence of minor components: deep green from $\approx 1\%$ each of Cu and Pb, violet from Fe (1%) with Mn (0.2%), turquoise blue from Cu (1–2%) with Fe (0.6%), yellow–orange from Pb(1%). Opacity is probably caused by antimonates, either Pb (yellow) or Ca (white). The Cu^{2+} in the turquoise blue material was characterized in the x-ray photoelectron spectrum by the low intensity of signals, the high electron binding energies, and the presence of shake-up satellites. The effect of bombardment of the sample with argon ions is illustrated by the Egyptian glasses, a Luristan bronze, and a Portuguese "bronze" manilla.

Elemental composition of artifacts yields a wide range of information. A recent study (1) of Egyptian glass used atomic absorption in conjunction with x-ray diffraction to determine the elements responsible for color and opacity. Atomic absorption (AA) is possibly the most accurate instrumental method for determining elemental composition, although it is insensitive to certain elements and each element must be determined separately (2). X-ray photoelectron spectroscopy (XPS) (3) is a technique developed relatively recently that can go beyond simple elemental constitution and examine the oxidation state of particular elements. We have extended the atomic absorption study of Cowell and Werner (1) to include further Egyptian glass materials, and we have

0-8412-0397-0/78/33-171-**189**$05.00/1

used some of their samples for a detailed examination of copper oxidation states by XPS. Color in glass depends not only on the specific trace elements present but also on their oxidation states. The very elements (Fe, Mn, Cr, Cu, Co, etc.) usually found to convey color to glass are those that can exist commonly in more than one oxidation state. Since XPS is one of the few techniques that can determine oxidation state (3), we felt a study of its applicability to this area would be worthwhile.

Results of Analysis

The fragments we have examined are all unregistered from the British Museum, Department of Egyptian Antiquities, and are believed to come from the XVIIIth Dynasty. In appearance they resemble closely

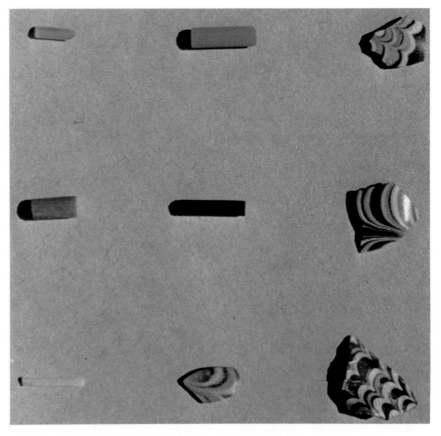

Figure 1. Egyptian glass fragments from the XVIIIth Dynasty. Objects 1–3 are in the first column (top to bottom), 4–6 in the second column, and 7–9 in the third column.

Table I. Description of Egyptian Glasses

Object No.	Type	Color	Opaque or Translucent
1	Trailing stick	light green	T
2	Trailing stick	dark green	O
3	Trailing stick	yellow–orange	O
4	Trailing stick	turquoise blue	O
5	Trailing stick	violet	T
6	Fragment with surface colors	turquoise blue with white streaks	T
7	Fragment with surface colors	violet with white, yellow, and blue streaks	T
8	Fragment with surface colors	violet with white, yellow, and blue streaks	T
9	Fragment with surface colors	turquoise blue with white and yellow streaks	T

those described recently by Cowell and Werner (*1*), which are also XVIIIth Dynasty and were originally collected by Howard Carter. The objects take the form either of a trailing stick (used to decorate glass surfaces) or a fragment of a larger object with polychrome surface designs (Figure 1). In the latter case we examined only the base glass free of surface designs. The samples are 2–4 cm in length. A description of the glasses is given in Table I.

XPS provides an excellent visual presentation of the qualitative composition of an artifact (*3*). The top spectrum in Figure 1 of Ref. *3* shows a survey spectrum for a blue Egyptian glass fragment. At a glance, one can detect Na, Ca, Mg, Cu, Cl, Si, and, of course, O. The lead peaks are from the sample holder, and the carbon peaks are from the organic vacuum pump oil. The spectrum shows that the glass is of the soda lime type, probably high in magnesium, as is characteristic of glass from the ancient Middle East (*4*).

Quantitative analysis for elemental composition was carried out by atomic absorption and (for silicon only) by colorimetry. Samples were removed from each fragment by sanding the glass with a diamond-impregnated abrading tool. The same method was used for XPS samples. In addition, the green glass (object 2) was sampled by crushing a chip in a percussion mortar to detect possible contamination from chromium in the abrading tool. The 10-mg samples were dissolved in $HF/HClO_4$ according to the method of M. R. Cowell, and 15 elements were determined by AA (Na, Ca, Mg, Al, K, Fe, Cu, Pb, Sb, Mn, Cr, Co, Ni, Li, and Sn). A separate 8-mg sample from each object was fused with lithium metaborate and was dissolved in nitric acid to determine silicon by colorimetry. The analytical results are given in Table II as percentages based on oxide content.

About 98% of each sample is composed of the following general matrix: SiO_2, 63%; Na_2O, 18%; CaO, 9%; MgO, 4%; K_2O, 2%; Al_2O_3, 1%; Fe_2O_3, 0.5%; CuO, 0.1%; Sb_2O_5, 0.1%; MnO_2, 0.03%; Li_2O, 0.004%.

Thus colorants comprise only about 2% of the total. The matrix is typical of soda lime glasses (high magnesium, high potassium) of this period and region (4).

The nearly colorless glass fragment (object 1) has essentially the matrix composition. The slight green tint probably comes from the very small amounts of iron, copper, and manganese oxides present.

There was no difference between the two analyses of the green glass (object 2), so that the abrading tool did not introduce chromium. The high nonmatrix constituents are copper and lead. The color is probably caused by copper (in the presence of iron), and the sole function of the lead (in conjunction with antimony) may be to render the sample opaque.

The yellow color of object 3 is perplexing, because lead is the only significant nonmatrix element. Normally such a color results from lead antimonate, $Pb_2Sb_2O_7$. Powder x-ray studies by M. Bimson (British Museum Research Laboratory) showed that the sample did contain lead antimonate. Thus some antimony must have escaped during dissolution and evaporation, possibly as SbF_5. We conclude that the yellow color and the opacity of this object results from the lead antimonate. This

Table II. Percentage Composition

	SiO_2[a]	Na_2O	CaO	MgO	Al_2O_3	K_2O	Fe_2O_3	CuO	PbO_2	Sb_2O_5
1	61.5	18.4	11.6	4.43	0.51	1.92	0.61	0.092	<0.05	0.08
2s[c]	64.0	17.5	7.6	3.44	0.61	2.91	0.41	0.93	0.95	0.12
2p[c]	65.0	17.7	7.9	3.53	0.63	2.96	0.53	0.95	0.85	0.10
3	62.0	18.4	9.3	4.57	0.53	2.11	0.42	0.059	1.03	0.12[f]
4	63.0	18.3	10.7	4.09	0.76	2.58	0.67	0.78	<0.05	0.63
5	65.0	20.2	6.7	3.10	1.66	0.54	1.01	0.34	<0.05	<0.06
6	61.5	15.9	10.7	5.65	0.97	2.20	0.55	0.78	<0.05	0.07
7	63.5	16.7	8.9	4.77	2.88	1.13	0.70	0.16	<0.05	0.16
8	59.0	17.6	10.9	4.99	3.68	0.50	0.98	0.081	<0.1	0.16
9	68.0	19.3	7.2	2.16	0.87	0.87	0.74	1.92	<0.05	0.20
Red[d]	64.0	16.0	7.4	4.4	0.9	1.95	0.43	3.9[h]	<0.1	0.5
Blue[d]	61.0	17.2	7.05	3.1	0.8	1.45	0.37	3.5	<0.1	2.1

[a] Determined by colorimetry; all other elements determined by atomic absorption.
[b] The average sum for samples 1–9 is 99.6%. Most of the error derives from the silicon measurement.
[c] Weight in mg for atomic absorption analyses.
[d] Weight in mg for colorimetric analyses.

material may also be present in object 2, giving it its opacity, but the copper component overcomes the coloring effect to produce the green hue.

The turquoise blue color of object 4 results from the high proportion of copper, in the presence of iron, which in turn is slightly higher than normal. The large amount of antimony again is probably associated with the sample's opacity (none of the remaining objects are opaque). Lead is not present, so the opacifying agent may be calcium antimonate, $Ca_2Sb_2O_7$. No x-ray studies were done to confirm this hypothesis, although calcium antimonate has been found in another opaque blue Egyptian glass (*1*).

Object 5 is high in iron, and to a less extent in copper and in manganese. In addition, there are small amounts of cobalt and nickel. The iron component is probably most important in determining the violet color, but the remaining elements must affect the exact hue.

The color of object 6 is similar to that of object 4 but without the opacity. Again, copper probably is dominant in causing a turquoise blue color. At 0.11%, chromium may not contribute to the color. The effect of 0.52% nickel, if any, is not known.

The color and constitution of object 7 are similar to those of the violet trailing stick, object 5. The color is caused by iron in the presence of manganese and cobalt. There is half as much copper (0.16%) in this object as in object 5 (0.34%), so that copper at this level must not influence coloring, or its effect is masked by the higher proportion of iron.

Object 8 is quite like object 7, with the violet color produced by iron in the presence of manganese, chromium, and cobalt. The relatively high

of Egyptian Glass Fragments

MnO_2	Cr_2O_3	CoO	NiO	Li_2O	SnO	Sum^b	Wt^c	Wt^d
0.031	<0.015	<0.007	<0.015	0.004	<0.9	99.2	10.18	7.92
0.025	<0.015	<0.007	<0.015	0.004	<0.9	98.5	10.60	8.37
0.026	<0.015	<0.007	<0.015	0.004	<0.9	100.2	10.73	8.37
0.023	<0.015	<0.007	<0.015	0.004	<0.9	98.6	9.97	8.14
0.026	<0.015	<0.007	<0.02	0.004	<0.9	101.5	9.97	8.01
0.104	<0.015	0.058	0.072	0.004	<0.9	98.8	9.81	8.15
0.031	0.11	<0.007	0.52	0.004	<0.9	99.0	9.76	7.75
0.316	0.02	0.069	0.11	0.004	<0.9	99.4	10.02	8.02
0.321	0.10	0.078	0.27	0.004	<0.9	98.7	10.11	7.81
0.023	0.07	<0.007	0.55	0.0025	<0.9	101.9	10.11	8.22
<0.04	0.02	<0.01	—	—	—	99.5	—	—
<0.04	0.04	<0.01	—	—	—	96.6	—	—

[e] Sample 2s, like 1 and 3–9, was obtained by sanding with the abrading tool. Sample 2p was prepared by crushing a chip in a steel percussion mortar.

[f] This figure must be low, since $Pb_2Sb_2O_7$ was observed by x-ray diffraction. Possibly SbF_5 was lost during the evaporation step.

[g] Analyzed by M. R. Cowell. *See Ref. 1.*

[h] Expressed as percentage Cu_2O; all others are CuO.

amount of nickel is probably not important. It is interesting that all three violet glasses (5, 7, 8) are high in aluminum (1.7–3.7%, rather than 0.5–0.9%).

The composition of object 9 is similar to that of the other two turquoise blue objects, 4 and 6. The color results from copper in the presence of iron.

In summary, the violet color (objects 5, 7, 8) is caused by iron in the presence of manganese and cobalt. The turquoise blue color (objects 4, 6, 9) is caused by copper in the presence of iron. These same components (copper, iron) in the presence of lead (probably as lead antimonate) produce a green color (object 2). Lead antimonate alone causes a yellow-orange color (object 3). Cowell and Werner (1) have found also that cobalt at the 0.2% level can impart a deep blue color, that antimony at the 2% level (as calcium antimonate) can produce an opaque white glass, and that copper in a certain form (see below) at the 4% level can give a deep red color. These results are in general agreement with those found by other workers (5).

Analysis for Oxidation State of Copper

Copper is the critical colorant in four of our nine samples. In addition to these blues and greens, copper can also produce a red color (1). It has been presumed that this color is caused by Cu^0 or Cu^+, whereas the blues and greens are caused by Cu^{2+}. Since XPS can often differentiate oxidation states, we wanted to see whether the differences in color are reflected in the photoelectron spectra (3). Previous workers have found that the XPS properties of Cu^{2+} are different from those of Cu^+ and Cu^0, whose properties in turn are not distinguishable (6,7,8).

For a frame of reference, it is useful first to examine the spectrum of copper metal in a bronze, since the oxidation state should be zero. The 930–960 eV binding energy region for a Luristan bronze sample has been presented as Figure 2 of Ref. 3. In order to clean the surface of the metal, the sample was first bombarded with argon ions ("etching") for 30 min. The spectrum after this process contains only the two peaks of the spin-orbit doublet (3), both resulting from the zero oxidation state expected for copper metal. The same sample without etching gives the Cu^0 (Cu^+) spin-orbit doublet at 933 and 953 eV and the analogous doublet from Cu^{2+} at 935 and 955 eV. There are also two broad peaks at 941–946 and 961–966 eV, which are "shake-up satellites" (3) (two again because of spin-orbit splitting). For our purposes, the most important characteristic of shake-up satellites is that they can be caused only by Cu^{2+} (6,7,8). Thus the typical spectrum for a dirty bronze contains two

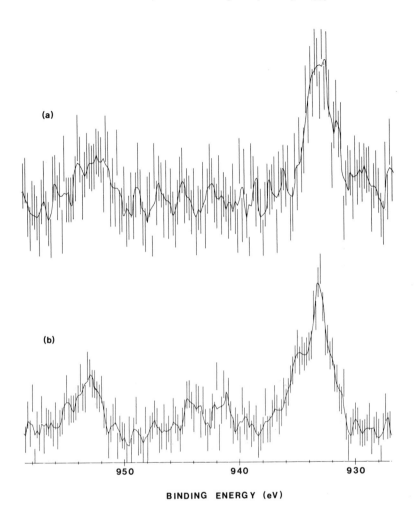

BINDING ENERGY (eV)

*Figure 2.   (a) The XPS copper region of the Red Egyptian glass object
in Table II after argon ion etching. (b) The same material without
etching.*

peaks from Cu^0 (Cu^+) and four peaks from Cu^{2+}, which is the expected
oxidation state for most corrosion and oxidation products.

The red glass from the study of Cowell and Werner (labeled "Red"
in Table II) provides an example of copper that is thought to be in the
0 or 1+ oxidation state (9). Without etching (Figure 2b), the sample
shows a broad peak (possibly a doublet) in the 930–937 eV region,
another peak in the 951–956 eV region, and a small peak in the 940–945
eV region. The last region is characteristic of Cu^{2+} shake-up satellites.
On etching (Figure 2a), however, the shake-up satellite disappears

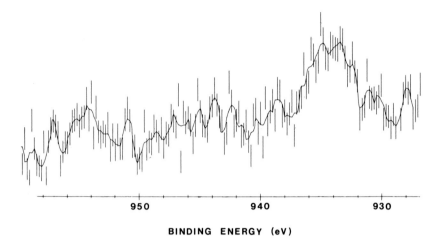

BINDING ENERGY (eV)

Figure 3. The XPS copper region of the Blue Egyptian glass object in Table II without argon ion etching. Etching brought about no spectral change.

entirely, and the peaks at 933 and 953 remain essentially unchanged, quite analogous to the behavior of copper metal in the bronze. These spectral characteristics exclude the presence of Cu^{2+} in the red glass, except as a possible surface contaminant and are consistent with either Cu^0 or Cu^+, in agreement with a previous supposition (9).

The turquoise blue glass most amenable to XPS is also taken from the samples of Cowell and Werner (labeled "Blue" in Table II), since its 3.5% Cu is the highest among those examined. The sensitivity of the copper peaks in all the blue/green glasses was much lower than that of the various red glasses that have been examined. Jorgensen (10, 11) has reported that the sensitivity of Cu^{2+} is almost half that of Cu^+. It is possible that the presence of shake-up satellites for Cu^{2+} dilutes the intensity, whereas no such dilution occurs for Cu^+, which has no shake-up satellites. The copper region of the photoelectron spectrum of the blue sample (Table II) is given in Figure 3. Although the spectrum is of poor sensitivity, the shake-up satellite is clearly visible in the 940–946 eV range, and it is almost as intense as the smaller spin-orbit Cu^{2+} doublet at 954 eV. Most significantly, the spectrum does not change on argon etching, the only alteration being a slight sharpening of the peaks. In particular, the shake-up satellite does not diminish in intensity, in contrast to the situation in Figure 2. The spin-orbit doublets of Figure 3 are at slightly higher binding energy than those in Figure 2a (934 and 954 vs. 933 and 952 eV). A 1–2 eV difference is normal for Cu^{2+} vs. Cu^0 or Cu^+ (6, 7, 8). These small differences were accurately measured by use of a small "internal standard"—a few particles of copper metal. The lower intensity of the peaks, the presence of the shake-up satellites, and the higher binding energies are all consistent with the Cu^{2+} oxidation state in the blue glass, as compared with the Cu^0 or Cu^+ oxidation state in the red glass.

These experiments point out the strengths and weaknesses of the XPS method. Its primary limitation is poor sensitivity, although newer instrument than ours (an AEI ES-200) can attain sensitivity an order of magnitude or so higher. In the present study, colorants below the 1% level simply could not be examined. It is hoped that the newer instrumentation will remedy this deficiency. Although the primary approach to differentiating oxidation states is by the detection of binding energy differences, this illustration points out that other properties of the photoelectron spectrum (in particular the shake-up satellites) are useful and sometimes necessary.

The use of the technique of argon etching was critical in the above examination of copper oxidation states. XPS examines only about the top 20–50 Å of the surface of an object. Thus it is extremely sensitive to surface impurities, which may be removed by the etching process. By the same token, of course, XPS can provide analysis of surface features that would be essentially impossible by other techniques. To illustrate the overall effect of etching, we offer the entire survey spectrum (50–1000 eV) for a Portuguese manilla, the form in which copper was traded from Europe to West Africa (Figure 4a). A typical manilla (*12*) contains copper in the 59–69% range and lead in the 25–37% range as the major

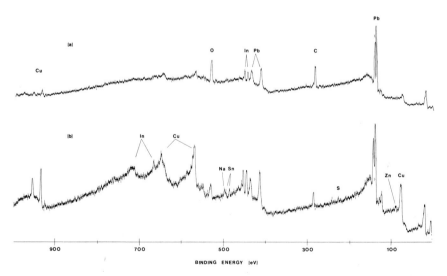

Figure 4. (a) The survey x-ray photoelectron spectrum of a Portuguese manilla. (b) Spectrum of the same object after argon ion etching.

elements. Figure 4a shows the unetched spectrum of a manilla mounted as a powder on indium rather than on lead, since lead is a major component of the matrix. The lead peaks are dominant, and the copper peaks are weak. The carbon comes as usual from the spectrometer pump oil, the indium comes from the mount, and the oxygen is a surface impurity. The spectrum in Figure 4b is obtained after the sample is bombarded for 30 min with argon ions. The oxygen peak has essentially disappeared, and the carbon peak is substantially lower in relation to the other peaks. The lead peaks are still quite large, and the copper peaks have become prominent. In addition, peaks are visible for Zn, S, Sn, and Na. Expected peaks from arsenic and antimony, however, are not detectable (the main antimony peak coincides with that of oxygen). Thus ion etching removes surface impurities and provides enhancement of peaks from the matrix. Etching is not always successful in improving the quality of the spectrum. In at least one example (a pottery matrix), we observed a slight reduction in the intensity of the potassium peak. The technique is clearly useful in the examination of metals, and it can have considerable utility with other materials, such as the glasses in the present study.

Summary

The AA and XPS techniques provide complementary analytical information. From AA we have learned that these XVIIIth Dynasty Egyptian glass fragments have the soda lime silica matrix with high MgO and K_2O that is typical of the Middle East during this period. The prime colorants are Fe, Cu, Pb, Mn, and Co, all in less than 2% (total). The prime opacifiers are probably lead and calcium antimonate. Copper (1–4%) can cause blue, green, or red colors. XPS can distinguish the Cu^{2+} oxidation state from Cu^+ or Cu^0 by relative binding energies and peak intensities and by the presence of shake-up satellites. In this manner the red glasses were shown to contain Cu^+ or Cu^0 and the blue/green glasses Cu^{2+}. The identification of oxidation state is assisted by etching the sample with an argon ion beam. This process cleans the surface of the sample, thereby removing corrosion products and enhancing the sensitivity of the spectrum.

Acknowledgments

The atomic absorption analyses were carried out by the author at the British Museum Research Laboratory. We are deeply grateful for the assistance by the staff (M. R. Cowell in particular) and for permission by the Keeper (M. S. Tite) to publish the data. We are also grateful to A. E. A. Werner, M. J. Hughes, and P. T. Craddock for advice during the course of the study; to the Department of Egyptian Antiquities of the British Museum (I. E. S. Edwards, Keeper) for the loan of the Egyptian glass fragments; to the Guggenheim Foundation for a fellowship to JBL during his tenure at the British Museum; to F. Willett (formerly of the Northwestern University Department of Anthropology and currently Director of the Hunterian Gallery, Glasgow) for the loan of the Portuguese manilla; and to the National Science Foundation, the Chevron Research Company, and the Research Corporation for financial support of the XPS studies.

Literature Cited

1. Cowell, M. R., Werner, A. E., "Analysis of Some Egyptian Glass," Annales du 6th Congrès de L'Association Internationale pour L'Histoire du Vere, pp. 295–298, Cologne, 1973.
2. Hughes, M. J., Cowell, M. R., Craddock, P. T., "Atomic Absorption Techniques in Archaeology," *Archaeometry* (1976) **18**, 19–37.
3. Lambert, J. B., McLaughlin, C. D., "X-Ray Photoelectron Spectroscopy: A New Analytical Method for the Examination of Archaeological Artifacts," *Archaeometry* (1976) **18**, 169–180.
4. Sayre, E. V., Smith, R. W., "Some Materials of Glass-Manufacturing in Antiquity," *in* "Archaeological Chemistry: A Symposium," M. Levey, Ed., pp. 279–311, University of Pennsylvania, Philadelphia, 1967.
5. Caley, E. R., "Analyses of Ancient Glasses 1790–1957," Corning Museum of Glass, Corning, 1962.
6. Frost, D. C., Ishitani, A., McDowell, C. A., "X-Ray Photoelectron Spectroscopy of Copper Compounds," *Mol. Phys.* (1972) **24**, 861–877.
7. Larson, P. E., "X-Ray Induced Photoelectron and Auger Spectra of Cu, CuO, Cu_2O, and Cu_2S Thin Films," *J. Electron Spectrosc. Relat. Phenom.* (1974) **4**, 213–218.
8. Brisk, M. A., Baker, A. D., "Shake-up Satellites in X-Ray Photoelectron Spectroscopy," *J. Electron Spectrosc. Relat. Phenom.* (1975) **7**, 197–213.
9. Hughes, M. J., "A Technical Study of Opaque Red Glass of the Iron Age in Britain," *Proc. Prehistoric Soc.* (1972) **38**, 98–107.
10. Jorgensen, C. K., Berthou, H., "Relative Intensities and Widths of X-Ray Induced Photoelectron Signals from Different Shells in 72 Elements," *Discuss. Faraday Soc.* (1972) **54**, 269–276.
11. Berthou, H., Jorgensen, C. K., "Relative Photoelectron Signal Intensities Obtained from a Magnesium X-Ray Source," *Anal. Chem.* (1975) **47**, 482–488.
12. Willett, F., unpublished results.

RECEIVED September 19, 1977.

13

Elemental Compositions of Spanish and Spanish-Colonial Majolica Ceramics in the Identification of Provenience

JACQUELINE S. OLIN

Conservation-Analytical Laboratory, Smithsonian Institution, Washington, DC 20560

GARMAN HARBOTTLE and EDWARD V. SAYRE

Department of Chemistry, Brookhaven National Laboratory, Upton, NY 11973

On the basis of the compositions obtained by neutron activation analysis two distinctive groups of pottery have been identified from the majolica sherds excavated from Spanish sites in the New World. The principal sites yielding majolica sherds analyzed in this project include Isabela, La Vega Vieja, Juandolio, and Convento de San Francisco in the Dominican Republic; Nueva Cadiz in Venezuela; and excavations in Mexico City at the Metropolitan Cathedral and for the Mexico City Metro transportation system. Concentrations of Na_2O, K_2O, BaO, MnO, Fe_2O_3, Rb_2O, Cs_2O, La_2O_3, Sc_2O_3, CeO_2, Eu_2O_3, HfO_2, ThO_2, Cr_2O_3, and CoO are reported. Petrographic and x-ray diffraction results confirm the existence of two distinctive groups. Comparisons to known specimens indicate a Spanish source and a Mexican source.

The study of tin-glazed ceramics (majolica) associated with Spanish colonization of the New World has identified the need for methods to classify the various types of majolica which have been excavated. One of the major efforts in this regard has been made by John M. Goggin (1). Goggin studied majolica from public and private collections and from field study and/or excavation in Cuba, Jamaica, Haiti, the Dominican Republic, Puerto Rico, St. Croix, Trinidad, Venezuela, Columbia, Yucatan, and Mexico. The system of type names which he assigned to ceramics with specific glaze decorations is now the basis for establishing the chronology and provenience of the majolica.

0-8412-0397-0/78/33-171-**200**$07.50/1

In the course of colonizing the New World the Spanish first brought majolica from Spain to the colonies, then in due course manufactured it there. Records reveal that between 1504 and 1555 some 2,805 ships sailed for the New World from Seville and the Canary Islands carrying cargos of weapons, provisions, metal, and household goods including "loza blanca" or white tableware. One of the types of majolica which is found in most of the early sites is a plain, white-glazed ceramic tableware which Goggin has named Columbia Plain. The glaze is made by adding tin oxide to a lead glaze. We have identified lead and tin by x-ray fluorescence in the glaze of four of our Columbia Plain sherds. Majolica types other than Columbia Plain have a colored design applied to a thin opaque white glaze or have a colored opaque glaze. Examples of these and other 16th century types are shown in Figures 1–4.

The tradition of majolica production in Spain is well documented. Its presence there dates back to before the 11th century. The types of majolica most frequently found in museum collections in Europe and the United States, however, are generally more decorative types than those excavated at the early Spanish sites in the New World. Nevertheless, we know that the types shown in Figures 1 and 2, Columbia Plain and Yayal-Blue-on-White, were used in Spain. These two types appear in paintings of Velasquez, Zurbaran, and Murillo, all of whom painted in Seville early in the 17th century. We know, therefore, that the majolica types excavated at the early 16th century Spanish sites in the New World were being produced in Spain. We also know that white tableware was being shipped to the New World. It is reasonable to conclude that majolica found in early sites in the New World may have originated in Spain.

Sources of Specimens

The sites in the New World from which these sherds have been excavated are among the earliest to be inhabited by the Spanish. In the Dominican Republic there are four: Isabela, which was the first substantial settlement in the New World; La Vega Vieja, which appears to have been founded as early as 1495; the Convento de San Francisco, the first of several religious houses to be constructed in the city of Santa Domingo (for which the church was completed in 1555 and the monastery in 1556); and Juandolio, which was an early 16th century site. In Venezuela the site of Nueva Cadiz was the earliest Spanish settlement. It was founded about 1515 and was abruptly destroyed by an earthquake in 1545 and subsequently deserted. The sherds from Mexico come from excavations within Mexico City. These include archaeological excavations at the Metropolitan Cathedral and commercial excavation for the Metro

Figure 1. *Columbia Plain sherds* (left to right): (top) *SA2 from Convento de San Francisco, Dominican Republic and SA41 from Juandolio, Dominican Republic; (bottom) SA19 from Nueva Cadiz, Venezuela and SA59 from LaVeija, Dominican Republic*

Figure 2. *Yayal-Blue-on-White sherds* (left to right): *SA61 and SA62 from Nueva Cadiz, Venezuela*

Figure 3. Caparra Blue sherds (left to right): *SA28 and SA30 from Nueva Cadiz, Venezuela*

Figure 4. Isabela Polychrome (left to right): *SA47 and SA48 from Nueva Cadiz, Venezuela*

subway construction. We have analyzed a few additional sherds recovered from Guatemala, Panama, Peru, and Equador.

The sherds from the Dominican Republic and all but four of those from Venezuela were sent to us by Charles Fairbanks. Some of these sherds were excavated by John M. Goggin and are from the collection at the University of Florida. The sherds from Mexico City and from the sites in Panama, Guatemala, Ecuador, and Peru were provided by Florence and Robert Lister.

To determine whether any of the New World material was actually fabricated in Spain it is necessary, of course, to have material from Spain itself for comparison. However, little if any excavated majolica of the 16th or 17th century from Spain is available. According to F. Lister (2), there are no controlled excavations of 16th and 17th century sites in the area of Seville–Triana which she and others believe to have been the source of the majolica shipped to the New World. In 1975, F. Lister learned of some excavations which had taken place at the Carthusian monastery at Jerez, Spain. This monastery was constructed about the time of the Spanish penetration in the Caribbean. The sherds which we have analyzed are from among those found in a dump area outside of the monastery (2). They are of the types Columbia Plain and Yayal-Blue-on-White and therefore do correspond to types which are included with the excavated materials from the New World Sites. Admittedly this is a very small sampling of comparison material from Spain but the lack of archaeological excavations of sites which would have produced majolica of this period has made these sherds more valuable than they otherwise would be. It is hoped that future archaeological excavation in Spain will provide a more complete sampling of majolica of authentic Spanish origin which we can analyze.

Experimental Methods and Results

In this research we have attempted to obtain evidence of the origin of the sherds found in the New World through neutron activation analysis, x-ray diffraction, and petrographic analysis. Neutron activation analysis provided the concentrations of 15 constituent oxides for 178 sherds. These analyses were carried out in the Chemistry Department at Brookhaven National Laboratory using the standard procedures developed there which are reported by Abascal et al. (3). Six U.S. Geological Survey standard rocks designated AGV1, DTS-1, PCC-1, GSP-1, and G-2 were used as standards (4). The specific standards used for each element are reported by Bieber et al. (5).

Neutron activation analysis revealed that the compositions of all the specimens from the five early sites in the Dominican Republic and Venezuela closely agree and that they differ significantly from all of the other specimens from New World sites. The individual data for these sherds are given in Table I, and in Table II the mean compositions of the sherds from the five sites are compared. The elements which do show some discernible variation among sites are the alkali metals rubidium and sodium. The alkali compounds tend to be water soluble and hence are susceptible to variations during burial.

Using the data in Table I for 40 Columbia Plain and Yaval-Blue-on-White sherds from Convento de San Francisco, Dominican Republic and for 41 sherds from Nueva Cadiz, Venezuela, we have calculated a value of t by dividing the difference between the means of the concentrations by the standard error of the difference between the means. We have calculated t for the oxides of rubidium, sodium, and cesium, and the numbers obtained are 5.6, 4.1, and 1.8, respectively. With this sample size the significance levels for the values of t for rubidium and sodium are greater than 99.5% and for cesium are greater than 90%. There is, therefore, a statistically significant difference between the concentrations of each of these three oxides for all of the sherds from Convento de San Francisco compared with those from Nueva Cadiz.

The Columbia Plain sherds from four different sites shown in Figure 1 have a different physical appearance. The physical condition of the glaze is consistent among sherds from a given site but is readily distinguishable in appearance from sherds from other sites. It appears that the glaze is altered to different degrees depending on different burial conditions. These differences in burial conditions could also have contributed to the variations in the alkali concentrations in the sherds. The basic similarity in composition of the sherds from all of these five sites strongly suggests, however, that they were manufactured from a single or at most a few closely related clays; that is to say, the similarity in composition would argue against a hypothesis that these wares were manufactured in each of the separate sites in which they were found. Indeed Goggin concludes that they were all probably of Spanish origin, most likely the Triana section of Seville (1).

The results of our analysis of the sherds from Spain which were supplied to us by F. Lister reveal that the composition of these sherds basically conforms to the same compositional pattern which characterizes the sherds from the five sites in the Dominican Republic and Venezuela. The data for the sherds from Jerez are given in Table III, and the mean compositions are given in Table II with the mean compositions for the sherds from sites in the New World. The agreement between the sherds found in the Dominican Republic and Venezuela and those found in Spain is excellent for 14 or 15 elements, the only exception being the component barium oxide. However, we frequently have observed erratic barium concentrations in related groups of sherds, and we tend to ascribe

Table I.   Concentration of Oxides in Majolica from Spanish

Specimen No.	Na_2O (%)	K_2O (%)	BaO (%)	MnO (%)	Fe_2O_3 (%)	Rb_2O (ppm)	Cs_2O (ppm)
Columbia Plain, Convento de San Francisco, Dominican Republic							
SA01	0.84	1.96	0.040	0.073	4.8	109	5.7
SA02	0.85	2.03	0.039	0.067	5.0	216	6.6
SA03	0.93	1.75	0.041	0.075	4.9	107	6.0
SA04	1.20	1.08	0.035	0.077	5.2	108	6.8
SA5B	1.01	0.91	0.036	0.058	4.6	110	6.0
SO6A	0.79	1.72	0.050	0.068	4.9	132	5.4
SA07	0.87	1.52	0.042	0.063	4.9	142	6.2
SA08	1.05	1.32	0.041	0.069	5.0	109	6.8
SA9B	1.15	1.06	0.035	0.071	4.7	95	5.5
SA10	1.01	1.10	0.041	0.064	4.4	122	5.1
SB86	0.91	1.77	0.050	0.086	5.1	82	4.8
SB90	1.03	1.19	0.044	0.073	4.9	92	5.2
SB91	1.16	1.12	0.046	0.073	5.0	79	5.6
SB92	1.36	0.98	0.032	0.076	5.2	83	5.7
SB93	1.13	1.14	0.042	0.070	5.0	96	6.4
SB94	0.98	1.27	0.042	0.070	4.6	85	5.2
SB95	0.98	1.07	0.043	0.069	4.7	84	4.7
SB96	0.64	2.28	0.044	0.078	4.5	85	4.0
SB97	1.04	1.44	0.057	0.086	6.2	69	4.8
SB98	1.11	1.56	0.048	0.093	5.2	81	5.4
SB99	1.42	0.90	0.031	0.075	4.8	80	6.2
SC01	1.07	1.56	0.048	0.080	5.3	74	4.3
Columbia Plain, Nueva Cadiz, Venezuela							
SA11	1.00	1.83	0.051	0.083	5.0	76	7.4
SA12	0.67	2.05	0.063	0.077	4.9	72	6.1
SA13	1.12	1.62	0.038	0.078	5.0	64	4.9
SA14	1.06	1.14	0.054	0.083	5.3	63	7.7
SA15	1.14	1.20	0.112	0.081	5.1	62	7.2
SA16	1.09	1.62	0.048	0.077	5.0	79	8.1
SA17	1.38	1.36	0.038	0.067	4.5	49	5.1
SA18	1.00	1.28	0.055	0.077	5.1	48	7.2
SA19	1.01	1.41	0.043	0.070	5.4	70	6.3
SA20	1.29	1.02	0.048	0.076	5.3	48	6.6
SB59	1.62	1.07	0.033	0.090	5.0	56	5.8
SB60	1.77	1.05	0.062	0.081	5.5	44	6.1
SB61	1.54	1.62	0.042	0.083	4.6	77	5.3
SB63	1.56	1.78	0.042	0.081	4.7	75	5.9
SB64	1.28	1.38	0.042	0.092	5.4	72	7.2
SB65	1.58	1.06	0.055	0.080	4.9	70	5.1
SB66	1.94	0.76	0.037	0.083	4.9	51	5.8
SB67	1.84	0.99	0.053	0.085	4.4	48	4.1
SB68	1.54	0.96	0.046	0.083	5.2	52	5.8

American Sites in the Dominican Republic and Venezuela

La_2O_3 (ppm)	Sc_2O_3 (ppm)	CeO_2 (ppm)	Eu_2O_3 (ppm)	HfO_2 (ppm)	ThO_2 (ppm)	Cr_2O_3 (ppm)	CoO (ppm)
\multicolumn{8}{c}{Columbia Plain, Convento de San Francisco, Dominican Republic}

Columbia Plain, Convento de San Francisco, Dominican Republic

La_2O_3 (ppm)	Sc_2O_3 (ppm)	CeO_2 (ppm)	Eu_2O_3 (ppm)	HfO_2 (ppm)	ThO_2 (ppm)	Cr_2O_3 (ppm)	CoO (ppm)
39	19.2	79	1.52	5.9	11.3	107	17.5
40	22.0	83	1.49	6.5	11.9	112	15.8
40	20.0	81	1.60	6.1	12.1	123	15.8
44	21.1	90	1.74	6.5	12.6	126	16.7
39	18.4	80	1.45	5.8	11.6	113	14.7
39	19.8	83	1.52	5.8	11.3	93	16.8
42	19.9	85	1.41	5.9	11.9	121	16.2
41	20.0	86	1.58	6.0	12.5	112	16.3
39	19.5	81	1.52	6.5	11.4	117	18.3
37	17.7	75	1.38	6.2	11.1	107	15.0
47	20.6	133	1.55	6.2	12.5	126	17.0
36	19.2	61	1.38	5.4	11.4	114	15.1
37	20.0	86	1.51	6.3	12.2	124	13.8
36	21.2	59	1.58	5.8	11.3	129	16.2
37	20.5	69	1.56	6.2	11.8	131	14.9
35	18.6	49	1.35	5.7	11.1	111	14.3
34	19.1	90	1.35	5.4	10.1	110	14.9
35	18.4	63	1.37	6.0	10.8	99	15.0
39	22.0	88	1.46	5.9	11.8	131	17.7
38	21.0	42	1.72	6.6	12.7	122	17.1
35	19.5	73	1.43	5.4	11.1	122	15.1
40	20.9	85	1.61	6.9	12.9	116	17.8

Columbia Plain, Nueva Cadiz, Venezuela

La_2O_3	Sc_2O_3	CeO_2	Eu_2O_3	HfO_2	ThO_2	Cr_2O_3	CoO
37	20.4	80	1.66	5.5	11.8	111	17.1
38	19.8	84	1.56	6.3	11.7	113	17.0
37	20.0	82	1.47	5.6	11.5	113	15.4
39	21.0	83	1.47	5.6	11.2	113	16.1
40	20.7	81	1.64	5.5	11.9	117	16.0
37	20.1	82	1.66	5.6	11.9	110	16.7
40	18.0	79	1.41	5.7	10.5	98	13.3
39	20.5	88	1.50	5.9	11.8	111	15.6
38	20.0	82	1.59	5.6	11.0	111	14.1
36	22.0	87	1.67	6.0	12.6	125	16.0
36	20.0	79	1.60	6.1	11.8	120	16.7
40	22.4	89	1.71	5.9	12.7	134	16.2
35	19.5	77	1.65	5.5	10.9	115	18.3
38	19.6	86	1.68	6.3	13.0	115	16.9
38	21.9	83	1.64	5.2	12.1	132	17.5
36	20.2	—	1.42	5.9	11.4	118	18.8
38	19.3	—	1.61	6.2	11.8	123	15.3
32	17.4	—	1.19	6.3	10.2	114	15.5
40	20.8	—	1.74	6.2	13.0	133	14.7

Table I.

Specimen No.	Na_2O (%)	K_2O (%)	BaO (%)	MnO (%)	Fe_2O_3 (%)	Rb_2O (ppm)	Cs_2O (ppm)
Columbia Plain, Juandolio, Dominican Republic							
SA32	0.92	2.20	0.043	0.075	5.0	81	4.2
SA33	0.72	2.08	0.043	0.077	5.2	83	5.4
SA34	0.93	2.10	0.048	0.093	5.8	89	5.6
SA35	0.96	1.87	0.045	0.088	5.4	76	5.9
SA36	0.66	2.60	0.046	0.080	5.1	98	4.6
SA37	1.17	1.47	0.044	0.099	5.6	116	6.9
SA38	1.17	1.53	0.046	0.082	5.6	80	6.3
SA39	0.89	1.41	0.042	0.079	5.0	92	6.0
SA40	0.78	1.78	0.047	0.085	5.1	93	5.3
SA41	0.95	1.73	0.046	0.086	5.3	85	5.7
Columbia Plain, La Vega Vieja, Dominican Republic							
SA50	0.67	1.92	0.056	0.098	5.6	98	5.8
SA51	0.45	2.32	0.052	0.072	5.7	116	5.6
SA52	0.24	2.51	0.045	0.076	4.7	97	3.9
SA53	0.69	1.73	0.046	0.078	5.2	116	4.9
SA54	0.49	2.01	0.065	0.088	5.8	94	5.2
SA55	0.68	1.61	0.055	0.084	5.8	112	6.1
SA56	0.49	1.70	0.055	0.093	5.7	71	3.6
SA57	0.59	2.13	0.060	0.076	5.8	129	5.8
SA58	0.51	1.82	0.049	0.064	4.7	79	4.4
SA59	0.61	2.02	0.056	0.088	5.6	95	5.2
Yayal-Blue-on-White, Convento de San Francisco, Dominican Republic							
SA21	1.51	1.23	0.046	0.078	5.0	59	5.9
SA22	1.29	1.20	0.058	0.084	5.4	96	8.5
SA23	1.15	1.49	0.045	0.202	4.9	68	7.8
SB69	1.27	1.20	0.057	0.086	5.4	99	7.6
SB70	0.95	1.50	0.056	0.082	5.3	80	5.3
SB71	1.06	1.52	0.051	0.087	5.4	74	6.0
SB72	1.38	1.46	0.046	0.080	4.8	98	6.2
SB73	1.24	1.41	0.048	0.080	5.1	91	5.6
SB74	1.05	1.52	0.053	0.093	5.1	58	4.3
SB75	0.85	1.67	0.065	0.089	5.2	73	4.2
SB76	1.10	2.32	0.052	0.089	6.1	111	6.3
SB77	1.02	1.87	0.056	0.089	5.2	67	4.9
SB78	1.17	1.42	0.053	0.090	5.1	58	4.0
SB80	1.07	1.96	0.062	0.097	5.6	67	3.9
SB82	1.14	1.09	0.058	0.090	5.1	59	3.9
SB83	0.82	1.75	0.041	0.077	4.6	70	3.6
SB84	1.13	1.10	0.046	0.073	4.9	73	5.0
SB85	1.14	1.52	0.042	0.071	5.2	98	6.0

Continued

La_2O_3 (ppm)	Sc_2O_3 (ppm)	CeO_2 (ppm)	Eu_2O_3 (ppm)	HfO_2 (ppm)	ThO_2 (ppm)	Cr_2O_3 (ppm)	CoO (ppm)
Columbia Plain, Juandolio, Dominican Republic							
47	20.4	84	1.71	5.7	12.6	112	15.4
41	21.0	86	1.67	6.0	12.4	110	15.2
46	22.9	95	1.78	7.0	14.9	123	17.5
43	22.0	88	1.71	6.2	12.8	124	16.2
34	21.0	85	1.67	5.8	11.6	110	15.8
42	23.2	88	1.61	5.4	13.0	124	16.8
44	23.0	93	1.80	7.6	14.1	129	15.0
46	20.6	78	1.62	5.7	11.4	114	14.6
40	20.4	82	1.62	6.5	11.9	109	18.8
41	22.0	87	1.67	5.5	12.7	121	16.5
Columbia Plain, La Vega Vieja, Dominican Republic							
46	22.0	96	1.82	6.9	13.5	125	18.1
42	21.7	91	1.69	6.1	13.2	126	17.3
44	19.1	92	1.54	5.5	14.4	88	17.2
37	18.7	80	1.50	5.4	11.7	114	14.9
46	24.0	94	1.77	6.5	14.6	135	16.5
44	23.4	94	1.77	6.1	12.8	134	15.5
43	22.5	94	1.60	6.8	13.5	129	14.5
40	22.9	89	1.62	6.6	13.3	133	13.9
40	19.0	84	1.46	5.5	11.5	113	16.4
42	22.0	87	1.71	6.3	12.8	126	16.4
Yayal-Blue-on-White, Convento de San Francisco, Dominican Republic							
32	19.5	83	1.59	6.9	10.9	110	18.8
36	22.2	87	1.61	5.6	12.3	123	19.3
40	19.5	80	1.62	5.5	11.9	114	68.3
41	21.9	—	1.78	6.6	13.1	140	17.7
38	21.4	—	1.97	6.0	11.8	113	18.1
40	21.6	—	1.63	6.3	12.2	128	16.8
54	20.0	—	1.64	5.7	11.6	128	19.9
39	21.1	—	1.72	6.2	11.7	116	17.8
42	20.3	—	1.83	5.6	12.7	121	20.9
37	20.2	—	1.58	5.8	11.9	121	22.8
43	24.5	—	1.82	5.8	13.6	142	20.5
52	20.6	—	1.60	6.0	11.8	128	21.9
39	20.4	—	1.63	5.8	12.6	131	22.3
42	23.7	—	1.71	5.8	13.0	137	22.9
41	20.0	—	1.56	6.0	11.4	132	22.9
35	19.1	105	1.30	4.7	11.2	110	18.5
35	19.7	56	1.34	5.7	11.0	118	17.9
37	20.9	75	1.58	6.3	11.5	132	15.8

Table I.

Specimen No.	Na_2O (%)	K_2O (%)	BaO (%)	MnO (%)	Fe_2O_3 (%)	Rb_2O (ppm)	Cs_2O (ppm)
Yayal-Blue-on-White, Nueva Cadiz, Venezuela							
SA31	1.19	1.42	0.040	0.097	5.0	65	6.0
SA60	1.01	1.30	0.039	0.082	4.7	56	4.5
SA61	1.47	1.59	0.043	0.088	5.0	57	4.1
SA62	1.18	1.36	0.045	0.083	4.8	52	4.4
SA63	1.50	1.06	0.042	0.081	5.2	58	6.1
SA64	1.15	1.23	0.038	0.075	4.8	52	5.2
SB46	1.28	1.38	0.044	0.079	5.0	70	6.3
SB47	1.47	1.47	0.044	0.077	5.1	72	5.8
SB48	0.73	1.41	0.042	0.079	4.8	77	5.7
SB49	1.16	1.54	0.037	0.074	5.1	58	5.7
SB50	1.20	1.34	0.039	0.082	5.1	63	6.4
SB51	1.13	1.55	0.050	0.077	5.2	82	6.4
SB52	1.10	1.35	0.067	0.083	6.3	52	6.1
SB53	1.56	1.55	0.045	0.081	4.9	61	5.9
SB54	1.08	2.00	0.037	0.082	5.3	73	5.7
SB55	1.01	1.39	0.041	0.068	4.5	71	5.5
SB56	1.11	1.14	0.050	0.078	5.0	67	6.0
SB57	1.38	1.21	0.041	0.070	4.6	64	5.5
Yayal-Blue-on-White, Juandolio, Dominican Republic							
SA46	0.94	1.63	0.046	0.090	5.7	84	5.3
Caparra Blue, Nueva Cadiz, Venezuela							
SA27	0.88	2.21	0.037	0.034	5.2	110	6.1
SA28	1.48	1.31	0.037	0.077	5.6	63	6.0
SA29	1.52	1.04	0.044	0.084	5.2	57	7.0
SA30	1.29	1.24	0.040	0.076	5.1	52	6.1
Isabella Polychrome, Nueva Cadiz, Venezuela							
SA47	1.30	1.64	0.037	0.072	4.9	85	5.4
SA48	1.44	1.59	0.044	0.076	5.3	76	5.8
SA49	1.25	1.54	0.039	0.076	5.4	70	5.7
Isabella Polychrome, Juandolio, Dominican Republic							
SA42	1.10	1.21	0.046	0.087	5.3	123	6.3
SA43	0.76	1.80	0.039	0.081	4.8	78	4.9
SA44	0.91	1.79	0.051	0.089	5.7	106	6.1
SA45	1.05	1.63	0.046	0.088	5.4	122	6.7
Isabella Polychrome, Isabella, Dominican Republic							
SA24	1.08	1.72	0.041	0.086	5.4	62	4.9
SA25	1.00	2.11	0.063	0.091	5.6	67	7.5
SA26	1.00	1.69	0.058	0.079	5.3	103	6.4

Continued

La_2O_3 (ppm)	Sc_2O_3 (ppm)	CeO_2 (ppm)	Eu_2O_3 (ppm)	HfO_2 (ppm)	ThO_2 (ppm)	Cr_2O_3 (ppm)	CoO (ppm)
\multicolumn{8}{c}{}							

Yayal-Blue-on-White, Nueva Cadiz, Venezuela

La_2O_3 (ppm)	Sc_2O_3 (ppm)	CeO_2 (ppm)	Eu_2O_3 (ppm)	HfO_2 (ppm)	ThO_2 (ppm)	Cr_2O_3 (ppm)	CoO (ppm)
38	19.9	83	1.70	5.8	12.2	111	26.7
39	18.9	74	1.44	4.6	10.4	109	14.6
46	20.3	98	1.59	6.5	14.5	114	19.0
39	19.4	83	1.51	6.4	11.7	112	14.7
38	20.8	82	1.53	6.0	12.3	118	16.4
36	19.4	80	1.39	4.8	11.2	114	14.6
36	20.5	76	1.55	5.9	11.6	119	20.7
36	20.3	81	1.67	5.8	11.8	120	17.1
35	19.7	74	1.61	5.1	11.5	117	16.6
36	20.5	78	1.51	5.9	11.5	117	20.2
38	20.6	85	1.65	5.2	12.6	122	20.3
38	20.4	83	1.69	5.2	12.2	130	17.1
38	22.3	84	1.62	5.5	12.6	134	20.6
37	19.6	79	1.40	5.5	11.8	126	16.2
40	21.7	82	1.68	5.6	12.0	131	19.9
34	18.2	73	1.38	5.1	10.8	108	17.8
38	20.9	83	1.69	5.7	12.1	121	16.5
35	18.3	77	1.25	5.9	11.1	111	16.8

Yayal-Blue-on-White, Juandolio, Dominican Republic

46	23.9	99	1.79	6.4	14.0	131	17.3

Caparra Blue, Nueva Cadiz, Venezuela

45	22.8	86	1.74	5.4	12.4	147	71.0
43	21.3	90	1.64	6.8	12.0	124	29.8
41	21.2	83	1.55	5.8	12.5	119	72.6
39	20.5	80	1.66	6.4	12.1	121	36.3

Isabella Polychrome, Nueva Cadiz, Venezuela

38	19.9	84	1.47	5.6	11.6	114	14.9
43	20.7	92	1.59	6.0	13.5	118	17.1
40	20.4	84	1.39	5.6	11.7	115	23.6

Isabella Polychrome, Juandolio, Dominican Republic

40	21.7	81	1.51	5.4	12.1	119	17.4
41	19.4	83	1.36	5.7	12.0	109	17.7
51	23.4	90	1.70	6.2	12.9	126	16.5
44	22.2	89	1.78	6.2	12.4	124	17.3

Isabella Polychrome, Isabella, Dominican Republic

38	20.7	83	1.70	5.9	12.4	111	17.0
41	22.9	90	2.13	6.9	13.9	127	20.6
36	20.9	82	1.58	6.4	11.5	119	18.0

Table II.   Mean Compositions of Majolica from Five Sites
Spain; and from

Average Concentrations

Oxides Determined	Convento de S.F.	Juandolio	La Vega Vieja
Sodium (Na$_2$O) (%)	1.06 ± 0.20	0.92 ± 0.16	0.57 ± 0.09
Potassium (K$_2$O) (%)	1.42 ± 0.35	1.76 ± 0.34	1.91 ± 0.22
Barium (BaO) (%)	0.046 ± 0.008	0.045 ± 0.002	0.054 ± 0.006
Manganese (MnO) (%)	0.080 ± 0.016	0.085 ± 0.006	0.082 ± 0.04
Iron (Fe$_2$O$_3$) (%)	5.1 ± 0.36	5.3 ± 0.30	5.5 ± 0.38
Rubidium (Rb$_2$O) ppm	87 ± 24	92 ± 15	99 ± 20
Cesium (Cs$_2$O) (ppm)	5.4 ± 1.1	5.6 ± 0.8	5.1 ± 0.9
Lanthanum (La$_2$O$_3$) (ppm)	39 ± 4.0	43 ± 4.0	42 ± 3.0
Scandium (Sc$_2$O$_3$) (ppm)	20 ± 1.3	22 ± 1.3	22 ± 1.9
Cerium (CeO$_2$) (ppm)	77 ± 18	87 ± 5.4	90 ± 5.5
Europium (Eu$_2$O$_3$) (ppm)	1.56 ± 0.15	1.66 ± 0.12	1.66 ± 0.13
Hafnium (HfO$_2$) (ppm)	5.9 ± 0.44	6.0 ± 0.6	6.2 ± 0.5
Thorium (ThO$_2$) (ppm)	11.8 ± 0.73	12.7 ± 1.0	13.0 ± 1.0
Chromium (Cr$_2$O$_3$) (ppm)	121 ± 10	119 ± 8	126 ± 8
Cobalt (CoO) (ppm)	18.0 ± 4.7	16.5 ± 1.2	15.9 ± 1.4

[a] Since we believe the data are logarithmically distributed, the geometric means
are used, and for convenience of presentation the standard deviations are expressed

Table III.   Concentration of Oxides in

Specimen No.	Na$_2$O (%)	K$_2$O (%)	BaO (%)	MnO (%)	Fe$_2$O$_3$ (%)	Rb$_2$O (ppm)	Cs$_2$O (ppm)
Columbia Plain							
SC08	0.60	2.19	0.078	0.066	4.7	99	5.7
SC09	0.68	1.82	0.060	0.050	4.9	101	7.7
Yayal-Blue-on-White							
SC03	0.53	2.37	0.058	0.056	4.4	90	6.0
SC04	0.80	2.12	0.088	0.068	4.9	106	7.9
SC05	0.61	1.69	0.069	0.057	4.9	100	7.7
SC06	0.97	1.68	0.095	0.059	5.3	80	4.5
SC07	0.60	2.13	0.101	0.063	5.0	84	4.9

in the Dominican Republic and Venezuela; from Jerez, Mexico City, Mexico[a]

Average Concentrations

Isabela	Nueva Cadiz	Jerez	Mexico City
1.12 ± 0.05	1.26 ± 0.28	0.67 ± 0.14	1.69 ± 0.42
2.08 ± 0.25	1.35 ± 0.29	1.98 ± 0.27	1.13 ± 0.40
0.053 ± 0.012	0.045 ± 0.011	0.077 ± 0.017	0.053 ± 0.010
0.085 ± 0.007	0.077 ± 0.011	0.060 ± 0.015	0.054 ± 0.019
5.5 ± 0.15	5.0 ± 0.34	4.9 ± 0.26	4.1 ± 0.59
76 ± 21	63 ± 12	94 ± 10	54 ± 11
6.2 ± 1.4	5.9 ± 0.9	6.2 ± 1.4	3.7 ± 1.3
43 ± 2.7	38 ± 2.7	45 ± 1.9	22 ± 3.9
21 ± 1.2	20 ± 1.2	20 ± 0.9	16.5 ± 2.8
85 ± 4.6	82 ± 4.8	80 ± 3.3	40 ± 8.6
1.79 ± 0.28	1.56 ± 0.13	1.47 ± 0.05	1.22 ± 0.21
6.4 ± 0.5	5.7 ± 0.5	5.5 ± 0.4	4.5 ± 0.8
12.6 ± 1.2	11.8 ± 0.8	11.7 ± 0.7	5.6 ± 0.7
119 ± 8	119 ± 10	122 ± 24	98 ± 27
18.5 ± 1.8	18.8 ± 6.7	15.4 ± 1.8	16.6 ± 14

as plus or minus one half of the total standard deviation spread of the groups as calculated logarithmically.

Majolica Sherds from Jerez, Spain

La_2O_3 (ppm)	Sc_2O_3 (ppm)	CeO_2 (ppm)	Eu_2O_3 (ppm)	HfO_2 (ppm)	ThO_2 (ppm)	Cr_2O_3 (ppm)	CoO (ppm)
			Columbia Plain				
45	19.4	80	1.55	5.0	11.5	111	15.2
44	19.0	81	1.48	5.4	11.0	103	16.9
			Yayal-Blue-on-White				
42	18.4	76	1.37	5.1	11.4	101	15.1
46	20.5	82	1.49	6.1	12.5	127	14.9
44	20.0	80	1.49	5.7	12.1	129	13.0
44	20.8	77	1.46	5.3	10.9	180	14.8
48	20.6	86	1.47	5.7	12.5	116	18.8

Figure 5.   Columbia Plain (left to right): (top) SC11, SC12, and SC14; (bottom) SC15, SC18, SC20, and SC21 from excavations from the Metropolitan Cathedral in Mexico City

Figure 6.   Abo Polychrome (left to right): (top) ST78–SA80; (bottom) SA81, San Luis Polychrome; SA82 and SA83 from the subway excavations in Mexico City

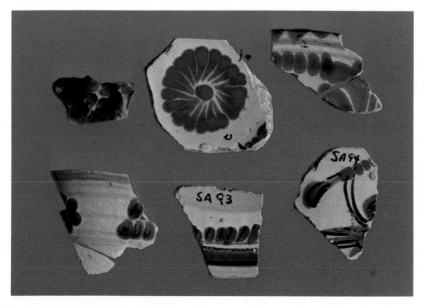

Figure 7. Puebla-Blue-on-White (left to right): (top) SA89–SA91; (bottom) SA92–SA94; San Elizario Polychrome from the subway excavations in Mexico City

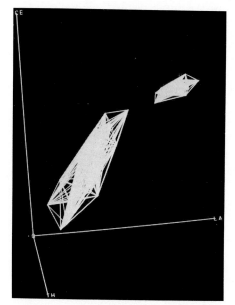

Bulletin of the American Institute for Conservation

Figure 8. Computer projection of three-dimensional plot of cerium, lanthanum, and thorium oxide concentrations for Spanish Colonial majolica sherds. The data divides into a group having its source in Spain and a group whose source is Mexican, as discussed in the text.

these divergencies to the high probability that barium carbonate might be either deposited in or leached from sherds during burial along with calcium carbonate. Accordingly, we usually do not regard deviant barium values as being very significant. This agreement strongly indicates that the matching group of New World sherds and those found at Jerez came from related sources. Since it is historically and archaeologically very probable that these specimens found in the New World came from Spain and very improbable that the Jerez specimens were imported from the New World, a Spanish source, as postulated by Goggin and later authors, gains additional support from these data.

The data in Table IV are for sherds from the excavations of the Metro and the Metropolitan Cathedral in Mexico City. Approximately 100,000 majolica sherds were uncovered during the excavations for the Metro subway, and we have analyzed a small group of these. Recent excavations have been carried out at the Metropolitan Cathedral in mine-like tunnels and shafts as much as 40 ft below the cathedral floor. These excavations are below the central cathedral foundation which church records state was laid in 1573. There is no evidence of intrusions through this floor so that the excavated sherds can be dated as earlier than 1573. The compositions of these sherds can be distinguished readily from those of the sherds in Tables I and III. The lanthanum oxide, cerium oxide, and thorium oxide concentrations are the most distinctly different. Selected examples of the sherds in Table IV are shown in Figures 5–7. The sherds in Figure 5 are examples of Columbia Plain from the excavations of the Metropolitan Cathedral and those in Figures 6 and 7 are from the subway excavations.

In Figure 8 we have plotted the lanthanum oxide, cerium oxide, and thorium oxide concentrations for sherds excavated in the Dominican Republic and Venezuela and sherds from the Metro excavations using a computer system developed for this purpose at Brookhaven National Laboratory (8). On the basis of these three oxides there is a distinct separation between the sherds from the Dominican Republic and Venezuela and those from Mexico City. Unlike the sherds from the Dominican Republic and Venezuela, the sherds from Mexico City appear not to have originated in Spain, at least at that specific source. There is further evidence of this distinction between the two sets of sherds. X-ray diffraction analysis of the samples from Jerez and from the New World showed that the sherds from Jerez, the Dominican Republic, and Venezuela had intense quartz peaks whereas the sherds from Mexico City did not. This constitutes additional evidence that the majolica from Mexico City came from a different source than the majolica from the Caribbean sites.

The development of the ceramic industry in Mexico and its precedents in Italy and Spain have been discussed by Florence and Robert

Lister (*9, 10*). They and Gonzalo Lopez Cervantes (*11*) also refer to records of early pottery production in Mexico. There is, however, no complete record of the early history of the manufacture of ceramics by the Spanish in the New World, and the analyses discussed here will be used to better understand that history.

It is logical to consider whether the majolica sherds which were found in Mexico City could have been fabricated of local clay. Fortunately data on clays and related pottery from the Valley of Mexico has been collected at Brookhaven National Laboratory over many years. The ceramic material, which had previously been anlyzed by Harbottle and Sayre in collaboration with other investigators, consisted of Precolumbian artifacts. The pottery and the clays from two archaeological sites within the Valley, Teotihuacan, and Tlatilco were all basically similar in composition, although the clays and pottery from the two separate sites could be differentiated through a subtle multivariate statistical analysis. It is likely that the entire Valley of Mexico is underlain with clay bed of moderately uniform trace impurity composition, and hence if the composition of the Mexico City majolica sherds was similar to that of ceramics and clay from Teotihuacan or Tlatilco, it would be probable that the majolica was fabricated from clays originating somewhere within the Valley of Mexico.

The data accumulated on clays and pottery from Teotihuacan are extensive, but only a few specimens from Tlatilco have been anlyzed. Accordingly comparison was first made between the majolica and the Teotihuacan statistical group of analyses. The mean concentrations for these two groups of pottery are compared in Table V. Except for only two components, the calcium and cesium compounds, the concentrations of all measured components in the majolica were about three quarters of the levels in the Teotihuacan specimens. The components cesium oxide and calcium carbonate are among those that can be most readily affected during burial; cesium compounds tend to be water soluble and hence susceptible to migration through the action of soil water, and calcium carbonate can either be dissolved from or deposited into burial sherds depending upon the levels of carbon dioxide in the ground water to which they are exposed. Another cause of aberrant calcium concentrations in pottery is the occasional addition to the clay of crushed marble or other calcareous material as a temper. Therefore the difference between the calcium and cesium concentrations in the majolica and Precolumbian specimens does not rule out the possibility that they may have been formed from related clays, and the parallelism in the concentration of all other components in both sets of specimens suggests that they indeed might have had related common origins.

Table IV. Concentrations of Oxides in Matching

Specimen No.	Na_2O (%)	K_2O (%)	BaO (%)	MnO (%)	Fe_2O_3 (%)	Rb_2O (ppm)	Cs_2O (ppm)
Columbia Plain, Metropolitan Cathedral							
SC11	1.93	1.61	0.054	0.027	3.3	59	4.0
SC12	1.76	1.60	0.058	0.055	3.8	51	5.3
SC13	1.96	1.26	0.046	0.040	3.1	44	2.0
SC14	2.29	2.02	0.058	0.054	4.0	80	5.5
SC15	1.80	0.72	0.045	0.055	4.0	50	4.1
SC17	1.14	0.98	0.050	0.039	3.0	42	3.0
SC18	1.81	2.00	0.057	0.042	3.8	73	6.5
SC20	3.10	1.75	0.058	0.025	3.2	62	3.0
SC21	1.35	1.26	0.047	0.035	4.3	67	5.0
Fig Springs, Polychrome, Metro Excavations, Mexico City							
SA65	1.64	1.04	0.058	0.038	3.8	43	2.7
SA66	1.46	1.57	0.043	0.039	5.3	55	4.6
SA67	1.38	1.66	0.052	0.047	3.6	75	3.9
Los Angeles Polychrome, Metro Excavations, Mexico City							
SA68	1.90	0.93	0.042	0.050	5.1	57	7.7
SA69	1.33	1.67	0.043	0.038	4.6	80	—
SA70	1.12	1.11	0.052	0.038	3.9	53	7.0
San-Luis-Blue-on-White, Metro Excavations, Mexico City							
SA71	1.65	—	0.042	0.052	5.0	82	4.0
SA72	2.40	0.98	0.062	0.099	4.5	52	3.0
SA73	1.20	1.60	0.044	0.052	4.5	59	5.4
Puebla Polychrome, Metro Excavations, Mexico City							
SA74	2.08	0.75	0.058	0.072	4.3	43	2.8
SA75	1.77	1.62	0.057	0.077	4.0	53	2.8
SA76	1.85	2.05	0.054	0.075	4.2	40	2.2
SA77	1.99	1.06	0.095	0.055	4.3	57	4.1
Abo Polychrome, Metro Excavations, Mexico City							
SA78	1.12	1.14	0.046	0.062	4.6	55	—
SA79	1.44	0.71	0.061	0.071	3.8	44	3.2
SA80	1.73	0.91	0.058	0.083	4.1	58	2.6
SA81	1.60	1.33	0.064	0.086	4.0	48	2.2
San Luis Polychrome, Metro Excavations, Mexico City							
SA82	1.97	1.20	0.052	0.069	4.4	53	2.9

Specimens of Majolica from Mexico City Sites

La_2O_3 (ppm)	Sc_2O_3 (ppm)	CeO_2 (ppm)	Eu_2O_3 (ppm)	HfO_2 (ppm)	ThO_2 (ppm)	Cr_2O_3 (ppm)	CoO (ppm)
Columbia Plain, Metropolitan Cathedral							
36	19.5	57	1.68	4.3	8.3	81	8.8
26	15.3	42	1.16	4.1	6.2	71	12.0
18	11.3	27	0.82	3.4	4.9	62	7.4
29	17.1	45	1.26	4.2	6.4	76	12.4
23	15.9	37	1.20	3.9	5.2	87	12.7
17	11.3	27	0.86	2.9	3.7	60	10.3
28	15.7	43	1.32	4.3	6.4	77	13.3
22	11.9	34	0.96	3.9	6.0	59	9.2
25	12.3	39	1.09	4.3	6.6	72	13.5
Fig Springs, Polychrome, Metro Excavations, Mexico City							
28	17.9	48	1.51	4.7	6.3	78	9.7
25	22.0	41	1.25	6.3	5.5	135	11.0
18	16.3	26	1.07	4.7	5.0	86	36.0
Los Angeles Polychrome, Metro Excavations, Mexico City							
22	20.7	35	1.37	5.3	5.9	153	13.4
22	20.4	35	1.22	5.7	6.6	123	15.7
19	17.8	31	1.10	4.7	5.1	96	10.7
San-Luis-Blue-on-White, Metro Excavations, Mexico City							
23	21.9	38	1.31	5.8	6.1	112	14.4
26	16.6	62	1.42	5.7	7.0	82	21.9
19	19.2	31	1.10	4.9	5.4	114	12.3
Puebla Polychrome, Metro Excavations, Mexico City							
123	16.0	50	1.42	5.7	5.5	92	15.3
24	15.0	52	1.32	5.1	6.0	82	13.8
23	15.4	50	1.38	5.8	5.6	102	16.0
21	17.8	44	1.33	5.0	5.4	133	20.8
Abo Polychrome, Metro Excavations, Mexico City							
17.5	17.8	26	1.01	4.9	5.6	116	11.3
21.9	14.0	43	1.21	4.8	5.0	85	13.6
21.2	16.4	44	1.33	4.8	5.6	85	13.0
21.6	15.1	45	1.33	4.9	5.4	83	13.5
San Luis Polychrome, Metro Excavations, Mexico City							
24.2	17.2	52	1.47	5.6	5.2	95	14.5

Table IV.

Specimen No.	Na_2O (%)	K_2O (%)	BaO (%)	MnO (%)	Fe_2O_3 (%)	Rb_2O (ppm)	Cs_2O (ppm)
Castillo Polychrome, Metro Excavations, Mexico City							
SA84	1.15	1.26	0.047	0.052	3.7	51	6.4
SA85	1.95	0.47	0.048	0.036	5.4	50	4.1
Aucilla Polychrome, Metro Excavations, Mexico City							
SA86	1.75	0.76	0.066	0.092	4.8	44	3.2
SA87	1.83	0.87	0.064	0.103	4.2	47	2.9
SA88	2.13	0.72	0.062	0.095	4.5	45	2.7
Puebla-Blue-on-White, Metro Excavations, Mexico City							
SA90	1.63	1.42	0.038	0.049	3.7	52	3.5
SA91	2.04	1.66	0.042	0.048	3.2	59	3.0
San Elizario Polychrome, Metro Excavations, Mexico City							
SA93	1.77	0.70	0.059	0.046	4.7	51	3.4
SA94	1.85	0.95	0.057	0.066	4.5	53	3.6
San-Agustin-Blue-on-White, Metro Excavations, Mexico City							
SA95	2.14	0.56	0.062	0.058	4.9	30	2.5
SA96	2.14	0.99	0.060	0.028	4.4	54	3.3
Huejotzingo-Blue-on-White, Metro Excavations, Mexico City							
SA99	2.09	0.94	0.062	0.070	4.3	59	4.1
SB01	1.77	1.06	0.042	0.053	3.9	55	4.9
Unidentified Types, Metro Excavations, Mexico City							
SA97	1.94	1.36	0.035	0.060	3.7	56	3.6
SA98	1.77	1.11	0.055	0.056	4.3	64	3.2
SB26	1.60	1.22	0.075	0.070	3.3	—	2.9
SB32	0.74	1.24	0.055	0.095	3.7	63	3.5
SB36	1.62	0.92	0.038	0.055	4.9	59	9.4
SB37	1.49	1.09	0.056	0.040	3.5	35	—

The high calcium content in the majolica found in Mexico City—21.4% calculated as pure calcium carbonate compared with 5.9% in sherds of Teotihuacan—suggests that a calcium compound such as calcium or calcium magnesium carbonate may have been added to the majolica either as a temper or through deposition during burial. Petrographic examination of cross sections of representative Mexico City majolica sherds show heavy deposits of birefringent material with structures

Continued

La_2O_3 (ppm)	Sc_2O_3 (ppm)	CeO_2 (ppm)	Eu_2O_3 (ppm)	HfO_2 (ppm)	ThO_2 (ppm)	Cr_2O_3 (ppm)	CoO (ppm)
\multicolumn{8}{c}{*Castillo Polychrome, Metro Excavations, Mexico City*}

La_2O_3 (ppm)	Sc_2O_3 (ppm)	CeO_2 (ppm)	Eu_2O_3 (ppm)	HfO_2 (ppm)	ThO_2 (ppm)	Cr_2O_3 (ppm)	CoO (ppm)

Castillo Polychrome, Metro Excavations, Mexico City

La_2O_3	Sc_2O_3	CeO_2	Eu_2O_3	HfO_2	ThO_2	Cr_2O_3	CoO
16.2	15.3	31	1.02	3.6	5.2	93	10.8
23.3	21.4	34	1.36	5.0	6.4	156	12.4

Aucilla Polychrome, Metro Excavations, Mexico City

La_2O_3	Sc_2O_3	CeO_2	Eu_2O_3	HfO_2	ThO_2	Cr_2O_3	CoO
26.8	18.1	56	1.60	5.1	6.5	96	16.5
24.8	15.8	50	1.41	4.6	5.5	92	14.9
25.9	17.2	54	1.50	5.1	5.7	105	16.0

Puebla-Blue-on-White, Metro Excavations, Mexico City

La_2O_3	Sc_2O_3	CeO_2	Eu_2O_3	HfO_2	ThO_2	Cr_2O_3	CoO
18.0	14.2	36	1.04	3.5	5.2	94	12.2
17.2	11.5	28	0.94	3.5	4.6	79	119.0

San Elizario Polychrome, Metro Excavations, Mexico City

La_2O_3	Sc_2O_3	CeO_2	Eu_2O_3	HfO_2	ThO_2	Cr_2O_3	CoO
20.7	18.3	40	1.30	4.7	5.2	138	17.3
21.9	18.2	38	1.36	4.2	5.5	129	16.5

San-Agustin-Blue-on-White, Metro Excavations, Mexico City

La_2O_3	Sc_2O_3	CeO_2	Eu_2O_3	HfO_2	ThO_2	Cr_2O_3	CoO
20.3	20.1	38	1.47	4.7	5.4	168	16.7
24.5	19.4	40	1.37	4.8	6.4	148	38.0

Huejotzingo-Blue-on-White, Metro Excavations, Mexico City

La_2O_3	Sc_2O_3	CeO_2	Eu_2O_3	HfO_2	ThO_2	Cr_2O_3	CoO
21.6	17.8	38	1.36	4.3	5.9	118	25.5
19.4	15.8	36	1.07	4.3	5.7	92	13.5

Unidentified Types, Metro Excavations, Mexico City

La_2O_3	Sc_2O_3	CeO_2	Eu_2O_3	HfO_2	ThO_2	Cr_2O_3	CoO
16.2	15.7	31	1.04	3.4	5.0	105	190.0
18.1	17.6	36	1.11	3.9	5.6	120	451.0
16.3	13.8	39	1.07	3.2	4.5	86	11.1
15.7	15.2	40	0.84	3.4	5.2	78	11.1
17.7	20.4	43	1.24	4.3	6.0	180	13.9
20.5	16.4	41	1.40	3.6	5.6	95	9.9

typical of secondary accumulations of carbonates lining the open spaces within the pottery structure. X-ray diffraction of samples from these specimens confirmed the presence of the mineral calcite within them. Inclusion of primary mineral calcite, that is, of crushed marble or the like added as temper, would have quite a different microscopic appearance, and little or no microscopic evidence of calcite added as a temper appears in the majolica specimens.

Table V. Comparison of Composition of Mexican Majolica with Precolumbian Ceramics

Oxide Concentration

Compounds	Majolica	Teotihuacan Clays and Pottery	Ratio Majolica/ Teotihuacan
	Major Components (%)		
Na_2O	1.69	2.42	0.70
K_2O	1.13	1.55	0.73
BaO	0.053	0.078	0.68
MnO	0.054	0.085	0.64
Fe_2O_3	4.09	5.36	0.76
CaO Expressed as $CaCo_3$	21.4	5.9	3.63
	Trace Components (ppm)		
Rb_2O	53.6	59.1	0.91
Sc_2O_3	16.5	20.4	0.81
La_2O_3	21.6	25.6	0.84
CeO_2	39.4	55.0	0.72
Eu_2O_3	1.2	1.7	0.71
HfO_2	4.5	5.1	0.88
ThO_2	5.6	6.9	0.81
Ta_2O_5	0.8	1.0	0.80
Cr_2O_3	97.9	109.2	0.90
CoO	16.6	19.5	0.85
Cs_2O	3.8	2.8	1.36

Mean ratio with $CaCo_3$ and Cs_2O values deleted 0.78 ± 0.08

Petrographic comparison of the Mexico City majolica with Teotihuacan sherds shows that except for the secondary deposition of carbonates, which is present in the majolica but absent in the Precolumbian sherds, the mineral composition of both sets of specimens is very similar. Both notably include hornblende and similar feldspars as inclusions, and both are low in quartz. Similarly, except for the calcite in the majolica, both sets of sherds show similar x-ray diffraction patterns. The mineralogical evidence, therefore, strongly suggests that both sets of sherds were made from closely related clays and that the compositional differences that exist between them are primarily the result of the accumulation of a secondary calcareous deposit within the majolica sherds during burial in the wet soil of Mexico City.

One would indeed expect the soil in central Mexico City, which to a great extent is the filled-in bed of Lake Texcoco, to be moist and hence conducive to carbonate deposition and the soil at Teotihuacan, which is situated on relatively high ground at the northern edge of the Valley of Mexico, to be dry and hence not favorable for such deposition. The neutron activation analytical data are consistent with this hypothesis. If the secondary carbonate deposit were relatively free of the other elements determined, then its presence would simply dilute the concentrations of these other components by a constant factor. Table V indicates that the concentrations of 15 components in the majolica are related on the average to those in the Teotihuacan specimens by the nearly constant factor 0.78 ± 0.08. Therefore on the basis of relative rather than absolute concentrations there should be good agreement between the two groups of specimens.

The Brookhaven computer program ADSTAT can adjust sets of specimens by factors which bring them into closest relative agreement on the basis of a least squares fit in logarithms of concentrations. Using this program all specimens of both the Mexico City majolica and Teotihuacan were adjusted into best-relative-fit agreement with the mean concentrations of the majolica. In this adjustment the elements calcium, cesium, and cobalt were eliminated; calcium and cesium because of their previously noted inconsistency with other components, and cobalt because some majolica specimens were decorated with cobalt-colored glazes and there was some evidence of occasional contamination from these glazes. In this process, a fitting constant was calculated for each specimen which was used to adjust all components for that specimen. Hence the relative values for each specimen are left unaltered in the adjustment. If one then assumes that the adjusted Teotihuacan specimens constitute a log-normally distributed statistical group, it is possible to calculate the multivariate probability that each of the adjusted specimens might belong to this group. This is a so-called Mahalanobis distance calculation corrected for a finite group of specimens through Hotelling's T^2 parameter.

Such a calculation showed that about half of the majolica specimens had a significant probability (i.e., greater than 5%) of belonging to the Teotihuacan group. This is a very sensitive test of agreement of specimens with a group which demonstrates that on a relative basis, for all elements other than calcium, cesium, and cobalt, there is a close agreement between the compositions of the two sets of specimens. We do not feel that this agreement proves that any or all of the majolica specimens were necessarily fabricated at Teotihuacan itself. It is historically unlikely that this would have occurred. It is more likely that the majolica was made from clays of a geological origin similar to those at Teotihuacan which in some instances are indistinguishable from Teotihuacan clays in their trace impurity patterns.

The usefulness of neutron activation analysis in assisting the archaeologist to establish the provenience of potsherd material is illustrated by a consideration of the sherds in Figure 9. SB30 and SB31 were among the majolica sherds excavated in the Mexico City subway excavations. They have been identified by F. Lister (2) as late 16th century Italian. In Table VI the concentration of several oxides in these sherds are markedly different from their mean values in the majolica now considered to be of Mexican origin. (SB30 and SB31 have much higher concentrations of CeO_2, ThO_2, and Cr_2O_3 than most of the pottery from Mexico City.) Although F. Lister thought that SB33, also in Figure 9, was of Mexican origin, it also has a distinctive composition, having a much higher concentration of HfO_2.

Table VI.   Nonmatching Majolica Sherds from

Specimen No.	Na_2O (%)	K_2O (%)	BaO (%)	MnO (%)	Fe_2O_3 (%)	Rb_2O (ppm)	Cs_2O (ppm)
Subway Excavations, Mexico City							
SB30	1.20	1.97	0.038	0.101	6.8	108	7.9
SB31	1.16	2.31	0.053	0.113	6.7	88	6.8
SB33	1.39	2.41	0.065	0.044	4.3	88	7.9
Santiago de los Cabelleros (Antiqua), Guatemala							
SB38	1.51	0.92	0.058	0.198	11.4	44	3.4
SB39	1.57	0.58	0.062	0.213	11.5	40	2.6
Panama Viejo							
SB40	2.09	2.94	0.082	0.142	7.3	151	33.2
SB41	2.07	2.93	0.062	0.104	7.7	146	40.1
SB42	2.09	2.40	0.074	0.137	7.5	193	127.0
Cuzco, Peru							
SB43	0.54	3.97	0.075	0.092	7.4	128	9.2
SB44	0.66	4.46	0.060	0.072	6.7	137	8.3
Quito, Ecuador							
SB45	3.33	1.49	0.098	0.114	7.0	54	2.8
Metropolitan Cathedral, Mexico City							
SC10	0.90	2.57	0.048	0.054	4.1	146	11.0
SC19	0.93	3.35	0.069	0.037	4.2	167	11.4
Means and Means + and −95% Confidence Limit for Mexico City Majolica							
Means	1.69	1.13	0.054	0.054	4.1	54	3.7
+0.05 limit	2.74	2.24	0.077	0.109	5.4	81	7.41
−0.05 limit	1.04	0.57	0.036	0.027	3.1	36	1.84

We have analyzed two sherds from Guatemala, three sherds from Panama Vieja, two sherds from Cuzco, Peru, and one sherd from Quito, Ecuador. These sherds are shown in Figures 10 and 11. In all cases the sherds from each geographic location had matching compositions which were different from the compositions of the sherds from other geographic locations. These data are also given in Table VI. The fact that the elements cerium, europium, thorium, and chromium show major distinctions strongly suggests that these differences are not caused by burial. These elements are not subject to leaching in the manner that the alkali elements are. Finally, included in Table VI are sherds SC10 and SC19 (Figure

Mexico, Guatemala, Panama, Peru, and Ecuador

La_2O_3 (ppm)	Sc_2O_3 (ppm)	CeO_2 (ppm)	Eu_2O_3 (ppm)	HfO_2 (ppm)	ThO_2 (ppm)	Cr_2O_3 (ppm)	CoO (ppm)
		Subway Excavations, Mexico City					
30	30.2	76	1.39	3.8	11.5	665	38
31	29.7	73	1.49	3.6	10.9	711	41
21	20.9	41	1.24	12.7	6.9	136	165
		Santiago de los Cabelleros (Antigua), Guatemala					
19	40.1	53	1.70	5.4	5.9	41	40
20	39.8	56	1.91	5.5	5.9	41	40
		Panama Viejo					
46	28.6	96	1.69	5.0	16.8	50	24
34	34.1	78	1.56	5.2	14.0	61	29
31	32.0	73	1.77	4.0	16.8	51	29
		Cuzco, Peru					
46	32.3	99	2.11	6.3	16.9	138	29
46	31.4	99	2.10	6.4	16.5	137	30
		Quito, Ecuador					
42	21.0	85	1.73	4.9	11.6	84	23
		Metropolitan Cathedral, Mexico City					
58	18.0	104	1.43	7.9	22.1	68	13.4
52	18.3	158	1.42	6.6	19.1	70	12.7
		Means and Means + and −95% Confidence Limit					
		for Mexico City Majolica					
22	16.5	39	1.22	4.5	5.6	98	16.6
30.9	23.2	61	1.72	6.4	7.3	167	76.5
15.1	11.8	26	0.87	3.1	4.3	57	3.6

Figure 9.   (left to right): (top) SB30, SB31, and SB32; (bottom) SB33 from the subway excavations in Mexico City; SB34 and SB35 from Maurica, Venezuela

Figure 10.   (left to right): (top) SB36–SB37 from the subway excavations in Mexico City; SB38 and (bottom) SB39 from Santiago de los Cabelleros (Antiqua), Guatemala

Figure 11. (left to right): (top) SB40, SB41, and SB42 from Panama Vieja; (bottom) SB34 and SB44 from Cuzco, Peru, SB45 from Quito, Ecuador

Figure 12. Columbia Plain (left to right): (top) SC10 and SC13; (bottom) SC19 and SC17 from excavations at the Metropolitan Cathedral in Mexico City

12) which are from the excavations at the Metropolitan Cathedral in Mexico City and are shown together with two additional sherds from that excavation. The compositions of sherds SC10 and SC19 do not match those of the other cathedral sherds. The appearance of the glaze is also not characteristic of the other cathedral sherds shown in Figure 5 and the two sherds to the right of Figure 12.

Conclusions

We have been able to distinguish two distinctive groups of pottery among the majolica sherds excavated from Spanish sites in the New World. These distinctions are based on the examination and analysis of the paste portions of the sherds and have involved the combined use of neutron activation analysis, x-ray diffraction analysis, and petrographic examination. Preliminary investigations of the relationships of each of these two groups of sherds to sherds of known origin have also been undertaken. There is evidence to support a Spanish source for the sherds from sites in the Dominican Republic and Venezuela and a Mexican source for the sherds excavated in Mexico City.

We have been able to compare our samples to a small group of majolica sherds from Spain and to a reasonably large group of Precolumbian sherds from Teotihuacan. The majolica sherds from Caribbean sites agree in composition with the Spanish specimens, and those from sites in Mexico City have compositions sufficiently similar to the sherds from Teotihuacan, considering the secondary deposits of carbonates of calcium which are in the majolica sherds and not in the Precolumbian sherds. The presence of these deposits of carbonates of calcium in the majolica and their absence in the Precolumbian sherds was determined by petrographic examination and x-ray diffraction as well as by elemental analysis.

Acknowledgments

We acknowledge the interest and encouragement of Richard Ahlborn (Smithsonian Institution) who brought this project to our attention and the further encouragement and generosity of our archaeological colleagues, Charles Fairbanks (University of Florida) and Florence and Robert Lister (Corrales, NM) who supplied well selected sherds for analysis. We also acknowledge the use of the data obtained from the analysis of the sherds from Teotihuacan provided to Brookhaven National Laboratory by Evelyn Rattray. We sincerely appreciate the assistance of Helen Warren (Santa Fe, NM) and William Melson, Martha Goodway, and James Blackman (Smithsonian Institution) for their important assistance in making the petrographic analyses. The contribution of Grover Moreland (Department of Mineral Sciences, Smithsonian Institution), who prepared the thin-sections for petrographic examination, was an invaluable part of this investigation. We are grateful to Joan Mishara (Smithsonian Institution) for carrying out x-ray diffraction analysis and to Barbara Miller (now at the National Gallery of Art) for x-ray fluorescence analysis of the glazes. Finally, we acknowledge the helpfulness and willingness of the Brookhaven Reactor Group and the very capable technical assistance of Elaine Rowland. The research performed at Brookhaven National Laboratory was under contract with the U.S. Department of Energy and was supported by its Division of Basic Energy Sciences.

Literature Cited

1. Goggin, J. M., "Spanish Majolica in the New World: Types of the Sixteenth to Eighteenth Centuries," Yale University Publications in Anthropology, No. 72, New Haven, 1968.
2. Lister, F., personal communication.
3. Abascal-M., R., Harbottle, G., Sayre, E. V., "Correlation between Terra Cotta Figurines and Pottery from the Valley of Mexico and Source Clays by Activation Analysis," in "Archaeological Chemistry," ADV. CHEM. SER. (1974) 138, 86–87.
4. Flanagan, F. J., *Geochim. Cosmochim. Acta* (1969) 33, 81.
5. Bieber, A. M., Brooks, D. W., Harbottle, G., Sayre, E. V., "Application of Multivariate Techniques to Analytical Data on Aegean Ceramics," *Archaeometry* (1976) 18(1), 62.
6. Olin, J. S., Harbottle, G., Sayre, E. V., presented at Symposium on the Application of the Physical Sciences to the Study of Medieval Ceramics, University of California, March 1975.
7. Olin, J., Sayre, E., *Bull. Am. Inst. Conserv.*
8. Bieber, A. M., Jr., "Neutron Activation Analysis of Archaeological Ceramics from Cyprus," Ph.D. thesis, University of Connecticut, 1977.
9. Lister, F. C., Lister, R. H., "Majolica in Colonial Spanish America," *Historical Archaeology* (1974) VIII, 17–52.
10. Lister, F. C., Lister, R. H., "Non-Indian Ceramics from the Mexico City Subway," *El Palacio* (1975) 81(2), Summer, 25–48.
11. Cervantes, G. L., "Colonial Ceramics in Mexico City," National Institute of Anthropology and History, Science Collection: Archaeology, No. 38, Mexico, 1976.

RECEIVED December 2, 1977.

14

Rare Earth Element Distribution Patterns to Characterize Soapstone Artifacts

RALPH O. ALLEN and S. EDITH PENNELL

Department of Chemistry, University of Virginia, Charlottesville, VA 22901

The rare earth element (REE) concentrations measured in soapstone artifacts can be used as "fingerprints" to match the artifacts to the geological source of the soapstone. Instrumental neutron activation analysis was used to measure 20 trace elements in over 700 samples of soapstone from artifacts and geological outcrops. The variations between samples from the same geological formation were small even though they differed considerably in mineralogy. Different outcrops or formations could be distinguished based upon their trace element contents, especially by using the relative distributions of the REE. Matching artifacts and quarries in eastern North America suggests that much of the prehistoric use of soapstone was from nearby (75–100 km away) sources. There are also cases of resource procurement from distant (> 200 km) quarries.

While knowledge of the age of an artifact is very important, knowing the source of the material used to produce the object is of equal value since it can be used to establish patterns of resource procurement and utilization. In order to identify the sources of any particular type of object, one must first identify some property of the material which is characteristic of the processes by which the original materials or the artifacts were formed. In addition, any property used to characterize a source must be shown to vary more between sources than for objects coming from the same source. For some objects such as glasses or metals the chemical composition as well as the shape may be altered during manufacture and either may be characteristic of the production or the object. For others such as carved steatite or soapstone objects only the shape is changed, and the chemical composition is determined by natural

0-8412-0397-0/78/33-171-**230**$07.00/1

processes. In the case of lithic materials, if the composition is charcateristic of a particular geologic formation or region, the composition could be used to identify the geographic source of that material.

One important group of lithic artifacts are those made of the softer stones such as steatite. These cultural objects are in general symbolic or decorative objects or are used as containers. These objects reflect an expansion of activities beyond the immediate quest for food. Soapstone is particularly interesting because it occurs in relatively few areas and yet was widely used. Its utility comes from its soft, easily carvable texture and its coefficient of thermal expansion which allows heating without cracking.

Geochemistry of Soapstone

Soapstone or steatite is a hydrous magnesium silicate material which is composed essentially of the mineral talc, with varying quantities of carbonate minerals, chlorite, amphibole magnetite, and various other minerals (1). Steatite is very soft, easily cut, and has a somewhat greasy or soapy feel from which it derives its common name.

All soapstone deposits originated during episodes of regional or contact metamorphism by the movement of hydrothermal solutions through older rock types. The original rocks may have been ultrabasic (such as pyroxenites, serpentinites, peridotites, and dunites) or sedimentary, with each type containing different amounts of trace elements. Since the starting materials vary in trace elements and since these can be removed or concentrated by the interaction of metamorphic solutions, one might expect that soapstone deposits would show substantial differences in their trace element contents.

The most common chemical process leading to the formation of soapstone begins with the formation of the mineral serpentine from olivine. The serpentine rocks or serpentines are closely associated with areas of high tectonic activity such as geosynclinal mountain chains and continental margins.

Soapstone is produced by alteration of serpentine in several possible ways. In some cases, the serpentine is directly replaced by carbonate (magnesite) and talc leaving a rock consisting chiefly of talc and magnesite with some minor magnetite.

$$2Mg_6Si_4O_{10}(OH)_8 + 6CO_2 \rightarrow Mg_6Si_8O_{20}(OH)_4 + 6MgCO_3 + 6H_2O$$

Serpentine   Talc   Magnesite

The addition of carbon dioxide and the removal of water are the chief chemical changes. If, however, the silica is added as well, the end result would be a soapstone which is predominantly talc instead of magnesite.

$$Mg_6Si_4O_{10}(OH)_8 + 4SiO_2 \rightarrow Mg_6Si_8O_{20}(OH)_4 + 2H_2O$$

Another common alteration involves the replacement of serpentine by chlorite, resulting in a rock composed solely of chlorite and magnetite. Further replacement by talc or talc–carbonate will result in a talc–chlorite (–carbonate) soapstone. Other alterations can result in the formation of tremolite–talc or chlorite–tremolite–talc soapstones.

On the basis of the different origins and mineralogies of the materials called soapstone, one might expect that the different formations could be classified in terms of their chemical and mineral contents. This has not been the case, and the reason can be explained in geochemical terms. First of all, as pointed out above, there are several distinct processes by which talcose rocks like soapstone can be formed. Although the initial reactants may be very different, the products can be very similar in mineralogy and composition. On the other hand, a single metamorphic episode may cause a series of rocks varying in mineralogy but which can still be classified as soapstone. This kind of variation is seen in a series of zones surrounding serpentinite blocks found as inclusions within a hornblende diorite dike in the Tamarack Lake region near Trinity Co., CA. The zones around these inclusions are the same as those along the original contact between the serpentinite rocks which were intruded by the hornblende diorite. It is convenient to use the inclusions as a scale model for the metamorphic reaction which resulted from the contact between these two chemically dissimilar rock types. The reaction zone sequence resulting is typical of the steatization process and is depicted in Figure 1. Details of this geochemical process are outside the scope of this chapter, but the important point is that in one reaction many of the minerals typical of soapstone (talc, chlorite, tremolite, and antigorite) are formed. The composition of a piece of soft carvable rock taken from this series of zones would depend upon exactly where it came from in relation to the two original rock types.

Detailed geochemical studies of the metasomatic reactions at Tamarack Lake (2) suggested that certain trace elements could be used to characterize soapstone. Using instrumental neutron activation analysis, precise trace element measurements were made for the original rocks and for the altered zones of several inclusions. The variations of several elements are shown as a function of distance from the center of one of the inclusions in Figure 1. For this discussion the most important factor is that the trace element levels vary little from the center of the inclusion out to near the original contact even though the mineralogy and major element composition vary considerably. These results have been explained in terms of a diffusion model with major and trace elements being transported in aqueous solutions through the pore spaces (2). For the trace elements

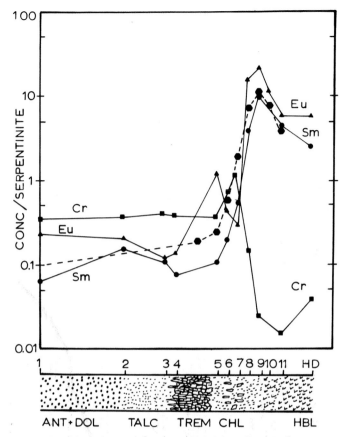

Figure 1. *Variations in the concentrations of trace elements*
as a function of distance from the center of a metamorphosed
serpentinite block (Tamarack Lake, CA).

The original serpentinite was altered to antigorite–dolomite, talc,
and tremolite, and the hornblende diorite (H.D.) was altered to
chlorite with some tremolite. The numbers refer to samples taken
along the alteration zone (the distance between samples 1 and 5
was about 2 ft), and the concentration of the Cr (■), Eu (▲), or
Sm (●) in each sample is shown normalized to the levels found
in unmetamorphosed serpentinite. The theoretical concentration
profile for Sm is shown by the dashed line and was calculated
assuming similar partition coefficients (mineral/solution) for all of
the minerals formed from the serpentinite.

the concentration of the element free to diffuse depends upon how
effectively it is removed by adsorption or reaction (3). This distribution
of a trace element between a solid phase (mineral) and a liquid phase
(where it is mobile) can be described in terms of a partition coefficient
which depends upon the affinity of the trace element for a particular
mineral. This partition coefficient:

$$P = \frac{\text{equilibrium concentration of element in mineral}}{\text{equilibrium concentration of element in solution}}$$

modifies the diffusion coefficient. Thus, Fick's second law of diffusion for radial flow can be written as:

$$\frac{\partial C}{\partial t} = D^* \left(\frac{\partial^2 C}{\partial r^2} + \frac{2}{r} \frac{\partial C}{\partial r} \right)$$

where C = concentration of a trace element in solution, t = time, r = distance from center of the original inclusion, D^* = effective diffusion coefficient = $D/(P + 1)$, D = normal diffusion coefficient, and P = partition coefficient as defined above.

This D^* is important in terms of trace elements because it is modified by the partition coefficient P of the element. Actually D^* is not constant, but is a function of r since the solution passes through different mineral zones with different partition coefficients. The problem is very complex, but it can be simplified somewhat to consider only two zones. In this case we assume that the partitioning of a particular element between a solution and either antigorite, talc, or tremolite to be very similar (nearly the same value of p). The movement of the element in a round inclusion composed of these three minerals can be represented by the modified diffusion coefficient $D^*_<$. Outside of this inclusion the partition coefficients for the various minerals present are assumed to be more similar to each other than to the antigorite, talc, or tremolite. Thus the diffusion outside the inclusion can be described in terms of another coefficient $D^*_>$. A further simplification allows Fick's law to be solved by numerical methods (2). We assume that an inclusion having the same concentration of some trace element as was found in the original serpenite is surrounded by a body of finite dimensions which has the same trace element present in concentrations like those found for the unaltered hornblende dionite. If the inclusion is altered to minerals which have low partition coefficients such that $D^*_<$ is greater than $D^*_>$, then the trace element will diffuse out of the inclusion. The element can either be lost from the system or be deposited in the outer zone depending upon the relative D^* values (2). Although the exact time, diffusion, and partition coefficients are not known, it is possible to determine some relative values of D^* by fitting the experimental data with this theoretical diffusion model. By using a value of $D^*_<$ which is about an order of magnitude greater than $D^*_>$, the samarium results shown in Figure 1 can be generated. Although not complete in all of its details, this does show that samarium would be depleted inside the inclusion as well as to a smaller extent just outside the inclusion. The samarium lost forms a zone of enrichment like a wave very similar to that observed for most of the trace elements measured (2).

Two important factors are suggested by this model. First, the concentration of a trace element in a particular mineral (like talc) formed in this manner depends upon many factors including the size and charge of the ion (which effect the partition coefficient), the composition of the aqueous solution, the minerals present in the intrusive (determines D*), and the time or extent of reaction. Since talcose rocks of similar composition can be formed by various different processes, the differences in any of the factors noted above would result in different trace element contents. Second, since the assumption that the partition coefficients for antigorite, talc, and tremolite are similar appears to be true, one would expect their trace element contents to be similar. This means that a series of soapstone samples collected from a particular geologic body might contain different amounts of these minerals but that they all would probably be very similar in trace element content despite differences in the bulk element composition of the samples.

A well established principle of geochemistry is that the partition coefficient for a trace element is a function of its ionic radius and charge. For example, experimentally determined partition coefficients P for the rare earth elements (REE) vary as a function of the size of the $3+$ ion (e.g., 4). The result of this smooth variation of P as a function of ionic radius, which in the case of the REE is a function of atomic number, is that the contents of the rare earth elements in minerals also vary smoothly. To see the effects of the geochemical processes which are based upon ionic size, it is necessary to remove the effects of having higher relative abundance for elements with even atomic numbers than those having odd atomic numbers. This can be accomplished by normalizing the REE concentrations measured in any mineral or rock to the concentrations of the particular element in chondritic meteorites since they are considered to represent the original (primordial) relative abundances of the nonvolatile elements in our solar system (5). Thus the REE distribution curve for the hornblende diorite intrusive at Tamarack Lake shown in Figure 2 resulted from a series of geochemical reactions which concentrated the heavier (smaller radii) rare earths relative to the presumed original composition (like chondrites) and relative to the lighter rare earths. Since these REE distribution curves almost always vary smoothly, it is not necessary to measure every one of the elements to determine the pattern or shape of the curve. The major exception to this is the element europium which behaves anomalously because during many geochemical processes it can be partially reduced to Eu^{2+} which partitions differently than Eu^{3+}. Figure 2 shows the REE distribution curves for the two starting materials, the serpentinite and the hornblende diorite, which were determined using

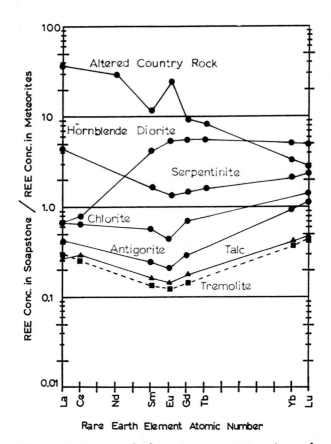

Figure 2. *Rare earth elements concentrations (normalized to chondrites) for the original rocks (serpentinite and hornblende diorite) and their alteration products antigorite, talc, tremolite (– – –), chlorite, and the altered country rock as a functions of the atomic number of the element. The smooth trends reflect the partitioning of the rare earth elements on the basis of their ionic radii.*

samples of these materials from the same areas but which did not show any signs of reaction between the two. When the two materials did react and the serpentinite was replaced by zones of antigorite, talc, and tremolite, the REE (especially the lighter ones) were lost as is also shown in Figure 2. Although different minerals were formed from the serpentinite, the patterns that result are very similar in shape. Although it is possible that the REE lost when the serpentinite is metamorphosed could be carried out of the region in the mobile solution, in this case most of the REE appear to have been redeposited in a reaction zone around the inclusion. In a region where there is a mixture of chlorite and hornblende diorite which is partially altered, the REE distribution (Altered Country Rock in Figure 2) shows an enrichment in most of the REE compared with the samples of unaltered hornblende diorite from the area. The anomalous europium in this altered rock (Figure 2) as well as the differences between samarium and europium concentration changes across the inclusion (Figure 1) suggest that some of the europium was present as Eu^{2+} which partitioned differently than the other REE ions. The REE distribution patterns observed for this set of samples does indicate that the REE are a particularly useful group of elements to use to characterize a geological formation or process as they provide an easily visualized pattern which can be used like a fingerprint as will be discussed later.

Characterization of Soapstone by INAA

To show that REE and other trace elements could be used to characterize soapstone formations, it was necessary to show that variations between geologic bodies were greater than those for soapstone samples found within the formation. For these studies instrumental neutron activation analysis (INAA) was used because of its sensitivity and preci-

Table I. Instrumental Neutron Activation Analysis of Soapstone Artifacts

Element	Isotope Analyzed	γ-Ray Energies	Count Set[a]	Concentration[b]
Ba	Ba-131	496	1, 2	83 ± 43 ppm
Ce	Ce-141	145	2	10.3 ± 0.7 ppm
Co	Co-60	1173, 1332	1, 2	78.1 ± 0.7 ppm
Cr	Cr-51	320	1, 2	4080 ± 100 ppm
Eu	Eu-152	122, 1408	2	55 ± 3 ppb
Fe	Fe-59	1099, 1292	1, 2	4.79 ± 0.05%
Gd	Gd-153	97, 103	2	0.43 ± 0.08 ppm
Hf	Hf-181	133, 482	2	0.25 ± 0.04 ppm
La	La-140	328, 487	1	7.42 ± 0.07 ppm
Lu	Lu-177	208	1	30 ± 1 ppb
Na	Na-24	1369	1	305 ± 4 ppm
Nd	Nd-147	91, 531	2	4.9 ± 3.0 ppm
Ni	Co-58	811	2	400 ± 100 ppm
Sc	Sc-46	889, 1120	1, 2	4.11 ± 0.02 ppm
Se	Se-75	264	2	50 ± 30 ppb
Sm	Sm-153	103	1	0.343 ± 0.003 ppm
Tb	Tb-160	298, 963	2	55 ± 20 ppb
Th	Pa-233	311	2	0.50 ± 0.11 ppm
U	Np-239	277	1	45 ± 15 ppb
Yb	Yb-175	282, 396	1	0.17 ± 0.02 ppm
	Yb-169	177, 199	2	

[a] Count set 1 was 4–10 days and count set 2 was 30–40 days after the irradiation.
[b] Concentrations in a soapstone plummet from Labrador along with the uncertainty in the analysis arse based upon counting statistics.

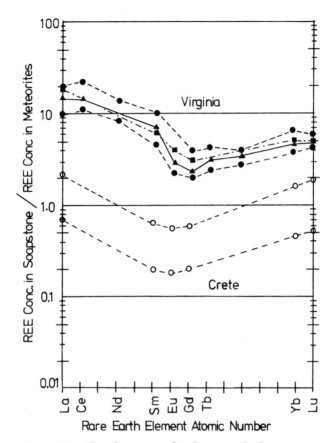

*Figure 3.   Chondrite-normalized rare earth element pat-
terns of soapstone samples from quarries in the Albemarle–Nelson County regions of Virginia and in Crete.
For the Albemarle–Nelson quarries (● – – ●) the two
lines give the range of 12 samples analyzed, and for the
Crete samples (○ – – ○) the two lines define the range
of 19 samples. Two artifacts made of soapstone from the
Albemarle–Nelson quarries are shown. One (■ – – ■)
is a pot from a habitation site in Cherokee County, NC
and the other (▲ – – ▲) a pot from a habitation site near
the quarry in Buckingham County, VA.*

sion. Normally .25–.60 g of soapstone was crushed with a hardened steel
mortar and was sealed into polyethylene tubes. After irradiation of 12–24
samples for 1 hr at a neutron flux of 3×10^{13} neutrons cm^{-2} sec^{-2} in the
University of Virginia Research Reactor, the samples were stored for about
three days to allow the short-lived radioactivity to decay. The samples
were then weighed into new glass vials for counting. At least one
standard was run with each group of samples. In the earliest experiments

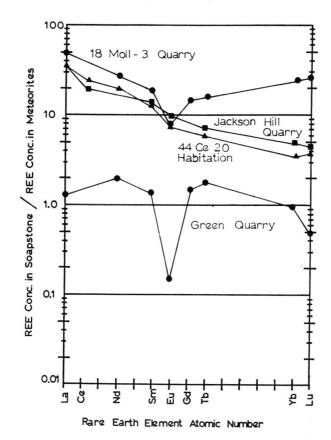

Figure 4. Chondrite-normalized REE patterns for three quarries: Green quarry, Watuga County, NY (●); Jackson Hill, Dekalb County, GA (■); and 18 Mo 11-3 quarry, Montgomery Co, MD (●). A soapstone pot found at a habitation site in Caroline County, VA is shown and is interpreted as being made of material from the Jackson Hill quarry (▲).

both aqueous solutions containing the elements of interest and the U.S.G.S. standard basalt BCR-1 were used as standards. In later experiments a soapstone standard prepared by crushing and homogenizing 2 kg of soapstone from an outcrop (Chula) in Amelia County, VA was analyzed extensively and then was used as a standard.

Samples and standards were counted twice; once 4–10 days and again 30–40 days after the irradiation. The elements analyzed in each count set are indicated in Table I. One of three high resolution Ge(Li) detectors coupled to 4996 channel pulse height analyzers was used to measure the γ-rays emitted from the activated samples and standards. Automated data reduction included the calculation of the analytical uncertainty based upon counting statistics as well as on the concentrations. Table I gives the concentrations found in a soapstone plummet from a Maritime Archaic (3700–4100 B.P.) habitation site at Rattler's Bight in Labrador along with the analytical precision.

The variations between a group of samples from the Albemarle–Nelson County (VA) quarries has been described elsewhere (6, 7). After analyzing 12 samples varying in physical properties from soft crumbly material to very hard soapstone, the REE patterns all fell within the region shown in Figure 3. More recently 19 samples of soapstone from the region of Gonies in Crete were analyzed. Although these samples varied considerably in color and texture, the REE patterns were all very similar in shape and all fall between the lines shown in Figure 3. These results as well as extensive analysis of other quarries suggest that differences between samples from the same formation or geologic setting are more a matter of magnitude rather than the shape of the REE distribution patterns. While the physical properties were different in the samples described above, the differences in mineralogy were not investigated. In Table II the petrographic estimate of the mineral contents of four soapstone samples are shown. While the Jackson Hill quarry sample from Dekalb Co., GA is quite different from the artifact from Caroline County, VA, the REE patterns (Figure 4) are nearly identical. On the basis of analyzing and comparing trace elements in more than 700 artifacts and quarry samples, we would contend that the Caroline County artifact originated in the vicinity of the Jackson Hill quarry. On the other hand, the other two samples (one from Green quarry in Watuga County,

Table II.   Petrographic Analysis of Thin Sections from Soapstone

	Color	Ave. Grain Size	Mineral Content	
			Talc	Chlorite
Jackson Hill 3 NAS 11 Quarry Dekalb Co., GA	green–gray, vitreous luster	1 mm	20%	30%
44CE2 Habitation Caroline Co., VA	medium gray with knots of dark green	1.5 mm (amphibole) 0.5 mm (talc and chlorite)	80%	7%
Green 007 Quarry Watuga Co., NY	Pearly greenish gray	1.5 mm	60%	35%
18MO 11-3 Quarry Montgomery Co., MD	Pearly light brown	1 mm	60%	25%

NY and a quarry in Montgomery County, MD) are similar in their mineralogy, but the REE patterns (Figure 4) clearly distinguish these two quarries. While other trace elements can be used for this type of fingerprinting of artifacts and quarries (as discussed later), the REE patterns have proven to be the easiest to use.

As was indicated above, when soapstone samples from the same formation were compared, the differences were in magnitude rather than in the shape of the REE patterns. Figure 5 shows that different formations have large variations in shape as well as in magnitude. These samples are all from some of the outcrops along the soapstone belt which runs along the Piedmont region of the eastern United States. The sizes of the outcrops where the soapstone is exposed are variable as, of course, are the sizes of the geologic body altered to soapstone. Often several geographically distinct outcrops are a part of the same formation. For instance, the samples falling in the range shown in Figure 3 came from several outcrops up to 700 m long 100 m wide along a 32-km region in Albemarle and Nelson Counties (VA). On the other hand, the Chula quarry (Amelia County, VA) appears to be a small body with limited outcrops which were nevertheless used extensively by the aboriginal inhabitants of eastern Virginia (Figures 6, 7). In some cases even soapstone boulders brought into a region by glaciers were used (9).

Often samples from outcrops are collected with little knowledge of the geologic settings. While a geographical description of a location is important, it can be misleading. For example a series of soapstone samples taken from Lancaster and Chester Counties in southeastern Pennsylvania

Samples Taken from Three Quarries and One Artifact (Bowl)

Amphibole (Tremolite Actinolite)	Accessory	Notes
48%	2% magnetite	hard, but curvable from well developed cleavage of amphibole
13%	trace	dark green knots are amphibole
trace	2% goethite	components intergrown
5%	7% goethite	minerals occur in small clumps

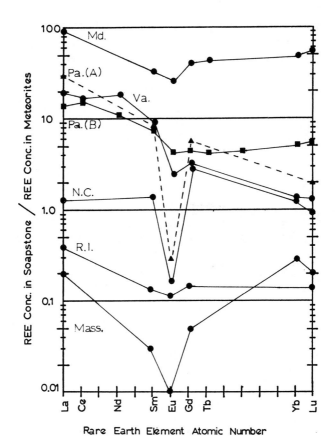

Figure 5.   Chondrite-normalized REE patterns for some soapstone outcrops along the eastern U.S. Piedmont. The source is indicated by the abbreviation of the state. The Md. sample is from Ednor quarry, Montgomery County, MD; Va. is from the Albemarle–Nelson County, VA quarry; Pa. (A) and (B) are representative of the two types of samples from outcrops in Lancaster and Chester Counties, PA; N.C. is from an outcrop in Watuga County, NC; R.I. is from Oaklawn, RI; and Mass. is from Westfield, Mass.

fell into two similar but distinct groups with REE patterns as shown in Figure 5. These results were forwarded to the late Davis Lapham who had made extensive studies of the geology in this region. He pointed out (*10*) that while there were too few samples for a good statistical treatment, the two groups did correspond to different geologic settings. The high europium samples are all from locations where a serpentinite pluton intruded granitic gneiss country rock. On the other hand the samples with the large europium anomalies (low europium) came from sites

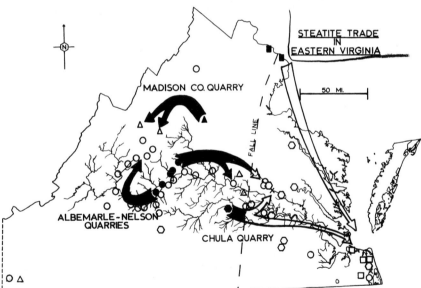

Figure 6. (top) *Soapstone boulders at the Chula quarry, Amelia County, VA showing evidence of the prehistoric manufacture of soapstone bowls (8)*

Figure 7. (bottom) *Utilization of soapstone or steatite trade in Virginia. Soapstone from extensive exposures in Albemarle–Nelson Counties (●) were found in over half of the habitation sites (○) in Virginia, most of them west of the fall line. The Madison County quarry (▲) was used less extensively but in the same area. A relatively small outcrop near Chula (⬤) was used extensively, being found primarily in habitation sites (⬡) east of the fall line. Other samples in this region came from northern Virginia (□) and Pennsylvania (×).*

where the country rock which was intruded was younger. In addition the low europium locations all had sodium-rich feldspar pegmatites associated with them while the others did not. Lapham concluded that since "the formation of talc was a process that occurs in place and thus affects both the chemistry of the ultramafic pluton and of the adjacent rocks, the REE contents might be expected to reflect these in-place reactions" (*10*).

Having shown by the analysis of soapstone samples from geologic outcrops that the variations in the REE patterns in a given region are smaller than between regions, the limitations of matching archaeological artifacts to these outcrops can be discussed. One approach is to use computers for the detection, perception, and recognition of regularities among sets of trace element measurements (*11*). An adaptation of the aggregative heirarchial clustering program AGCLUS was used to match about 100 samples. This program takes as its input an N x N symmetric matrix of dissimilarity measures for a set of items to be clustered. The matrix is produced by the program NADIST which compares the trace element concentrations of all the samples being considered. The dissimilarity index is essentially the distance between each sample in n space, where each of the n values is the concentration of a particular element. It is necessary to do some scaling since the elemental concentrations measured for the soapstone samples varied in magnitude from ppb to %. Since the size of a cluster is defined by the maximum of the dissimilarity scores, scaling of the concentrations is important as well as the precision of the data. While the program offers several choices of how samples can be clustered, they all start with N single item clusters and merge pairs of clusters repeatedly until only a single N-item cluster remains. For each cluster there is a measure of the cluster distance (in n space) related to the dissimilarity of the samples. By appropriate choice of the cluster distance, the clusters obtained were nearly identical to those determined by visually matching REE patterns.

Our experience has been that it is difficult to determine beforehand the most desirable clustering distance for this program, especially since large differences from a single element can influence the results in a disproportionate manner. The same difficulty exists for visually matching the REE patterns; however in this case one in effect compares ratios as well as the absolute magnitude. There are five important parameters to compare:

(1) the slope of the light REE from lanthanum to samarium,

(2) the slope of the heavy REE from gadolinium to lutetium,

(3) the slope between samarium and gadolinium,

(4) the europium anomaly (difference between the normalized europium value and the average of the normalized samarium and gadolinium values), and

(5) the magnitude of the normalized concentrations (only one element such as lanthanum need be used to indicate magnitude).

These five parameters, which are based on those REE measured with the greatest precision, were used rather than elemental concentrations. When these parameters were used with AGCLUS, the clusters were almost always the same as those determined visually. The magnitude was scaled by using the log of the concentration. There was, however, no way to take into account our observations that the magnitude of the variations within a quarry area were inversely proportional to the absolute magnitude. Thus when concentrations were about 10 times the chondritic values, the lanthanum concentrations differed by about a factor of two while they differed from each other by about a factor of six when the absolute concentrations fell below chondritic levels.

When possible quarry sources have been analyzed extensively (as in Virginia), matching artifacts to quarries is fairly straightforward. However, when an artifact matches no known quarry samples, the source cannot be identified. Since many of the soapstone outcrops are small, it is nearly impossible to locate and sample all of the possible sources that could have been used throughout prehistoric times. Valuable information is obtained from such artifacts if they can be grouped as coming from the same, although unknown, source. This clustering does not show the geographical and cultural distribution of resources. Similarly, when an artifact matches only very distant quarry regions or outcrops, the possibility exists that a nearby unsampled quarry also matches the artifact.

Although we have not yet found distant outcrops which can not be distinguished on the basis of REE patterns along with several other trace elements, some caution must be used in interpreting distant sources for an artifact. It is, therefore, necessary to continually enlarge our data base to include more quarry samples as well as artifacts. This makes the highly automated instrumental neutron activation analysis a desirable approach since it gives accuracy and precision with a relatively small amount of time required for the actual analysis.

Utilization of Soapstone in Eastern United States

Having established the validity of using REE distribution patterns to characterize soapstone deposits, we can discuss the patterns of resource procurement suggested by the analysis of soapstone artifacts. While many artifacts have been analyzed, the most complete picture of soapstone utilization we have established is in the eastern regions of North America. In particular, the following discussions will be limited to the distribution of soapstone from quarries in Virginia along with some generalities about the patterns of soapstone procurement in the southeastern United States

and to the use of soapstone over the past 4,000 years along the central
Labrador coast.

The series of soapstone outcrops in Albemarle and Nelson Counties
is probably the largest exploitable deposit and is still quarried. Since
the James River runs parallel to and within easy distance of this geological formation for nearly 60 km, it is not surprising that there was
extensive prehistoric use of this soapstone. Evidence exists of several
prehistoric quarries in this region (12). It was not surprising to find that
many of the soapstone artifacts along the upper portions of the James
River basin originated from the outcrops in this area. A preliminary
account of this work in Virginia was presented earlier (7), but further
analysis of artifacts has confirmed these trends. Many soapstone bowls
or fragments from habitation sites in Virginia have REE patterns which
match the Albemarle–Nelson outcrops (see Figure 3). The geographical
distribution of these sites in Figure 7 indicates clearly that soapstone
from this area was utilized extensively and that the James River may have
been an important means of access to the quarries. It is also clear that
the utilization of these quarries was limited except in rare instances in
the eastward direction by the fall line which separates Virginia's rolling
Piedmont area from the coastal plain. Whether this was a result of some

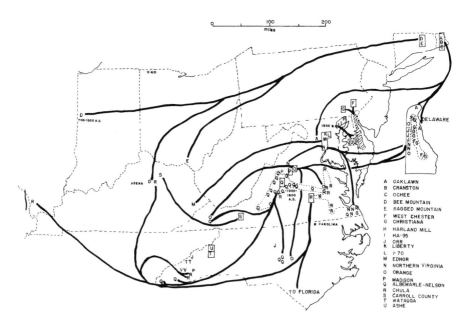

*Figure 8. Resource procurement and the utilization of soapstone in the eastern
U.S. Sources of soapstone are identified by a letter in a square (Ⓐ); soapstone
artifacts which have REE patterns matching a particular quarry are identified
by the letter assigned to the quarry placed on the map at the habitation site (8)*

cultural or political division is not clear. At the time of European occupation of Virginia there was a division at the fall line between the Algonquin-speaking Powhattan groups of the coastal plain and the Siouan-speaking Monacan and Mannahoacs of the Piedmont. Whether the patterns of soapstone procurement from the Albemarle–Nelson quarries suggest an earlier division is unclear since the dates for most of the habitation sites have not been determined. Thus, while we have the geographical distribution of materials in Virginia, the important temporal distribution is missing. While the commonly cited period of maximum utilization of soapstone is during the Transitional Archaic/Early Woodlands periods (roughly 3000–1500 B.P.), there is no evidence that all of these samples came from that period. Since the Algonquin entry into Virginia from the north is normally considered to be much later (perhaps 700 B.P.), the ages of these soapstone artifacts would be of great importance in resolving the discrepancies.

Before discussing other sources of soapstone in Virginia, it should be pointed out that the utilization of the Albemarle–Nelson soapstone

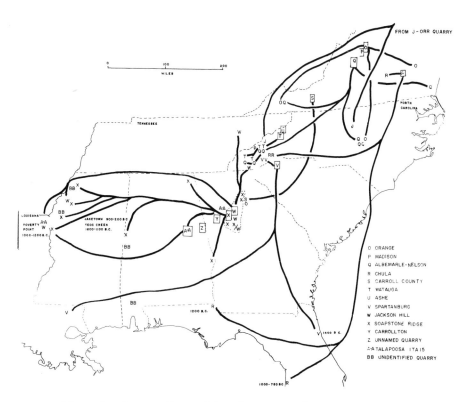

Figure 9. Same as Figure 8 but showing southeastern states (8)

was not limited to Virginia. Artifacts from five North Carolina habitation sites, most in the western mountainous regions were found to match this Virginia source (see, for example, Figure 3). Although the number of artifacts analyzed from sites northeast of this Virginia quarry area was about equal to the number from the southwest, none of the former came from Albemarle and Nelson Counties (Figure 8).

Evidence of even more distant utilization of the Albemarle–Nelson soapstone was found in northwestern Mississippi. While most of the 16 samples from this region originated from regions further south nearer the end of the Appalachian soapstone belt in the Carolinas and Georgia or from unknown quarries, three appear to have come from the Albemarle–Nelson quarries. Whether this material was traded (in its most general sense) or obtained directly by its users can not be established and is a matter of interpretation. However, the distribution of materials from this large Virginia soapstone formation does show some of the same characteristics we have found for all of the soapstone in the eastern United States (Figures 8 and 9). The main features are:

(1) artifacts from a particular quarry area are found south and west of the quarry at considerable distances but rarely are found more than 100 km in northerly or easterly directions;

(2) most artifacts in a given area come from soapstone formations within 150 km of the habitation site, although there are usually artifacts in the same site or region which came from more distant sources;

(3) the distributions of material from quarries overlap each other at least to some extent although separation of these patterns on a temporal basis and more extensive sampling of artifacts could resolve this overlap in utilization patterns.

An example of this last point is a comparison between the utilization of the Albemarle–Nelson quarries and of the Chula quarry (Figure 7). Although the Chula outcrop is small, it was utilized quite extensively east of the fall line.

In addition to the patterns of soapstone utilization in Virginia from these two regions, there are other interesting differences in the geographical distribution of the material outside the state (Figures 8 and 9). The Chula material is found more along the seacoast as far south as Florida. One Florida sample comes from a site dated at about 3200 B.P., suggesting very early utilization of the quarry. However, while the utilization of the Chula soapstone was distinctly different from that of the larger outcrops further west, there is considerable overlap in the southwesterly direction. Although Chula samples are not as common as those from the Albemarle–Nelson quarries, both are found in North Carolina.

The other quarry regions in Virginia were utilized in a similar manner, but less extensively. Quarries in Orange County and in Carroll

County (the former north and the latter south of the Albemarle–Nelson outcrops but along the Blue Ridge Mountains) appear to have been utilized in a manner very similar to that of the Albemarle–Nelson quarries. In other words the artifacts were found south and west of the quarries. On the other hand, artifacts from quarries in Northern Virginia near Washington, D.C. were distributed more like the Chula materials (Figure 8).

These studies of artifacts from the eastern United States shed some light on the "minimum effort" hypothesis that the nearest exploitable resources were utilized. While we have found that many artifacts came from nearby quarries (75–100 km away), there are numerous examples where this is not true. For example two platform pipes from the Elk Garden site in Russell County (VA) were not made from soapstone in the region of Carroll County (VA) about 90 km from the habitation site. Instead the materials came from the Albemarle–Nelson and the Orange County regions over 250 km to the northeast. A similar but distinct situation exists for artifacts found in habitation sites which are great distances (over 200 km) from any soapstone source. For instance, in the midwestern region of the United States where the Hopewellian trade is being studied, there are no nearby sources of soapstone. Most of the limited number of artifacts analyzed from the midwest do come from great distances, but usually from one of the nearest soapstone sources east of the habitation site. Artifacts from Kentucky originated in Maryland (Ednor quarry) and Virginia (Chula quarry) though the latter is not quite as near as the Albemarle–Nelson quarries. The Chula quarry was also the the source of a pipe bowl found at a site in Union County, IL. As in the case of regions near quarries, there are also exceptions to the "minimum effort" hypothesis for these regions far from any soapstone outcrops. For example, one artifact from southern Ohio (Scioto Co.) and one from central Kentucky (Bourbon Co.) did not originate in the nearest quarries of Virginia or Maryland. Rather they originated from the much more distant soapstone formations in Connecticut. Examples such as these, where the nearest resources were not utilized, could point to some special cultural contacts. Again it would be important to know the temporal relationships in these patterns of resource procurement.

Soapstone Utilization in Labrador

For one group of artifacts from Labrador the temporal relationships between samples are known. These samples are particularly intriguing because they come from the juncture of the arctic and subarctic zones. The cultural boundaries along the central Labrador coast have shifted through time (e.g., 13). Indians have been present in the southern and central parts of Labrador for the past 7,500 years. The first Eskimos appeared from the north about 4,000 years ago, after which the two groups advanced and retreated across the central Labrador coast as many as six times (*13*).

A third ethnic group, namely the Norse Vikings, also settled briefly in this area about 1,000 B.P. While soapstone was used by all three of these groups, it was utilized differently by each one. Norse utilization is reflected by a soapstone spindle whorl found in Newfoundland. Soapstone plummets, perhaps used as fishing sinkers, have been analyzed from several Maritime Archaic (Indian) sites occupied during 4000–3500 B.P. (Figure 10). Soapstone oil lamps, cooking pots, and amulets from Dorset Eskimo sites (2000–1000 B.P.) have also been analyzed. In fact, nearly 100 artifacts from about 20 different sites in Labrador and Newfoundland have been analyzed. The dates of occupation of the various sites ranged from 4000 to 150 B.P. The REE patterns of this collection of soapstone artifacts can be separated into several groups, each presumably coming from separate soapstone formations. Most of these quarry areas were

Figure 10.   Soapstone plummets from the Maritime Archaic (4000–3500 B.P.) site at Rattler's Bight in the Hamilton Inlet region of Labrador (14)

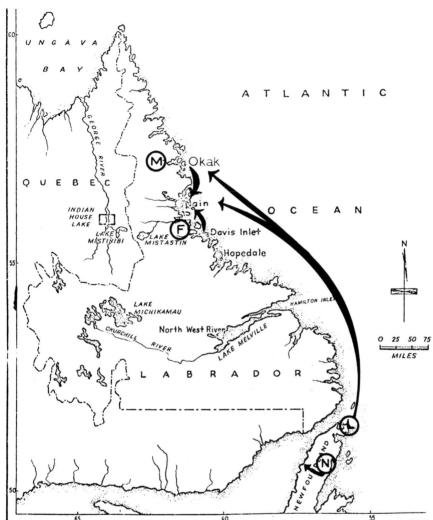

Figure 11. Utilization of soapstone by Eskimos on the central Labrador coast during the Middle Dorset period (200–600AD). The letters in circles (○) are the four quarries from which samples have been analyzed. The letters refer to the letters used in Figure 12 where the characteristic REE patterns are shown. The quarries near Okak and Davis Inlet were utilized locally and in the Nain region. In addition material from northern Newfoundland (L) is found in this region as well as many samples for which the quarries have not been identified.

apparently used over a considerable period of time since most quarries are represented by samples from each of the four major cultural periods as is discussed below. The problem is that while the REE patterns in the artifacts suggest at least 10 different source regions, only four of these

sources have been identified. The four sites from which soapstone outcrops have been analyzed are shown in Figure 11. Part of the problem is
that much of the Eskimo utilization of soapstone disappeared after contact with Europeans, and the locations of the soapstone sources have
been lost in many cases (14).

While it is possible to draw some conclusions with regard to the
utilization of soapstone from the four quarries sampled, less than half of
the artifacts analyzed originated from these four sources. The similarities

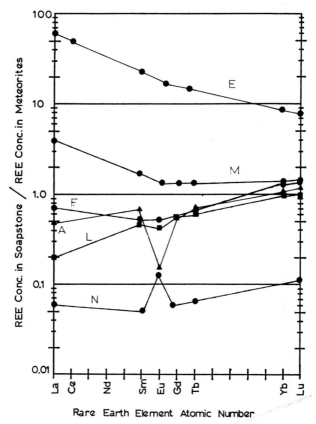

*Figure 12.   Chondrite-normalized REE patterns from
quarries and artifacts in Labrador and Newfoundland.
Sample N is from the Fleur-de-Lys quarry on the east
coast of Newfoundland; L from an outcrop near the
Viking habitation site at L'anse aux Meadows in northern Newfoundland; F from soapstone outcrops in the
Freestone Harbor region; and M from outcrops of
Moore's and Okak Islands on the central Labrador coast.
Both E and A are artifacts which represent a group of
artifacts having similar REE patterns but for which no
quarry of origin has been found.*

in the REE patterns for groups of the other artifacts may suggest that all of the artifacts with similar patterns originated in the same quarry. But until more sources are actually sampled, the interpretation of these groups must remain tentative. The REE patterns for the four source locations which are actually known are shown in Figure 12 along with two artifacts which represent two of the other groups (A and E) for which the source is unknown. Figure 13 shows the range of three samples from outcrops in the Okak region along with two artifacts which we content came

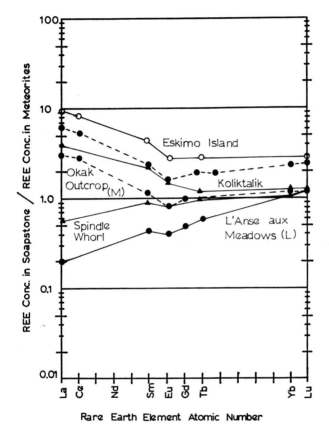

Figure 13. Chondrite-normalized REE patterns for soapstone from the central Labrador coast. Dashed curves (● – – ●) outline the range of three samples from the Moore's Island, Okak outcrops (M) and two artifacts made of this soapstone. The Eskimo Island (○) artifact is from a Labrador Eskimo site (historic period) in Hamilton Inlet and the Koliktakik (▲) artifact is from a soapstone pot from a late Dorset Eskimo site in the Nain region. A sample of the soapstone outcrop near L'anse aux Meadows (L) is shown (●) along with the Viking spindle whorl (▲) made from this material.

from this same soapstone formation. While artifacts can be easily matched to quarry samples, the grouping of artifacts is not always as clear. For instance, some of the groups of samples (for example E and M in Figure 12) are really fairly similar to each other in REE patterns. It is possible that they are from the same area, but until the quarry is found and the variations between samples in that quarry have been determined, this question cannot be resolved. Most of the groups are, however, quite distinctive.

Using this classification scheme to designate the source of the artifacts, the geographical distribution of material from a quarry can be examined as a function of time. In Table III the quarry for a particular artifact is shown (by a letter for each object measured) as a function of time and of a general geographical region. The cultural context (Indians and Eskimos) is indicated along with the dates. Two factors are obvious from this table. First, some of the same sources have been used by the different cultural groups over a very long period. Second, some quarries appear to have been utilized locally while others were the source of soapstone that was distributed widely. The distribution of soapstone during the Middle Dorset period (200–600 A.D.) is shown in Figure 11. The arrows indicate the apparent movement of material from the known quaries to habitatioin sites where artifacts with REE patterns matching those of the quarry were found. Further statistical validity to the trends observed.

Table III. Geographical Distribution of Soapstone Periods of Major

Location (North–South)	Maritime Archaic Indian (3000– 2000 BC)	Early Dorset Eskimo (700–300 BC)
Okak Island	—	L F F D
Nain	—	F F M C
Hopedale/Davis Inlet	FF	—
Hamilton Inlet	L M M A B H H H	C

ᵃ Each letter represents one artifact which originated in the quarry area indicated by the letter code. F = Freestone Harbor quarry, Davis Inlet; L = L'anse aux Meadows quarry, Northern Newfoundland; M = Moore's Island and Okak quarries;

In addition to general trends there are a couple of interesting samples from this region. One is a sample of a Maritime Archaic plummet from Rattler's Bight (Hamilton Inlet) which has a REE pattern which matches very closely a surface-collected sherd from Merrimac, MA. The other trace elements also match very well for these two samples, but does this represent contact between Indians in New England and along the Labrador coast? It is unfortunate that more information about the age and cultural context of the Massachusetts bowl sherd is not available. Fowler, who provided the sample, did note that the color and soft texture of this sherd differed from any he had found in his excavations of New England habitation sites. No quarry sample analyzed to date has a similar pattern.

The other sample of particular interest was a spindle whorl from what is considered to be a Viking settlement at L'anse aux Meadows in Newfoundland (*16*). A small (20 mg) sample of this unique artifact was provided by Parks Canada. After duplicate analysis of 10-mg samples, the material could be returned to Canada for further testing since INAA is a nondestructive technique. As can be seen in Figure 13, the spindle whorl has a very similar REE distribution pattern to a sample of soapstone collected from an outcrop about a mile from the habitation site (*17*).

While the L'anse aux Meadows spindle whorl is similar to those found in Greenland and Norway, its origin may be enigmatic. It is possible that the Norse settlers at L'anse aux Meadows discovered the

Artifacts along the Central Labrador Coast during Cultural Tradition[a]

Middle Dorset (200–600 AD)	Late Dorset (1000 AD)	Labrador Eskimo (Historic Period)
N B B B	CCCC BB	B
M M M M F F F L L A A A A D D D D	—	F
—	—	—
—	—	F M M N N N N D D C A A

N = Fleur-de-Lys quarry, Northeastern Newfoundland. A,B,C,D are artifacts grouped on the basis of REE patterns, quarry locations unknown.

nearby outcrop. Fitzhugh (14) has suggested another possible origin of the spindle whorl based upon its unusual thinness as well as on its soot-blackened concave base. A Dorset Eskimo soapstone fragment of about the same wall thickness as the whorl was found below the turf and above the Norse level at L'anse aux Meadows (16). The presence of Dorset implements which predate the Norse settlement at the site suggest to Fitzhugh that the Norsemen located and scavenged soapstone from the Dorset site (14).

Conclusion

While some caution must still be used in interpreting how closely samples must match, it is clear that analysis of the REE in soapstone can be used to obtain new information about the utilization of lithic resources. The reasons that this group of elements can identify the source of soap-stone are geochemical. Soapstone is a relatively rare lithic resource formed by fairly complex metamorphic processes which leave a "fingerprint" in terms of the trace element content. The distribution of material with this "fingerprint" will help archaeologists to understand the process and pattern patterns of resource procurement. However, the examples dis-cussed above show that it is also important to know age of the artifacts to interpret the geographical distribution patterns.

It is possible that still other lithic materials could be analyzed and their sources determined on the basis of trace element "fingerprints." On the basis of the experience with soapstone it seems that it would be best to use a systematic geochemical approach. After evaluating the processes responsible for producing a particular geologic material, those trace elements effected by the process can be determined. This is an area which with the cooperation and interaction of analytical chemists, geologists, and the archaeologists can continue to grow in importance.

Acknowledgments

A study of this magnitude requires the cooperation of a large number of archaeologists and geologists to provide samples, and we express our appreciation to each of them although they are too numerous to list. We especially acknowledge the cooperation of C. G. Holland who originally brought this problem to our attention, helped collect many of the samples, and provided great help in the archaeological interpretation of the data. W. W. Fitzhugh provided the Labrador samples, and M. Becker collected the samples from Crete. J. Chernosky provided samples and his interpre-tation of the metamorphic changes for the study of the Tamarack Lake samples. L. Thomas helped model the trace element redistribution. The assistance of W. A. Williams in computer programming as well as in the analysis of samples was especially important. K. K. Allen, J. Burton, A. H. Luckenbach, and M. D. Baker also assisted in the analysis of some samples. Finally, we gratefully acknowledge the financial support of NSF grant 41500 without which this study could not have been made.

Literature Cited

1. Smith, J., "Talc, Soapstone and Related Stone Deposits in Virginia," *Va. Miner.* (1961) **7**, 75.
2. Williams, W. A., "Trace Element Migration During Metamorphism," Ph.D. thesis, University of Virginia (1977).
3. Curtis, C. D., Brown, P. E., "The Metasomatic Development of Zoned Ultrabasic Bodies in Unst, Shetland," *Contrib. Mineral. Petrol.* (1969) **24**, 275.
4. Jensen, B., "Patterns of Trace Element Partitioning," *Geochim. Cosmochim. Acta* (1973) **37**, 2227.
5. Haskin, L. A., Frey, F. A., Schmitt, R. A., Smith, R. H., "Meteoritic, Solar, and Terrestrial Rare-Earth Distributions," *Phys. Chem. Earth* (1966) **7**, 167.
6. Allen, R. O., Luckenbach, A. H., Holland, C. G., "The Application of Instrumental Neutron Activation Analysis to a Study of Prehistoric Steatite Artifacts and Source Materials," *Archaeometry* (1975) **17**, 69.
7. Luckenbach, A. H., Holland, C. G., Allen, R. O., "Soapstone Artifacts: Tracing Prehistoric Trade Patterns in Virginia," *Science* (1975) **187**, 57.
8. Holland, C. G., Department of Anthropology, University of Virginia.
9. Fowler, W. S., "Stone Eating Utensils of Pre-historic New England," *Am. Antiq.* (1947) **13**, 146.
10. Lapham, D., personal communication.
11. Jurs, P. C., Isenhour, T. L., "Chemical Applications of Pattern Recognition," Wiley-Interscience, New York, 1975.
12. Bushell, D. I., "The Five Monacan Towns in Virginia—1607," *Smithson. Misc. Collect.* (19) **82**, 1.
13. Fitzhugh, W. W., "Preliminary Culture History of Nain, Labrador: Smithsonian Fieldwork, 1975," *J. Field Archaeol.* (1976) **3**, 123.
14. Fitzhugh, W. W., Smithsonian Institution, personal communication.
15. Fowler, W. S., personal communication.
16. Ingstad, H., "Westward to Vinland," St. Martins, New York, 1969.
17. Allen, R. O., Allen, K. K., Holland, C. G., Fitzhugh, W. W., "Utlization of Soapstone in Labrador by Indians, Eskimos and Norse," *Nature* (1978) **271**, 237.

RECEIVED September 19, 1977.

15

Atomic Absorption Spectroscopy of Archaeological Ceramic Materials

VANCE GRITTON

Department of Chemistry, Chapman College, 333 N. Glassell,
Orange, CA 92666

NICHOLAS M. MAGALOUSIS

Departments of Archaeology and Anthropology, Chapman College, 333 N.
Glassell, Orange, CA 92666 and Institute of Archaeology,
University of California, Los Angeles

Atomic absorption sample preparation procedures applied to archaeological samples can be streamlined. Important factors include sample preparation, sample size, sample decomposition, standards, instrumentation, and practical and conceptual applications of atomic absorption analysis. On a comparative basis the sample preparation procedure reported was convenient and rapid; the AAS instrumentation proved to be flexible, sensitive, rapid, and inexpensive in the analysis of archaeological materials.

The theory, technique, and applications of atomic absorption spectroscopy (AAS) are well known to chemists and are described extensively in the literature, which is estimated at nearly 1,000 papers a year. Excellent sources of information are the Annual Review issues (April) of *Analytical Chemistry* and the *Annual Reports on Analytical Atomic Absorption* from the Society for Analytical Chemistry, London. Several pertinent references are listed by Hughes et al. (*1*). Some books found generally useful are listed in the bibliography (*2, 3, 4*).

Two reports directly concern the application of AAS to ceramics of archaeological interest and serve as a basis for the procedures recommended here. The first report is by Magalousis and contains the method used prior to the current study (*5*). The second is by Hughes et al. (*1*). These will be referred to as M procedure and H procedure, respectively.

Although flameless AAS is a valuable technique, this report will be limited to a discussion of conventional flame methods. The problems associated with AAS may be grouped into four somewhat interrelated areas: sampling and sample preparation; sample decomposition; standards; and instrumentation.

Much of the following discussion is in the nature of a preliminary report. Considerable research remains to be done, especially in the areas of sampling and standards. This research is currently in progress. Since this report is concerned with the methodology of the analysis, results will be reported in a separate publication. The elements determined were: Ag, Al, Ca, Cd, Co, Cr, Cu, Fe, K, Mg, Mn, Na, Ni, Pb, Sr, and Zn. Of these, potassium, sodium, and strontium were determined by flame emission. The choice of elements determined by AAS was dictated almost entirely by the hollow cathode lamps available to the investigators. Concentrations of the elements determined ranged from about 10% to undetectable.

Samples analyzed for this research were part of a larger study incorporating petrographic, botanical, optical emission, and advanced computer analyses to maximize the data base in regard to sourcing. The long-ranged goal of this sourcing, as with many other studies of this nature, was to determine social, political, and economic trends of ancient peoples. The archaeological materials sampled and analyzed in this study centered on ceramics, bricks, soils, and vessel contents from two sites in the Near East—Tell Terqa and Tell Dilbat.

Sampling and Sample Preparation

Obtaining the Gross Sample. It is difficult to simply assess the problem of obtaining a representative sample from an archaeological specimen. In many, if not most cases, the specific technique used must be determined on a specimen-to-specimen basis. If the entire shard cannot be used, the difficulties become compounded.

The sample obtained is very important—the results of analysis can be no better than the quality of the sample. If the sample used for analysis is not truly representative of the item under investigation, the results are worse than useless.

Several methods of sampling have been proposed. From the point of view of the analyst the methods, in decreasing value, are crushing an entire shard; drilling through representative portions; taking a small chip; and abrading a portion of the shard.

A casual examination of the shards currently being studied reveals that many are visibly nonhomogenous. Layers and striations are easily seen. If the entire shard is crushed, finely ground and well mixed, then the analytical sample taken will represent the best estimate possible of the composition of the entire object. If a drilling technique is used, the drill should pass through the entire piece and, if possible, in several locations. Abrading tends to remove the analytical sample from only one small location. Again, this should at least be taken from the entire thickness of the shard.

Drilling is usually performed using a sapphire or tungsten carbide tipped bit. Wear of the sapphire tip as well as wear of the metal shank in the bore hole may introduce some slight contamination.

Abrading is generally done with a diamond-charged metal spatula. Again, some slight contamination is possible from the base metal (1).

Glazes or other surface treatments must also be considered. Glazes may be very high in some elements compared with the pottery substrate. Whether or not this surface is to be included in the sample must be decided for each shard or set of shards. If the glaze is not to be included, it must be carefully ground off. Since the specimens currently being analyzed are not slipped, glazed, or painted, these problems did not occur.

Sample Comminution. However the sample is obtained, it must be reduced to a fine powder before use. Decomposition of the sample is greatly abetted by small particle size; if not adequately comminuted, part of the sample may not be converted to a soluble form. If the entire sample is not used, as will generally be the case when an entire shard comprises the gross sample, then a finely ground material is needed to ensure taking a sufficient number of particles to give a representative sample. Material passing through a 100 mesh sieve (150 μm) should be adequate; 200 mesh (75 μm) would be preferable. It is recommended that individual samples not be screened unless the larger-than-desired material is recovered, ground (an agate mortar will be very useful here), and combined with the rest of the material. Rejection of part of a comminuted sample on the basis of particle size will tend to classify the sample on the basis of hardness (or at least resistance to grinding) and thus will bias the results. Once a comminution procedure is chosen, a typical specimen should be treated and the resultant powder be subjected to screening or to estimation of particle size by microscopy. If the procedure is judged to be adequate, other specimens may then be similarly treated and used without screening.

In the M procedure, the shards were ground by hand with a mortar and pestle. There are several objections to this technique. First, it is time-consuming and tedious. Second, it is not practical to reduce the several grams of material to a sufficiently fine powder. Third, the common mortars and pestles are not hard enough to prevent contamination from the porcelain itself. Mortars made of agate or mullite are much superior in those instances when hand grinding is used. In the H procedures, an agate mortar was used.

There are many types of mechanical grinding equipment available: ball mills, ball and mortar sets, planetary mortars, impact mills, etc. These are available in a variety of materials: steels, porcelain, agate, mullite, silicon carbide, tungsten carbide and others. Prices range from $200 to about $2,000. Unfortunately, the cost increases rather markedly as one goes to the harder materials, especially the carbides.

In the procedure currently under study, a CRC micro-Mill is used. This device is somewhat like a blender, consisting of a grinding camber and a blade driven at high speed. The capacity is about 50 mL of sample, and pieces of ¼ in. or so can be accommodated. The grinding chamber and blade are interchangeable units. The chamber and blade now in use are made of Stellite 6, a very hard cobalt-based alloy. A carbide-tipped blade is available, but not a carbide-lined chamber. While performance of the unit has not yet been evaluated completely, no problems have been encountered. The samples currently being studied were ground in the mill for approximately 2 min. Examination by microscopy of the powder produced from a typical sample indicates that at least 90% of the material has a particle size of less than 150 μ, and at least 50% is smaller than 75μ. Further studies of grinding times and particle size distribution are part of the research in progress.

A drawback to this device is that it is not possible to regrind a minor fraction of larger-than-desired particles since it takes a reasonable amount of sample to provide a minimum charge. Hence the grinding must be adequate the first time through.

Any method of sample comminution is a potential source of contamination, and a systematic study of analysis procedures should make some effort to ascertain the degree of contamination thus introduced. It is proposed that this be attempted through the analysis of a standard material, to be mentioned later. Preliminary observations indicate that grinding with the Micro-Mill is not a serious source of contamination. Visual examination of the blade after grinding about 100 specimens indicated very little wear. The cobalt analysis of the samples are generally very low (0–100 ppm) while 10 are greater than 100 ppm. These samples will be reanalyzed to better judge the possibility of contamination from the grinding process.

Sample Size. In general, the larger the size of the analytical sample, within reasonable limits, the better. The number of particles which should comprise a sample is a function of both the degree of homogeneity of the sample and the concentration of the material in the sample (6). Since we are frequently concerned with trace elements in materials which may be rather nonhomogenous, a large number of particles is needed. Large samples, however, are more tedious and expensive to work up, so some compromise must be made. The present study uses 0.25-g samples which are economical with regard to both time and materials. The estimated particle size is sufficiently small to indicate that portions of this size will provide a representative sample of the material under study.

Drying. The powdered gross sample should be dried for at least 4 hr at 110°C and subsequently stored in a desiccator before weighing out the analytical sample. This step is not mentioned in the M procedure but is in the H procedures.

Studies are underway, but by no means complete, to determine loss on drying, weight loss as a function of drying time, water regain in normal laboratory environment, water regain in nearly saturated air, and water regain in a silica-gel-charged desiccator. Typical samples have shown a loss on drying of 0.5–1.5%. Short-term water regain is about 0.2 mg/min. This was measured during a rainy period, hence high humidity, and on large (7–8 g) samples. Even if this entire weight gain were to be applied to the analytical sample, normal technique would hold the error to less than 1 ppt on a 0.25-g sample. Thus, no extraordinary precautions need to be observed in weighing the analytical sample.

Preliminary observations indicate that the dried sample is, not surprisingly, an efficient dessicant in its own right, although of low capacity and low speed. The sample will regain some moisture from partially spent silica gel, and hence it is recommended that anhydrous magnesium perchlorate (anhydrone) be used as the dessicant for sample storage. A 3.5-g sample stored over partially depleted silica gel for 5 weeks regained 15.6 mg while 7–8 g samples stored over anhydrone for 16 weeks regained only 1–2 mg.

Sample Decomposition

In the M procedure, the sample is fused with sodium carbonate, the cooled fusion mixture dissolved in HCl, filtered, diluted to volume, and stored in plastic screw-cap bottle. Two procedures are reported by the H procedures. One uses a lithium metaborate fusion followed by HNO_3 dissolution. No filtration is indicated (this procedure is used only when silicon is to be determined). In the other H procedure, the sample is treated with HF, evaporated completely, and the residue taken up with $HClO_4$.

While fusion techniques allow silica to be determined by AAS, they have certain disadvantages in AA work. The fluxes used will introduce a certain amount of contaminants. For example, consider the Na_2CO_3

fusion. Reagent grade sodium carbonate contains about 50 μg of potassium per gram of Na_2CO_3. Since the weight of flux used is five times the weight of the sample, this will add 250 μg of potassium per gram of sample. This is close to the level of potassium which is found in the specimen under study. Other trace elements in sodium carbonate may also be contaminants, of course. The situation is similar with $LiBO_2$, which is considered to be a superior flux for silicate materials (7). The use of blanks compensates for this to a large degree, but never completely. The precision and accuracy of the analyses will be poorer when large amounts of contaminants are added.

The use of fluxes produces a final solution having a high salt concentration which leads to deposits and clogging of the aspirator and burner as well as to bothersome memory effects. The high salt content also leads to nonatomic absorption which is especially troublesome for those elements which are determined using the UV portion of the spectrum. The silica which remains in solution interferes with the determination of some elements and requires that blanks have an approximately equal silica concentration. The silica also tends to precipitate slowly from the solution, which can lead to scavenging of some trace elements as well as exacerbating the problems of clogging.

The techniques which rely on the destruction of silica by HF produce solutions which behave much better during the analysis, and this technique is the one recommended and the one currently being used. If silica must be determined, and if the amount of sample permits, it is suggested that a separate sample be used. The silica content of most ceramic samples is sufficiently high that a classical gravimetric determination may be made at not much, if any, more bother than an AA determination. Two of the current specimens were analyzed for silica by a classical method and indicate that the ceramics may be expected to have a silica content of about 50%. A rapid gravimetric procedure for silica is in preliminary tests and has produced results within 1% of the classical methods and with a considerable saving of operator time. It is also possible to use an HF decomposition technique which retains silicon by carrying out the decomposition in a Teflon bomb (8).

In the method currently in use, a 0.25-g sample is placed in a 50-mL Teflon beaker. Ten mL of HF (40%) and 3 mL of $HClO_4$ (70%) are added. The beaker is placed on a hot plate, the mixture is evaporated to fumes of $HClO_4$, and fuming is continued to dryness or near dryness. The mixture is cooled, 3 mL of $HClO_4$ are added, and again it is fumed to dryness. The fuming is repeated once more. One mL of $HClO_4$ and 10–15 mL of DI water are added, and the mixture is heated to effect solution. The samples are then diluted to 100 mL and are stored in plastic screw-top bottles. A blank is prepared by fuming 10 mL HF with 9 mL $HClO_4$, adding 1 mL $HClO_4$, and diluting to volume.

Reagents and Equipment Details. Regular reagent grade acids are being used. The water is distilled water passed through a mixed bed deionizer. Ideally, the very highly purified reagents such as Ultrex (J. T. Baker) would be used, but their high cost precludes their use in production analyses. A double-distilled perchloric acid available from the G. Frederick Smith Chemical Co. is not much more expensive than the regular acid and is a good compromise. An all-Teflon apparatus is available for the sub-boiling distillation of acids which might prove economical in the long run.

The use of Teflon beakers in place of platinum crucibles (which are used in the M and H techniques) is a great savings since the beakers are less than $7.00 each. The fuming goes very smoothly in these beakers; no bumping, spattering, or decrepitation was observed. They rinse easily and wash quickly. An added advantage is that they may be removed from the hot plate with bare hands. A supply of Teflon stirring rods aids in breaking up the residues. A Labindustries Repipet or similar device is very handy for dispensing $HClO_4$ and costs approximately $80. The HF is dispensed by plastic graduated cylinders.

The plastic bottles used for sample storage should have all-plastic caps; paper cap liners must be avoided. Prior to use, the bottles should be soaked in 1:15 nitric acid for at least a day; longer soaking is fine. Trace metal contamination from plastic ware is well documented.

Procedural Details. Triple fuming with $HClO_4$ is most probably unnecessary, but at least one fuming to dryness or near dryness does seem to be necessary. This converts the fluorides to perchlorates and destroys any organic matter. Since all perchlorates are soluble, this is a necessary step. Studies will be run using one, two, and three fumings and the results compared. Each fuming takes 2–3 hr, so sample preparation time is high. However, the operator time involved is low since the fumings proceed unattended.

In general, the procedure as outlined works well and yields a clean solution for analysis. The normal precautions used with HF and $HClO_4$ must be followed. All fuming must, of course, be done in a hood, preferably one which has been constructed especially for $HClO_4$ work.

Standards and Interferences

The preparation of suitable standards depends on the composition of the sample, the method of decomposition, and the type of flame used in the determination of a given element. It is felt that many workers pay inadequate attention to the problems of standards preparation and interelement interferences. A concise and thorough, although rather dated, compilation of interferences arranged by element is presented by Angino and Billings (2).

Concentrated aqueous standards are readily available from laboratory supply companies and are also easily prepared from pure elements or compounds. Detailed directions for such standards are found in the manuals furnished by the various instrument manufacturers (9).

Simple aqueous working standards prepared by dilution are often not suitable because of the effect of one element on another in the actual samples. Ideally, the standards used would closely approximate the composition of the sample solution itself. However, such a situation is not practical when many and diverse samples are to be analyzed, and compromises must be made.

Many of the interelement interferences result from the formation of refractory compounds such as the interference of phosphorous, sulfate, and aluminum with the determination of calcium and the interference of silicon with the determination of aluminum, calcium, and many other elements. Usually these interferences can be overcome by using an acetylene–nitrous oxide flame rather than an acetylene–air flame, although silicon still interferes with the determination of aluminum. Since the use of the nitrous oxide flame usually results in lower sensitivity, releasing agents such as lanthanum and strontium and complexing agents such as EDTA are used frequently to overcome many of the interferences of this type. Details may be found in the manuals and standard reference works on AAS. Since silicon is one of the worst offenders, the use of an HF procedure is preferable when at all possible.

The other major source of interelement interference is related to ionization. Since AAS depends on the absorbance of light by atoms, any change in the degree of ionization of an element will be reflected by a change in apparent concentration. Since the ionization of atoms in a flame represents an equilibrium with electrons within the flame, the ionization of one element affects the degree of ionization of another. The extent of ionization increases with flame temperature, so that these effects become exacerbated when the higher temperature acetylene–nitrous oxide flame is used to overcome the interference mentioned above. This ionization also leads to a lower sensitivity since a smaller proportion of the element is present in the atomic form.

The worst offenders, as far as ionization interference goes, are the easily ionizable alkali metals. For example, in the determination of calcium, the absorbance is increased when sodium is added to the mixture since the ionization of sodium represses the ionization of calcium. Obviously, the apparent concentration of calcium will increase as the sodium content increases. The customary method of minimizing this effect is to swamp the system with an easily ionizable element. Typically, both samples and standards are prepared so that the final solutions will contain 1–5 g/L potassium. This technique is equally useful for flame emission methods.

In the samples under consideration, calcium, sodium, and potassium are all rather high, in the vicinity of 1800 ppm, 600 ppm, and 300 ppm respectively. Since the addition of a large amount of potassium would prevent its determination, as well as introducing to a moderate extent the problems connected with higher salt content which were mentioned earlier, it is suggested that an aliquot of the original sample solution be withdrawn, KCl added, the mixture diluted to volume, and used to determine calcium and other elements so affected. For example, 25 mL of the original solution is placed in a 50 mL volumetric flask, 1 mL of a KCl solution containing 50 g/L potassium added, and the new mixture diluted to volume. This will result in a solution containing 1000 ppm potassium. (Twice this amount of potassium was actually used with the present series of specimens, but in a few cases led to the formation of a precipitate of $KClO_4$.) The degree of dilution of the original sample may be varied easily to bring the concentration of the elements being determined into a convenient range.

To assess the effects arising from variations in sample communition, sample decomposition, analysis methods and conditions, and the effects of interelement interferences, some sort of a standard material is needed; for example, a quantity of good quality clean bisque. This material is the basis of a program which is to investigate more carefully the variables noted above. In addition, synthetic samples prepared from reagent grade silicic acid and other materials will be studied. It is hoped these studies will eventually lead to modifications of procedures and, if necessary, to techniques for correcting raw data so that future analyses may most realistically represent the true composition of the specimens under investigation.

Another standardization technique which is widely used in AAS is standard addition. It is especially useful for samples where the matrix is difficult to reproduce, such as samples decomposed by fusion. In this method, a known quantity of the element to be determined is added to a portion of the sample, preferably at the start of the decomposition procedure. This known amount then serves as the standard when the absorbances of the two solutions are compared. Because of the rather prevalent curvature of absorbance vs. concentration plots, it is usually necessary to run three standards as well as the unspiked sample. Thus, even if one could use a single multiple element spike as the solution for standard addition, this would quadruple the work load for sample decomposition and analysis. If such a solution were added after decomposition, it would require splitting the quantitatively diluted solution, spiking, and rediluting, and would quadruple the work load for analysts.

This technique, then, is not considered practical for routine analyses. However, it is of great value when studying interferences and in the development of good standard mixtures.

When producing working standards from the concentrated aqueous standards (which are usually 1000 μg/100 mL) certain routine precautions should be observed. The bottles should be pretreated as outlined in the section on sample decomposition. Working standards of less than 100 μg/mL should be prepared fresh daily. All working standards should contain 1 mL perchloric acid/mL. The availability of high quality micropipets makes the direct preparation of dilute standards more convenient and probably more reliable than preparation by serial dilution. Thus 100 μL of concentrated standard when diluted to 100 mL will provide a 1 μg/mL (1 ppm) standard. It is convenient to provide a secondary stock solution containing 100 μg/mL by 10-fold dilution of the original standard. If desired, the working standards may contain several elements.

Instrumentation and Analysis

Atomic absorption spectrometers are available from several instrument manufacturing firms. Since the technique is now well established, the instrumentation is generally high quality. Still, because it is such a widely used analytical method, each year sees improvements in design, flexibility, and ease of use.

The instrument being used in the current study is a Varian/Techtron model 1200. This is a rather compact unit with considerable flexibility and is very convenient to operate. The unit needs very little maintenance or repair, and personnel may be trained to use it in minimal time.

Whatever instrument is used, provision must be made for using both air and nitrous oxide-supported flames. A fume exhaust must be provided. If arsenic, selenium, or mercury are to be determined, an apparatus for vapor generation should be used. Such apparatus is usually available from the instrument manufacturer. Mercury is usually determined by a flameless or cold vapor technique.

A major expense is the purchase of the lamps needed for the number of elements to be determined. There are several multi-element lamps available, which helps to reduce this cost.

When determining those elements which absorb in the UV portion of the spectrum, corrections for nonatomic absorption are almost essential. This is especially true for solutions of high solids content, such as those which result from fusion techniques. This correction is generally made by using a hydrogen continuum lamp. Some of the newer instruments have provisions for automatic and continuous background correction. A field installable kit is available to retrofit the model 1200 now used.

In addition to the normal flame technique, AAS may be carried out using a graphite furnace. This technique is capable of greater sensitivity than flame operation but usually shows poorer accuracy. The furnace technique requires the same steps for sample decomposition as do the flame techniques.

One new instrument is worthy of note. This is the Hitachi Zeeman effect atomic absorption spectrometer, model 170-70. It provides background correction for nonatomic absorption at all wavelengths through use of the Zeeman effect. It is presently offered only in a carbon furnace configuration. The cost of the instrument is considerably higher than conventional instruments.

Little detail need be given on the actual analysis. Wavelengths and flame conditions are well detailed in the manuals. Blanks and standards should be checked often to keep the baseline constant. The burner height should be adjusted for each element to maximize absorption. In some cases the main sample solution may have to be quantitatively diluted to bring the component being determined into range. Often a satisfactory method is simply to rotate the burner head 90° to present a much shorter path length. This will decrease the sensitivity by a factor of 10 to 20, depending on the burner in use.

Sodium, potassium, and strontium were determined by flame emission. Most AAS instruments are capable of this mode of operation. Ideally, the sample to be analyzed for sodium would be heavily spiked with calcium and potassium; that for potassium, with sodium and calcium. This was not done in the current study; however, potassium was added before the determination of sodium, calcium, and strontium.

Practical and Conceptual Applications of Atomic Absorption Analysis

The application of AA analysis in archaeology must be looked upon conceptually and practically. It has been stated that systematic standardization is necessary if credible and reproducible results are to be obtained. AA is a proven technique; its credibility and sensitivity have for several years been depended upon in the fields of medicine, industry, soil science, and now archaeology. If the analytical circle in archaeology is to be completed, procedures and standards must be followed prior to AA analysis and after; e.g., during field, computer, and interpretive phases.

There are good reasons for acceptance of AA instrumentation. Archaeologists universally have found that funding is a major consideration in element analysis research. However, the purchase cost of necessary AA equipment, including support equipment, is minimal (approximately $8,000–$12,000). In addition the cost of analysis per sample is a large factor. One may expect to pay as much as $250.00 per sample using neutron activation whereas AA is dramatically below this cost. Only a minimal amount of maintenance and repair is required, and AA saves time since rapid and direct readout is obtained. These data may be immediately relayed to existing data storage banks for evaluation, updating, and further study.

Atomic absorption and atomic emission instrumentation have made dramatic advances in the past five years. Instrumentation may now be computerized internally and externally and in the future has the capability of multielement analysis.

The availability of AA instrumentation compared with many other forms of instrumentation is again a significant consideration. AA spectrophotometers are found on almost every campus and usually in various departments; on the other hand, few universities possess nuclear reactors.

AA innately possesses standardizing and organizational concepts not attainable by most other techniques in that it is flexible and mobile. This mobility allows analysis directly in the field in areas of high archaeological interest. Thus, compact, standardized laboratory units utilizing AA as well as other applicable instrumentation and procedures would be available on site. Laboratory expenses could be shared by several archaeological institutes, thus creating international standardization and an even more economical profile than suggested previously.

Laboratories of this nature would also alleviate the problems of transporting artifacts between international borders, which in the past has been time-consuming and expensive. Each laboratory could provide a systematic handling of artifactual data, from primary photography to cataloging and analysis while simultaneously using the laboratory as an international training unit for students of archaeology and chemistry.

Synopsis

To conclude, the element analysis of artifactual materials is in a formative yet positive phase. However, a great deal more communication and standardization is necessary if qualitative and quantitative results are to be experienced by archaeologists and chemists worldwide.

In our attempts to refine and standardize various phases of AAS analysis it has become quite apparent that scientific controls and standardization must be used throughout the entire spectrum of sample acquisition. This necessitates known and accepted sample collection procedures, beginning in the field (noting possible contamination factors) and continuing through the recording, laboratory, computation, and interpretive phases of analysis.

Acknowledgment

Appreciation is extended to Drs. Giorgio and Marilyn Buccellati, Institute of Archaeology, UCLA, and also to several individuals for their assistance and the use of laboratories at Chapman College, University of California at Riverside, University of California at Irvine, and University of California at Los Angeles.

Literature Cited

1. Hughes, M. J., Corvell, M. R., Craddock, P. T., "Atomic Absorption Techniques in Archaeology," *Archaeometry* (1976) **18**, 19.
2. Angino, E. E., Billings, G. R., "Atomic Absorption Spectroscopy in Geology," Elsevier, New York, 1967.
3. Price, W. J., "Analytical Atomic Absorption Spectrometry," Hayden and Sons, London, 1974.
4. Slavin, W., "Atomic Absorption Spectroscopy," Wiley–Interscience, New York, 1968.
5. Magalousis, N. M., "Atomic Absorption Spectrophotometric Analysis of Nabataean, Hellenistic, and Roman Ceramics in an Attempt to Reconstruct History," Master's thesis (1975).
6. Harris, W. E., Kratochovil, B., *Anal. Chem.* (1974) **48**, 313.
7. Suhr, N. H., Ingamells, C. O., "Solutions and Technique for Analysis of Silicate," *Anal. Chem.* (1966) **38**, 730.
8. Rantalla, R. T. T., Loring, D. H., "Multi-element Analysis of Silicate Rock and Trace Elements by Atomic Absorption Spectroscopy," *Atomic Absorption Newsletter* (1975) **14**, 117.
9. "Analytical Methods for Flame Spectrometry," Varian–Techtron, Brisbane, Australia, 1972.
10. Shepard, Anna O., "Ceramics for the Archaeologist," Carnegie Institution, Washington, DC, 1954; reprint ed., 1976.
11. Magalousis, N. M., "Atomic Absorption Spectrophotometric Analysis of Babylonian Ceramics: A Preliminary Evaluation," unpublished research, University of California at Los Angeles (1976).
12. Magalousis, N. M., lecture delivered at the International Symposium on Archaeometry and Archaeological Prospection, University of Pennsylvania and University Museum, Philadelphia, March 16–19, 1977.

RECEIVED September 19, 1977.

Metals

The Possible Change of Lead Isotope Ratios in the Manufacture of Pigments: A Fractionation Experiment

I. L. BARNES, J. W. GRAMLICH, and M. G. DIAZ

Center for Analytic Chemistry, National Bureau of Standards, Washington, DC 20234

R. H. BRILL

The Corning Museum of Glass, Corning, NY 14830

Because of the increasing use of lead isotope ratios obtained from artifacts to determine provenance and because sampling from artifacts is frequently from heavily oxidized or otherwise altered areas, concern has been expressed about the possible fractionation of lead isotopes either by the process of manufacture or through time. A pigmenting compound commonly used in ancient times was manufactured from a large sample of galena through four intermediate steps. As expected, the lead isotope ratios in the starting material, in the final pigments, and in each of the intermediate compounds were identical within experimental error. The lead isotope ratios in heavily weathered glass and in the underlying, protected material are compared.

For the past 10 years extensive effort has been devoted to studying the isotopic ratios of the element lead contained in glasses, glazes, and pigments and from bronze, gold, and silver items from the ancient world (1). A parallel study has been conducted to determine the isotopic ratios of lead in galenas (PbS) from known ancient mining sites with the hope that, by matching the sets of ratios, the provenance of the objects could be ascertained (2). The work has all been based on the fact that of the four isotopes of lead (^{204}Pb, ^{206}Pb, ^{207}Pb, and ^{208}Pb), the heaviest three are the stable end products resulting from the radioactive decay of ^{235}U, ^{238}U, and ^{232}Th.

0-8412-0397-0/78/33-171-**273**$05.00/1
© 1978 American Chemical Society

During the geological formation of major lead deposits the lead is separated from the parent uranium and thorium, and the isotopic composition of the lead so separated depends on the geochemical history of the source materials, specifically on the U/Pb and Th/Pb ratios and the timing of any variations in these ratios including the time of the final separation of lead. In theory, it is possible for each lead deposit in the world to have its own unique isotopic composition.

This uniqueness has been tested (3, 4), and differences are found in all examined deposits, although in many cases the differences are small and may require highly precise and accurate analytical techniques for identification. Fortunately, suitable analytical methodology is available (5, 6).

Because of the widespread use of lead in the metallurgical refining of brasses and bronzes and in the manufacture of glasses and glazes, there is nearly always enough lead in these products for an isotopic analysis without using inordinate amounts of sample. Most such materials contain from 0.1 to as much as 20% or more lead, and a modern mass spectrometric isotopic analysis requires approximately 1 μg of lead. This requirement may even be reduced to 1 ng or less if extreme cleanliness is practiced in the chemical processing and subsequent analysis.

Thus the uniqueness exists, the analysis in most cases feasible, and if a sufficient base of data from lead deposits of known ancient exploitation is obtained then the combinations may provide the archaeologist with vital information on the provenance of ancient objects. Even negative information may be important since it has been possible at times to state that it is unlikely that an object was made with materials from a particular area.

In the work to date a tacit assumption has always been made. This assumption is that nothing done in the use of lead in the manufacture of ancient objects would cause an isotopic fractionation of the lead and thus cause a difference between that measured from the original ore and that measured in the bronze, glass or pigment that was the subject of study. A further part of this assumption was that nothing in the subsequent history of the object would cause a change in the isotopic composition. Such things as the corrosion of brass and bronze and the heavy weathering or devitrification of glass might be considered.

To evaluate the possible effects of isotopic fractionation on the validity or usefulness of the lead isotope method, a fractionation experiment was devised to simulate the probable ancient production methods.

Experimental

To produce the samples used, a large (~200 g) sample of natural galena was obtained from the Corning Museum of Glass. After removing

Table I.

Material	Weight (g)	Theoretical (wt %)	
Corning 200 mesh sand	27.97	SiO_2	53.86
Potassium carbonate (anhydrous)	10.78	K_2O	15.72
Potassium nitrate	1.75		
Lead oxide	15.00	PbO	28.88
Calcined alumina	0.55	Al_2O_3	1.06
Antimony trioxide	0.25	Sb_2O_3	0.48

all extraneous material at the surface, this material was crushed in a steel mortar to pass through a 50 mesh screen. An 80-g sample of this ore was mixed with 20 g of coconut charcoal ($+50$, -200 mesh) and was heated in a laboratory furnace in a 100 cm³ alumina crucible with an alumina cover. The sample was heated at 900°C for approximately 18 hr and then for an additional 3 hr without the lid. At this point air was directed over the sample at 0.5 SCFH for 2 hr through an alumina tube placed directly over the sample. This reduced the galena to molten lead (98.6% yield) and removed the excess charcoal. A small sample (6.6 g) of this material was saved for later analysis, and the remaining lead was reheated (900°C) for 1 hr with an air flow of 0.8 SCFH passed onto the material. This procedure reoxidized the lead to lead oxide with a 95% yield. X-ray diffraction analysis indicated that this material was PbO with a very small amount of a second phase, probably PbO_2. A potassium lead silicate glass was made as shown in Table I.

These materials were mixed in a laboratory ball mill for 30 min and then were melted in a 60-cm³ platinum crucible at 1450°C for 4 hr. A clear but viscous glass was obtained.

An additional 15 g of the lead oxide was mixed with 10.9 g of Sb_2O_5 (molar ratio 2:1) in a ball mill for 30 min. This material was then heated in an alumina crucible at 950°C for 2 hr with an air flow of 0.8 SCFH directly on top of the melt. The resulting material was identified as $Pb_2Sb_2O_7$ by x-ray diffraction.

These materials, the original galena, the refined lead metal, lead oxide, glass, and pigment were put into solution. The lead was purified by anodic electrodeposition and was analyzed by a mass spectrometer using previously published procedures (6). The results of the measurements of the isotopic ratios of the above samples are shown in Table II.

Table II.

Sample No.	Material	Ratio		
		208/206	207/206	204/206
890	galena	1.8782	0.7281	0.04566
891	metallic lead	1.8789	0.7283	0.04567
892	litharge (PbO)	1.8783	0.7281	0.04564
893	$K_2O:PbO:SiO_2$ glass	1.8791	0.7283	0.04567
894	$Pb_2Sb_2O_7$ pigment	1.8786	0.7282	0.04564

Table III.   Lead-Containing Glasses and

Source	Sample	Description
Kenchreai	Pb-400	mixture of ~20 red opaque glasses
(Greece)	Pb-401	mixtures of weathered products from same glasses used for Pb-400, but possibly not in same proportions
	Pb-402	mixture of ≃ 10 yellow opaque glasses. Some lead present as $Pb_2Sb_2O_7$, some is in glass matrix.
	Pb-402	weathered products from same glasses used for Pb-402
Nimrud	Pb-423	single piece of red opaque glass
(Iraq)	Pb-424	weathered products from Pb-423
	Pb-425	single piece of red opaque glass
	Pb-426	weathered products from Pb-426
	Pb-427	single piece of red opaque glass
	Pb-428	weathered products from Pb-427

Discussion

All of the measured isotopic ratios agree within experimental error. While the preparation of these samples could not duplicate the processes used in early time, it is reasonable to assume that oxidizing conditions and chemical processes were severe enough to indicate that isotopic fractionation in measurable amounts would not occur during these procedures.

As mentioned in the introduction, however, there is an additional process which affects all ancient objects to some degree—that of weathering. This weathering is in some cases severe, and it is frequently true that the analytical work must be done on the weathered products. We have had an opportunity to examine five sets of samples where material was available both as "fresh" base material and as the overlaying weathered product. These samples are described in Table III.

The agreement between the isotopic ratios in each of these pairs is good within the experimental error. A sixth pair available gave a very poor agreement; however, subsequent reexamination of the sampling records indicates that these did not come from the same glass fragment but in fact are from different vessels. Thus even the severe weathering of materials buried or submerged for several thousand years does not appear to produce isotopic fractionation of lead.

Weathering Products from the Same Objects

Conditions	Total Time Buried or Submerged (years)	Agreement
submersion in brackish water	$\simeq$ 1600	good
submersion in brackish water	$\simeq$ 1600	good
buried in soil; annual rainfall $\simeq$ same as New York State	$\simeq$ 2600	good
buried in soil; annual rainfall $\simeq$ same as New York State	$\simeq$ 2600	good
buried in soil; annual rainfall $\simeq$ same as New York State	$\simeq$ 2600	good

Acknowledgments

The authors gratefully acknowledge the help of F. E. Woolley and E. H. Francis of the Corning Glass Works who prepared the various materials from the galena.

Literature Cited

1. Brill, R. H., Wampler, J. M., "Isotope Studies of Ancient Lead," *Am. J. Archaeol.* (1967) **71**, 63–77.
2. Brill, R. H., "Lead and Oxygen Isotopes in Ancient Objects," *Philos. Trans. R. Soc., London* (1970) **A269**, 143–164.
3. Brill, R. H., Shields, W. R., "Lead Isotope Studies of Ancient Coins," *in* "Methods of Chemical and Metallurgical Investigations of Ancient Coinage," E. T. Hall and D. M. Metcalf, Eds., Royal Numismatic Society, London, 1972.
4. Barnes, I. L., Shields, W. R., Murphy, T. J., Brill, R. H., "Isotope Analysis of Laurion Lead Ores," ADV. CHEM. SER. (1974) **138**, 1.
5. Catanzaro, E. J., Murphy, T. J., Shields, W. R., Garner, E. L., "Absolute Isotopic Ratios of Common, Equal Atom and Radiogenic Lead Isotopic Standards," *J. Res. Natl. Bur. Stand. Sec. A* (1968) **72A**, 261–267.
6. Barnes, I. L., Murphy, T. J., Gramlich, J. W., Shields, W. R., "Lead Separation by Anodic Deposition and Isotopic Ratio Mass Spectrometry of Microgram and Smaller Quantities," *Anal. Chem.* (1973) **45**, 1881–1884.

RECEIVED March 8, 1978.

17

Lead Isotope Analyses and Possible Metal Sources for Nigerian "Bronzes"

CANDICE L. GOUCHER—Departments of Chemistry and Visual Arts, University of California, San Diego, La Jolla, CA 92093

JEHANNE H. TEILHET—Department of Visual Arts, University of California, San Diego, La Jolla, CA 92093

KENT R. WILSON[1]—Department of Chemistry, University of California, San Diego, La Jolla, CA 92093

TSAIHWA J. CHOW—Scripps Institution of Oceanography, University of California, San Diego, La Jolla, CA 92093

A series of lead isotope analyses has been made of Nigerian copper alloy castings (∼ AD 850–1850). The three lead isotopic ratios from each analysis have been compared among the objects as well as with the isotopic ratios for known lead sources. Benin and Benin-related objects have a strong clustering of ratios while Igbo-Ukwu and Ife castings, which are believed to represent earlier cultures, fall outside this cluster. Comparison with published data on world lead sources eliminates most ore sites but reveals several possible matching sources in Africa, the Near East, and Europe. The results of these studies and information on mining and trading activities are analyzed to determine the possible sources of metal used in these Nigerian castings.

The beauty of the cast metal sculpture of Nigeria (Figure 1), has attracted the attention and admiration of both scholars and the general public. Despite this attention given to Nigerian artistic traditions, basic questions regarding chronology and relationships of styles, as well as technologies and sources of metals, remain unanswered. Benin art brought back to Europe by the British Punitive Expedition in 1897 was disseminated to the museums and collections throughout the world and aston-

---

[1] Author to whom correspondence should be addressed.

0-8412-0397-0/78/33-171-**278**$05.55/1

Figure 1. (upper left) *Huntsman, "Lower Niger Bronze Industry," or possibly Owo-style, Yoruba;* (upper right) *Ife head;* (lower left) *rectangular commemorative plaque by the "Master of the Circled Cross;"* (lower right) *Early Period head of a Queen Mother, Benin*

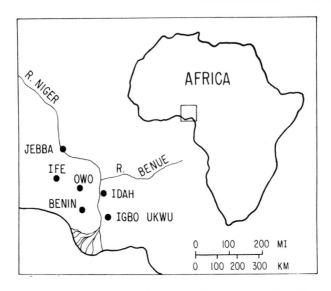

Figure 2.   Locations of major Nigerian metalworking traditions

ished viewers in spite of the fact that the kingdom of Benin had been in contact with one European power or another since the late 15th century. The aesthetic appeal of this captured treasure created an academic interest in Nigeria and subsequently led to the uncovering of an artistic tradition probably reaching back more than 10 centuries—hundreds of years before the first Portuguese contact.

Since the cultures which produced this art were preliterate, it is necessary to turn to analyses of the objects themselves to trace their background and meaning. Interest in the elemental composition of the metals from which these objects were cast began in the late 19th century shortly after the "discovery" of the Benin "bronzes" by the British. Many then found it intriguing to speculate on the origin of the technology as well as on the sources of the metals. The immediate suggestion was that the metalworking knowledge and skills had a European origin. However, this speculation was soon contradicted by archaeological discoveries which dated the metal arts and culture of Ife to around the 10th–12th century (*1*) and that of Igbo–Ukwu as early as the 8th or 9th century (*2*), although the radiocarbon dates have been disputed (*3*). The relationship between the metalworking centers (Figure 2) of Ife and Benin has been explained, in part, by oral traditions which reveal that the knowledge of the casting of memorial heads of past kings was brought to Benin by Ife artists (*4*). However, the Ife–Benin relationship remains problematical since the interpretation of the oral traditions is by no means certain. The art of Ife has been considered by some to be far more naturalistic than that of Benin, and this stylistic separation has been used to categorize Benin art into Early, Middle, and Late Periods (5, 6, 7). Within this interpretive scheme, the earlier, less stylized works are therefore positioned closer in time to the tradition of Ife. If the oral traditions are at all accurate, then their suggestion of the contemporaneity of Benin and Ife artists is very significant. Recent discoveries at Owo (8) seem to confirm this relationship, since Owo terracotta sculpture appears to be stylistically intermediate, sharing qualities and iconography with the Ife and Benin traditions. Therefore, it should be helpful to examine the isotopic composition of the copper alloys from Ife, Owo, and Early Benin in order to further clarify the Ife–Benin relationship.

The question of the source of metal used in the lost-wax castings has not been answered yet. The copper was certainly imported since there are no workable deposits in Nigeria (9). Although there are tin, lead, and zinc deposits being worked today, it is not known whether these were exploited in earlier times. It is also not known whether the trade which supplied the metal used in the bronze–brass castings brought metal in the form of an alloy or whether imported copper was melted together with lead, zinc, or tin already at hand. The trade between Portugal and the Kingdom of Benin in copper and brass manillas (bracelets) is well documented from the late 15th century onward (10). Since this trade, as well as the activities of the metalworking guild, were directly controlled

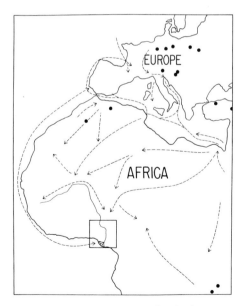

Figure 3. Patterns of trade. (●), possible Benin + lead sources as suggested by isotopic correlation with ore data.

by the Oba, the divine king of Benin, it has been suggested that the manillas were melted down and used for royal castings (10). Although the metal in later centuries was imported by the European traders, the original source of this traded metals is also not definitely known. The excavation of crucibles at a Benin site dated to the 13th century (11) extends the knowledge of casting to pre-European times. On the basis of stylistic analysis and of the documented observations of European travelers to Benin City, it is clear that metalworking at Benin spans several centuries (7). The Middle Period is designated as the period in the 16th century during which the rectangular plaques likely were produced (6). Several objects which fall into the Benin + group (defined below) are assigned to the Early Period which is thought, again on the basis of stylistic analysis, to correspond to a time prior to the 16th century and which may also pre-date European contact. Speculations as to pre-Portuguese trade routes with West Africa (Figure 3) include trade with North Africa across the Sahara and trade with Central Africa; both journeys (12, 13, 14) were viewed as heroic when first accomplished by Europeans. By learning more about the sources of metals we can thus add to our knowledge of West African cultural, artistic, and economic history.

On the basis of recent spectrographic analyses, Willett, Shaw, and others (Table I) have identified the elemental composition, and in some cases have grouped objects (2), but they have been unable to provide clear correlation with previously made stylistic divisions or to solve the problem of metal sources. While the compositions of the Nigerian alloys

Table I. Elemental Analyses of Nigerian

Source	Cu	Sn	Pb	As	Sb	Ni	Fe	Zn	Ag	Cd
Barker (1965)	x	x	x				x	x		
Connah (1975)	x	x	x	x	x	x	x	x	x	
Giauque (in Thompson 1971)	x	x	x	x	x	x	x	x	x	x
Gowland (in von Luschan 1919)	x	x	x	x	x	x	x	x		
Gray (in von Luschan 1919)	x	x	x	x	x	x	x	x		
Mauny (1962)	x		x			x	x	x	x	
Obst (in Frohlich 1966)	x	x	x				x	x		
Rathgen (in von Luschan 1919)	x	x	x	x	x	x	x	x		
Shaw (1965)	x	x	x	x	x	x	x	x	x	
Shaw (1966)	x	x	x	x	x	x	x	x		
Shaw (1969)	x	x	x	x	x	x	x	x		
Shaw (1970)	x	x	x	x	x	x	x	x	x	
Werner (1970)	x	x	x	x	x	x	x	x	x	x
Willett (1959)	x	x	x	x	x	x	x	x	x	
Willett (1964)	x	x	x	x	x	x	x	x	x	x
Willett & Werner (1975)	x	x	x	x	x	x	x	x	x	x
Wolf (1968)	x	x	x	x	x	x	x	x	x	

vary considerably, the elemental analyses have shown a consistently prominent percentage of lead, indicating that in most cases it was probably added intentionally.

The use of mass spectrometric lead isotope analysis as an archaeometric tool began with the initial efforts of Robert Brill and co-workers to distinguish archaeological materials from Greece, England, Spain, and Egypt on the basis of their lead isotope ratios and ultimately to determine the source of the metal itself (15–21). Lead isotope studies have, in addition, been applied to pigments (22). The method is based on the variability in isotopic compositions of lead ores, which is a result of differences in their varied geological histories. The concentrations of the isotopes of lead reflect the age of the deposit and the relative amounts of lead, uranium, and thorium which were originally in the material from which the lead ore was derived; the uranium and thorium provide lead isotopes as end products of radioactive decay. In particular, the isotopes ^{206}Pb, ^{207}Pb, and ^{208}Pb are the endproducts of the radioactive decay of ^{238}U, ^{235}U, and ^{232}Th. Another abundant isotope, ^{204}Pb, is not known to have been created by any radioactive decay process. Since each lead ore deposit has a set of characteristic (although not necessarily constant or unique) ratios among the isotopes, source distinctions can often be drawn clearly.

Several aspects of this method of analysis are important for archaeometric use. The technique has an important advantage over chemical analyses in that the ratios remain virtually constant throughout the chemical history of the object. That is, while variations in mining, smelting,

Copper Alloys Indicating Elements Determined

Bi	Se	In	Co	Al	Mg	Si	K	Na	Ba	Ti	Va	Mn	Ca	Au	P	Ref.
													x			36
x																37
	x	x														38
			x													39
			x													39
		x	x	x	x	x	x	x	x	x	x					40
																41
				x												39
																42
x																43
x																44
													x			45
x													x			46
x				x		x								x		47
x													x			48
x													x			49
x																50

refining, and casting methods, as well as environmental effects (such as corrosion), do alter an object's chemical composition, the lead isotope ratios remain essentially unchanged. The possibility exists, however, that the lead or lead-containing alloys were reused, mixing together leads from different sources. This would result in intermediate isotopic ratios for the objects involved.

A final point should be made in regard to the correlation between isotopic ratios of objects and their possible sources. Because the isotopic ratios for a mining area are not unique to that source, but may be identical to sources with a similar geological past, the matching of object to source may not be proven conclusively without other corroborative evidence. However, as Brill and Wampler (15) have pointed out, negative assertions may be made. One may say that the lead using in making an object could not have come from this or that particular mining area. This in fact may prove to be an equally significant statement.

Experimental

With the above mentioned artistic, historical, and economic questions in mind, we began to study the lead isotope ratios of Nigerian copper alloys, as briefly reported earlier (23). Small samples were removed by scraping objects in the collections of the British Museum (London), the Nigerian National Museum (Lagos), the Department of Archaeology of the University of Ibadan, and the University of Ife. Mass spectrometric analyses were made from the samples taken from 19 of these objects (see Table I and Figures 4, 5, and 6).

The lead is first isolated by a standard ion exchange technique followed by several dithizone extractions and then is converted to the sulfide. Mass spectrometric analyses are carried out on a 30-cm radius solid source instrument with an electron multiplier. Square root of mass ratio corrections are incorporated in the data to compensate for velocity discrimination in the electron multiplier. For purposes of comparing our data with other measurements, we have assigned the generous absolute error estimates of 0.3% (for the 206/204 ratio) and 0.5% (for the 207/204 and 208/204 ratios). The resulting isotopic ratios have been examined for clustering with interactive three-dimensional computer graphics, and then have been compared similarly with isotopic ratios from worldwide lead sources taken from the U.S. Geological Survey lead isotope data bank (24).

Absolute error estimates for these data are based on the recommendations of the data bank compiler (25) and (for one standard deviation) are 0.1% for normalized triple-filament, silica-gel, and lead tetramethyl methods and 0.3% (for the 206/204 ratio) and 0.5% (for the 207/204

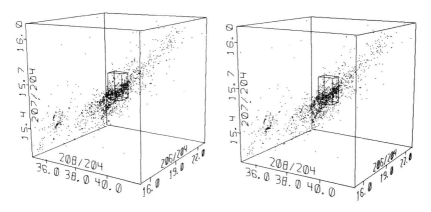

Figure 4. Isotopic ratios for world-wide distribution of lead ore samples (data from Ref. 23). The inner rectangular box corresponds to the cube containing our measured points, as shown in Figure 5, transformed into a rectangular box by the scale transformation. For clarity, only a subset consisting of the more accurately known points from the data bank are illustrated in Figures 4 and 5. Specifically, these are the normalized triple filament, silica gel, lead tetramethyl, and lead sulfide measurements. Each of Figures 4 and 5 is a stereo pair and may be perceived as a three-dimensional graph with depth by fusing the left and right images by slightly crossing the eyes.

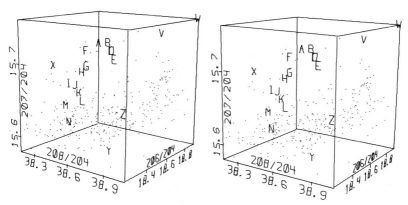

Figure 5. Three-dimensional graph of lead isotope ratios for the Nigerian copper alloy objects listed in Table II.

and 208/204 ratios) for the normalized lead sulfide method. Because of the difficulty of precisely estimating the errors for other categories of data, the error estimates have rather arbitrarily been taken as being the same as for the normalized lead sulfide method. The internal relative precision of our measurements, which is the criterion for cluster discrimination within our own set of 19 samples, is considerably better than the error estimates relative to absolute isotopic ratios. We estimate our own

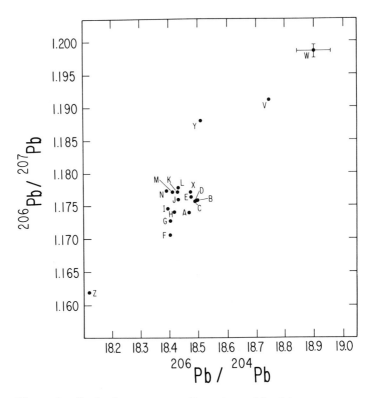

*Figure 6.   Projection onto two dimensions of lead isotope ratios
for the objects listed in Table II. The ratios chosen are different
than those shown in Figures 4 and 5, and the error bars, which
represent all the points, are for the relative internal precision at
our points, the proper error measure for judging clustering
within our own data set. The clear clustering of the Benin +
group is visible, as well as the separation from the Igloo–Ukwu
and Ife samples.*

relative errors as ±0.06 for the 206/204 ratio and ±0.001 for the 206/207
and 206/208 ratios in terms of which the data were originally expressed
and in terms of which Figure 6 is drawn.

Discussion of Results

The strong clustering shown in Figures 5 and 6 of the isotopic ratios
(A–N and X) forming what we will call the Benin + group is most con-
spicuous. Very similar ratios place all the Benin pieces, including a
manilla (a possible source material), in the same group, despite the
stylistic dating which separates, for example, the Queen Mother's head
(D) and the Late Period head (A) by several centuries. Several pieces
(the Huntsman, B, and the stool, G) now thought to be of Owo (Yoruba)

origin (26); the Jebba bowman (H), found hundreds of miles north of Benin City; the Udo-style head (J); and several unclassified Lower Niger style pieces also fall within this Benin + group. The Lower Niger pectoral (X), which lies at the outside of the cluster, is a possible exception. The Benin + cluster suggests a single source of lead for the entire time span and geographical area represented by that group of objects. Much has been written about the remelting of scrap metals, but it seems unlikely that the lead from the Benin + group is a mixture from isotopically different sources, since the mixing ratio would have had to have been nearly constant for all of the objects, unless the sources fortuitously had the same sets of isotopic ratios.

The remaining points represent the earlier traditions of Ife (represented by the one point Y) and Igbo–Ukwu (points V, W, and Z). Most

Table II.   Lead Isotope Ratios in Nigerian "Bronzes"[a]

Object[b]	$\dfrac{^{206}Pb}{^{204}Pb}$	$\dfrac{^{207}Pb}{^{204}Pb}$	$\dfrac{^{208}Pb}{^{204}Pb}$
Benin Late Period head (BM 1903.10-22.3)	18.47	15.73	38.59
"Lower Niger" huntsman (BM 1952.Af11.1)	18.50	15.73	38.64
Benin Early Period head (BM 1963.Af9.1)	18.49	15.73	38.68
Benin Early Period head of Queen Mother (BM 97.10-11.1)	18.49	15.73	38.68
Benin musician (BM 1949.Af46.156)	18.48	15.71	38.73
"Lower Niger" Igala bell-head (NML 72.16.26)	18.41	15.73	38.56
"Owo" stool (I 65.4.1)	18.41	15.70	38.56
"Jebba" bowman (NML exhibit) (0.80% Pb)[c]	18.42	15.69	38.50
Ke "manilla" (UI no reference number)	18.40	56.67	38.40
"Udo-style" head (BM 1948.Af9.1)	18.43	15.67	38.42
Benin plaque by Master of the Circled Cross (BM 98.1-15.34)	18.43	15.66	38.46
Benin plaque (BM 98.1-15.32)	18.43	15.65	38.48
"Lower Niger" head (BM 1956.Af27.224)	18.41	15.64	38.36
Benin Early Period head of Queen Mother (BM 97.10-11.1)	18.39	15.62	38.41
Igbo–Ukwu ritual shell shaped vessel (NML 39.1.13) (6.4% Pb)[d]	18.75	15.74	38.90
Igbo–Ukwu bowl (NML 39.1.2) (4.5%,9.8% Pb)[d]	18.90	15.77	39.07
"Lower Niger" ram's head pectoral (BM 1930.4-23.1)	18.47	15.69	38.17
Ife head (BM 1939 Af34.1) (11.4% Pb)[e]	18.51	15.58	38.65
Igbo–Ukwu "Horseman Hilt" (NML I-R-350) (13.8% Pb)[d]	18.21	15.66	39.04

[a] Numbers indicate museum catalog reference.
[b] BM = British Museum, London; NML = Nigerian National Museum, Lagos; I = University of Ife, Institute of African Studies; UI = University of Ibadan, Department of Archaeology.
[c] Ref. *36*.
[d] Ref. *45*.
[e] Ref. *47*.

significant is what appears to be the isotopic separation of the three major traditions—Igbo–Ukwu, Ife, and Benin—which correlates with the stylistic separation of the artistic traditions and with what may be their temporal separation.

Correlation with Source Data

In light of the strong cluster obtained for the Benin + group, we have compared these ratios with ratios for possible lead sources from a worldwide compilation of 3500 measurements (24), rejecting those sources whose separation in isotopic ratios from any of the Benin + objects indicates that a match is improbable. Specifically, we have eliminated those sources which could be rejected at the 90% confidence level, given the standard deviations quoted above and assuming uncorrelated errors between our measurements and those drawn from the data bank. As can be seen in Figures 3, 4, and 5, most known lead sources can be ruled out, including the local Nigerian sites. Several sources whose isotopic ratios match those for the Benin + group remain as possibilities. These possible sources occur in North Africa, Europe, the Near East, and Central Africa and are shown in Figure 3. In particular, matching lead sources consistently occur in Poland; Hungary; the Rammelsberg, Harz, and Meggen mines in Germany; Bleiberg in Austria; Turkey; Morocco; and Zaire. Occasional or perhaps insignificant matches with Chilean, Mexican, and Japanese sources also occur. It is possible, of course, that the lead for the Benin + group might have come from a source not represented by the isotopic data in the U.S. Geological Survey lead isotope data bank.

We do not have enough Igbo–Ukwu or Ife measurements to form defined clusters, so it is conceivable that the sampled objects contain a mixture of leads from different sources, caused by remelting and mixing. A variation in mixing ratios could partially explain, for example, the spread of the three Igbo–Ukwu points in isotopic space. Thus the matches with possible source data must be viewed with caution for the Igbo–Ukwu and Ife samples, as our measured isotopic ratios might represent intermediate values between components of a mixture and thus might correspond to no real source. With this caveat, the sources which cannot be rejected at the 90% confidence level for the three Igbo–Ukwu points are found in Chile, Peru, Mexico, Spain, Greece, Yugoslavia, Hungary, Turkey, Oman, Indonesia, and Japan. Similarly, the unrejected sources for Ife are found in Chile, England, Germany, Hungary, Bulgaria, Egypt, Turkey, Indonesia, and Japan. There is an absence of any known African sources except for one Egyptian source for Ife. Additional analyses are needed to provide enough data for possible clustering before definitive conclusions are drawn.

History of Sources and Trade

Having arrived at this list of possibilities, the next step involved searching the existing literature on trade patterns and ancient mining activities in these parts of the world. Up until the end of the 15th century, the great trading cities of Italy served as distributors of metals. About this time, Bruges and then Antwerp gained prominence as international entrepots. By the beginning of the 16th century, Central European mines controlled by the German Fugger family were producing lead, zinc, and copper for the Antwerp markets (27). This new direction in the movement of metals probably accommodated the interests of the Portuguese traders. One scholar suggests (28) that the southern and western German mining centers were exporting directly to Lisbon at the end of the 15th century. The trade network in metals must have been sufficiently complex to have accommodated the movement of lead from any one of the possible European sources to a market accessible to the Portuguese.

Arabic documents mention several particular African sources of copper known to the ancient world and the use of this copper as a medium of exchange (29). The exact beginnings of this movement in metals is not known, but a key point may be the period in the 8th century when an Arab movement south from Morocco took place (30). It is necessary to recognize the North African confluence of European and Saharan trading patterns as early as the 13th century. For example, at this time Moroccan leather (actually obtained in the Sudan) was supplied to Christian merchants, whereby it found its way to the markets of Normandy and England (31). The movement in metals could conceivably have traversed this trading network in the opposite direction. In fact, by this time North African cities were handling exports which included both lead and copper (32). These North African regions, rich in silver and lead, also contained copper deposits which were exploited. By the 15th century when Portuguese contact was established with Benin, Portugal was already well established in North Africa with the sanction of the Papal bull Romanus Pontifix (33). Thanks to the spectacular discovery of Monod (34), who stumbled onto the remains of a caravan which presumably had attempted to cross the Sahara with a cargo of 2000 brass bars, we have evidence for trade in metals in the 12th century. Certainly either North African or European lead (or both) could have found its way south of the Sahara. In addition, the Ottoman State (c. 1300–1919) which controlled the Turkish sources must not be disregarded as a metals supplier, considering its intimate connections with both the European trading community and the rest of the Arab world.

Pre-European trade between West Africa and the Congo can only be conjectured. However, early Portuguese documents record the exploitation of Congolese sources of copper as early as 1530, as well as the production of manillas which were then traded to the people of the Guinea coast (35).

Conclusions

A comparison of the lead isotope ratios among 19 Nigerian copper alloy objects, shown in Table II and in Figures 5 and 6, leads to two conclusions. The first is that since the Benin + group forms such a tight cluster in isotopic ratio space, it likely arises from a single source of lead, unless there is a fortuitous mixture of sources with essentially the same isotopic ratios. The isotopic evidence thus indicates a stable pattern of trade lasting over many centuries, linking the makers of these objects to their lead source. Second, the fact that the Igbo–Ukwu and Ife objects have isotopic ratios quite different from Benin indicates separate sources of lead and corroborates the stylistic and radiocarbon evidence that these are indeed separate centers from Benin.

Comparison of the lead isotope ratios which we have measured with isotopic measurements from worldwide lead sources allows the clear elimination of most sources. Given the uncertainties which must be assigned to our measurements and the range of uncertainties which must be assigned to other measurements, a number of source areas are close enough in isotopic ratio space to the Benin + cluster that they should be considered. In particular, the source areas of Central Europe, Turkey, North Africa, and Central Africa remain candidates. These possibilities indicate that a long-distance trade in metals reached West Africa, Nigeria in particular, at an early date. In sum, these objects, artistic metaphors for a divine kingship, also hold clues to the economic, technological, and cultural history of West Africa.

Acknowledgments

The authors gratefully acknowledge support received from the Chancellor's Club; the UCSD Foundation; the Project for Art/Science Studies; the Samuel H. Kress Foundation; the receipt of a University of California President's Undergraduate Fellowship; the cooperation of museums cited in Table II as well as the Museum of Cultural History, University of California, Los Angeles; and the use of computer systems supported by the National Science Foundation, Division of Computer Research, and the National Institutes of Health, Division of Research Resources. Thanks to Nathan Meyers and Peter Berens for programming help. Our gratitude also to William Fagg, Douglas Fraser, Thurstan Shaw, and Frank Willett for their encouragement and helpful discussions.

Literature Cited

1. Willett, F., "Ife in the History of West African Sculpture," McGraw-Hill, New York, 1967.
2. Shaw, T., "Igbo-Ukwu: An Account of Archaeological Discoveries in Eastern Nigeria," Northwestern University, Evanston, Illinois, 1970.
3. Lawal, B., *J. Afr. Hist.* (1973) **14**, 1–8.
4. Egharevba, J., "A Short History of Benin," 4th ed., Ibadan University, Ibadan, 1968.
5. Underwood, L., "Bronzes of West Africa," 2nd ed., Tiranti, London, 1968.
6. Fagg, W., "Nigerian Images," Praeger, New York, 1963.
7. Dark, P. C., "An Introduction to Benin Art and Technology," Clarendon, Oxford, 1973.
8. Eyo, E., *Afr. Arts* (1970) **3**, 44–47.
9. *Annual Reports of the Geological Survey Department,* Nigeria, 1949–50, 1950–51, 1956–57, 1958–59.
10. Ryder, A. F. C., "Benin and the Europeans: 1485–1897," Longmans, London, 1969.
11. Connah, G., "The Archaeology of Benin," Clarendon, Oxford, 1975.
12. Fage, J. D., "An Atlas of African History," Edward Arnold, London, 1970.
13. Rosenberger, B., *Rev. Geog. Maroc* (1970) **17**, 71–108.
14. Lombard, M., "Les Metaux dans l'ancien Monde du Ve au XIe Siècle," Mouton, Paris, 1974.
15. Brill, R. H., Wampler, J. M., in "Application of Science in Examination of Works of Art," W. J. Young, Ed., Museum of Fine Arts, Boston, 1965.
16. Brill, R. H., Wampler, J. M., *Am. J. Archaeol.* (1967) **71**, 63–77.
17. Brill, R. H., in "The Impact of the Natural Sciences on Archaeology," T. E. Allibone et al., Eds., The British Academy, London, 1970.
18. Brill, R. H., Shields, W. R., in "Methods of Chemical and Metallurgical Investigations of Ancient Coinage," E. T. Hall, D. M. Metcalf, Eds., Royal Numismatic Society, London, 1972.
19. Brill, R. H., Shields, W. R., Wampler, J. M., in "Application of Science in Examination of Works of Arts," W. J. Young, Ed., Museum of Fine Arts, Boston, 1970.
20. Brill, R. H., Barnes, I. L., Shields, W. R., Murphy, T. J., in "Archaeological Chemistry," *Adv. Chem. Ser.* (1974) **138**, 1–10.
21. Brill, R. H., Barnes, I. L., Adams, B., "Recent Advances in Science and Technology of Materials," Plenum, New York, 1974.
22. Keisch, B., Callahan, R. C., *Archaeometry* (1976) **18**, 181–193.
23. Goucher, C., et al., *Nature* (1976) **262**, 130–131.
24. Doe, B. R., Rohrbough, R., *U.S. Geological Survey Open File Report* (1977) **77–418**, 1–137.
25. Bruce Doe, U. S. Geological Survey, Denver, private communication.
26. Fraser, D., *Afr. Arts* (1975) **8**, 30–35, 91.
27. Jeannin, P., "Merchants of the Sixteenth Century," Harper & Row, New York, 1972.
28. Strieder, J., *Z. Ethnologie* (1932) **64**, 249–259.
29. Herbert, E., *J. Afr. Hist.* (1973) **14**, 179–194.
30. Posnansky, M., *World Archaeology* (1973) **5**, 149–162.
31. Bovill, E. W., "The Golden Trade of the Moors," 2nd ed., Oxford University, London, 1968.
32. Lopez, R. S., Raymond, I. W., "Medieval Trade in the Mediterranean World," Norton, New York, 1971.
33. Marques, A. H. de Oliveira, "History of Portugal," Columbia University, New York, 1972.
34. Monod, T., *Bull. Inst. Fr. Afr. Noire Ser. A* (1964) **36A**, 1392–1402.
35. Ryder, A. F. C., *J. Hist. Soc. Nigeria* (1965) **3**, 195–210.
36. Barker, H., *Man* (1965) **65**, 23–24.

37. Connah, G., "The Archaeology of Benin," Clarendon, Oxford, 1975.
38. Giauque, R., Appendix in "Black Gods and Kings: Yoruba Art at UCLA,"
    by R. F. Thompson, Museum and Laboratories of Ethnic Arts and
    Technologies, University of California, Los Angeles, 1971.
39. von Luschan, F., "Die Altertümer von Benin," Berlin, 1919.
40. Mauny, R., *J. Hist. Soc. Nigeria* (1962) **2**, 393–395.
41. Frohlich, W., *Ethnologica* (1966) **3 N. F.**, 231–310.
42. Shaw, T., *Archaeometry* (1965) **8**, 86–95.
43. Shaw, T., *Archaeometry* (1966) **9**, 149–154.
44. Shaw, T., *Archaeometry* (1969) **11**, 85–93.
45. Shaw, T., "Igbo-Ukwu: An Account of Archaeological Discoveries in East-
    ern Nigeria," Northwestern University, Evanston, Illinois, 1970.
46. Werner, O., *Baessler-Archiv.* (1970) **18 N. F.**, 51–53.
47. Willett, F., *Man* (1959) **59**, 189–193.
48. Willett, F., *Archaeometry* (1964) **7**, 81–83.
49. Willett, F., Werner, O., *Archaeometry* (1975) **17**, 141–156.
50. Wolf, S., *Abh. Ber. Staatl. Mus. Volkerkunde* (1968) **28**, 91–153.

RECEIVED September 19, 1977.

Ternary Representations of Ancient Chinese Bronze Compositions

W. T. CHASE

Freer Gallery of Art, Smithsonian Institution, Washington, DC 20560

THOMAS O. ZIEBOLD

Braddock Services, Inc., 17200 Longdraft Rd., Gaithersburg, MD 20760

The simple method of representing the percentages of copper, tin, and lead in an alloy on a three-component plot has helped to categorize ancient Chinese bronze compositions. While the representation method is straightforward and the plots can be drawn by computer, some of the problems involved, such as the methods of grouping values on the plot and the normalization procedure, are not trivial. The ternary representations give new insights into the evolution of bronze alloys in China, show clearly the great control over alloy composition exerted by ancient Chinese foundrymen, and reveal the caster's remarkable grasp of the metallurgy of the copper–tin–lead system. Ternary representations have helped to assess the date at which an object could have been made and to show, graphically and imemdiately, any inconsistencies between stylistic and technical attribution to period.

Metal compositions of Chinese bronze ceremonial vessels and other bronze objects have fascinated scholars for some time. It seems that Chinese bronze founders must have had good intuitive control of composition. They must have considered the final uses of these objects and must have adjusted the compositions accordingly. Economics, availability of ore sources, trade routes, and many other factors influenced the metals placed in the crucible. One has long thought that the compositions of ancient Chinese bronze objects should vary in a regular manner with the time and place of manufacture and with the object type.

0-8412-0397-0/78/33-171-**293**$11.25/1
© 1978 American Chemical Society

History of Bronze Production in China

Before examining what other scholars have found concerning ancient Chinese bronze compositions, let us first look briefly at the population of bronzes with which we are dealing. With a few exceptions, the chronology of Chinese history is clear. Table I is a chronological table of Chinese history adapted from a recent publication of the People's Republic of China (1). The information in this table is firmly based on Chinese historical records (2, 3).

Table I. Chronological Table of Chinese History[a]

Period	Date
Primitive Society	~ 500,00–over 4,000 years ago
Slave Society	~ 21st century–475 B.C.
Hsia	~ 21st–16th centuries B.C.
Shang	~ 16th–11th centuries B.C.
Western Chou	~ 11th century–770 B.C.
Spring and Autumn Period	770–475 B.C.
Feudal Society	475 B.C.–1840 A.D.
Warring States Period	475–221 B.C.
Ch'in	221–207 B.C.
Western Han	206 B.C.–9 A.D.
Interregnum of Wang Mang	9–24 A.D.
Eastern Han	25–220
Three Kingdoms	220–280
Western Tsin	265–316
Eastern Tsin	317–420
Southern and Northern Dynasties	420–589
Southern Dynasty	420–589
Sung	420–479
Ch'i	479–502
Liang	502–557
Ch'en	557–589
Northern Dynasty	386–581
Northern Wei	386–534
Eastern Wei	534–550
Western Wei	535–557
Northern Ch'i	550–577
Northern Chou	557–581
Sui	581–618
T'ang	618–907
Five Dynasties and Ten Kingdoms	907–979
Sung	960–1279
Yuan	1271–1368
Ming	1368–1644
Ch'ing	1644–1911

[a] Adapted from Ref. 1.

Bronze production began in China possibly in Kansu province in the early Shang dynasty or possibly in the late Neolithic period. These few isolated finds do not seem to have any relevance to later Chinese bronze production, and we have not included analyses of any of them in our tables. The mainstream of bronze production begins in the Homan area, probably about 1400 B.C. with workshops near Cheng-chou, which at that time probably was the capital of the Shang state. These early (pre-An-yang) bronzes have characteristic decoration, shapes, and thinness.

With the move of the Shang capital to An-yang, bronze production begins in earnest. The great Shang bronzes, some weighing a ton or more, were produced here. Production in various other regions started at the same time. With the Chou conquest, the An-yang foundries continued for a while; then the bronze styles change markedly to early Chou decoration schemes and vessel types.

In the mid- to late Chou, there seems to have been a great flowering of Chinese technology. Pattern stamps were widely adopted, facilitating the production of ornate bronzes with repetitive patterns. Inlay was first used in bronzes. Mercury was discovered as an element and then was used for mercury gilding. Casting by the lost-wax method may have begun at this time, and, probably the most important technological discovery, iron production started. This period from about 550 B.C. up until the Ch'in conquest in 221 B.C. seems to have been in some ways the equivalent of the European Renaissance. During the reign of Ch'in and into the Han bronze vessels were still produced, with a gradual change in style. More stress was put on inlay and coloristic effects including gilding and silvering in two and three colors. While the production of bronze weapons continued into the earlier Han, the later weapons seem to have been intended primarily for ceremonial or display purposes only. The majority of bronzes included in this study date from before the end of the Han Dynasty.

With the division of China into the Three Kingdoms, bronze production seems to have fallen off temporarily. Buddhism was introduced during this time, and some of the great Buddhist images were produced during the Sui and T'ang periods. Mirrors were produced from the late Chou up through the T'ang and even later. (For a more detailed overview of metal objects in ancient China, *see* Ref. *4*).

At the end of the T'ang Dynasty and in the Sung Dynasty, interest in Chinese archaeology began in China (*5*). Great numbers of vessels were made as reproductions or archaistic imitations of early bronzes; some of them were actually fakes or forgeries (*6, 7*). Even today reproductions of ancient vessels are being cast in the People's Republic of China and on Taiwan (*8*).

So we have a production of varied bronze types—ceremonial vessels, mirrors, weapons, belt hooks, horse trappings, etc.—for over 1000 years in ancient China. This is followed by 2000 years of both original production and imitation, leading us up to the present.

The bronze production in China was both long-lasting and extensive. A lot of bronzes were produced in these 3000 years! As an example, over 1295 mirrors have been excavated and listed in publications during 1923–1966 when Barnard compiled his tables (4). When we add the pieces in Western collections which have come out of China without provenance data and the numbers of objects lost, still undiscovered, or melted down as scrap, the totals are staggering. The story of Chinese bronze alloy compositions is a complex one.

Earlier Studies of Alloy Compositions

Earlier studies of Chinese bronze alloys have been compiled by Barnard and others (4, 9). Needham's book (10) contains a fine bibliography. Most of the authors come to the same conclusion as Gettens (7):

"The question is frequently asked: Do the vessels of different time periods show any difference in average composition in respect to the major elements, or even the minor elements? Close inspection of the analytical results show that, on the whole, they do not. There seems to be little or no relationship between composition and age."

Barnard looks closely at the lead compositions of bronzes and concludes that Shang bronze is for the most part a binary alloy (4). In Western Chou bronze was a ternary alloy with 2–7% lead, and in Eastern Chou the Chinese bronzes have higher lead. This, indeed, seems to go along with what we found. Barnard stresses, correctly, the problem of working with unprovenanced material—objects which have not come from archaeological excavations and which have no reliable data about their sources. In any case, the picture of ancient Chinese bronze alloying seems to be complicated and confused. No clear conclusions have been drawn, and no firm picture emerges.

In the past few years Chase has examined Chinese bronze compositions from various viewpoints to see if there was any consistency in the alloy formulas. He seemed to see a glimmer of hope when examining four Fang-ting vessels, one of which belongs to the Freer Gallery of Art (FGA 50.7). In this object and its three counterparts the proportions of tin-plus-lead add up to 20% quite consistently. Within a fairly wide range this seems to be true of several other Shang and early Chou vessels. With this in mind, Chase proceeded to graph these objects out on a tin-vs.-lead coordinate system with the tin + lead = 20% line drawn clearly on it (11).

Shortly after this Ziebold suggested that it would be more logical if these compositions were graphed out in a standard ternary representation, and this is the course we have been pursuing. The first plots produced were made simply by hand. Compositional figures were not normalized in these plots; the plots were intended to be histogramamtic and not to represent the individual points. The frequency of points in a given hexagon from the diagram was the quantity observed.

Figure 1 shows two histogrammatic plots used to prove a point about a specific vessel in a private collection. Definite trends seem to be evident. The centroid of the Shang and Early Chou bronzes is different from that of Late Chou, and a trend from about 15–20% tin down towards the 40% lead corner of the digram is clear. Areas of the plot with infrequent

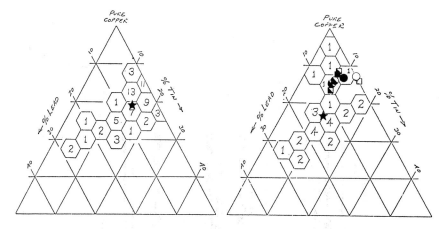

Figure 1.   Chemical composition of objects Z71.16 and Z76.17 compared with that of 92 Chinese bronze vessels in the Freer Gallery of Art (7). (left) Shang and Early Chou vessels; (right) Late Chou objects. Comparison of objects Z76.16 and Z76.17 with 92 Freer ceremonial vessels. The numbers in boxes show the number of vessels having the composition range enclosed. (★), center of gravity of each distribution. The reported compositions of objects Z76.16 and Z76.17 are shown as: Z76.16—O body; ◐ handle; ● cover. Z76.17— □ body; ◨ handle; ■ cover.

appearance of bronze compositions are also clear. These diagrams were simply made by plotting the three points tin-vs.-copper, lead-vs.-copper, and tin-vs.-lead on the diagram and tallying them in the individual hexagonal boxes. The boxes were sized to represent the stated analytical error from Gettens (7); i.e., ±2½%. On these ternary plots a hexagon gives equal edges of uncertainty from its center.

There are, however, certain problems with this approach. The boxes at the edges of the ternary diagram enclose only half of the area of the boxes in the center. The numerical frequencies in these boxes probably should be weighted in some way. The hexagonal network can be placed with either of two different origins. Ziebold plotted these diagrams twice with two different positionings of the hexagons, selected the net that maximized the most probable composition frequency from the Freer data, and has used this net throughout. This method emphasizes population differences but loses sight of individual objects. Chase has preferred to work simply with scatter diagrams in the same ternary representation. Representing each point enables one to see more easily individual differences between bronzes.

Computer Program

The computer program (TGPL78) is written to extract the copper, tin, and lead values for individual objects from a data file, to normalize the values, and to plot them on a modified teletype terminal. Details of the program can be seen in the Appendixes I–V. Various plotting parameters, such as the computation and plotting of centroids, partitioning of symbols, normalization procedure, and labeling of individual points can be adjusted by the operator.

All of the plots shown used the procedure of composition normalization by weight. The copper, tin, and lead percentages for an individual object are added. This total is then divided by 100% to obtain a normalization factor. The original percentages are then each multiplied by this factor to give the normalized percentages. The three normalized percentages sum to 100%; this satisfies the requirements of the ternary plot. Normalization does, in a way, constitute "fudging" of the original analytical data. In most cases, the adjustment of compositional percentages by normalization is rather small. Most bronzes have an analytical total of these three components which is close to 98–99%. There are, however, a few analyses where the total falls far short of 98%. The normalization procedure also ignores the presence of zinc. To add another component would require three-dimensional quaternary diagrams, a complication which we did not want to get into at this time.

Initial Test of the Program

When the program was functioning correctly, we fed in all the data we had in the computer, producing the total picture (Figure 2). On this plot, the overlap of ceremonial vessels and belt hooks is marvelously confused. This plot does not include any weapons, nor does it include any analyses not made in the Freer Gallery laboratory, or under our strict supervision, or by R. J. Gettens and his students before he came to the Freer. In other words, most of the analyses represented here were made by standard wet chemical techniques (7).

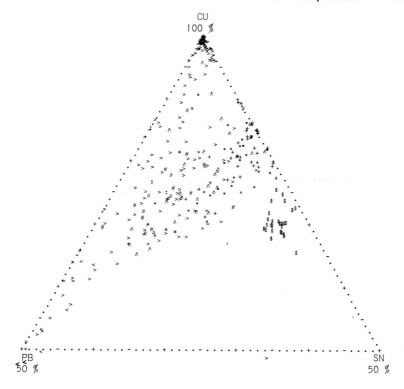

Figure 2. A ternary plot of Chinese bronzes analyzed at the Freer Gallery of Art. The mirrors ($) form a distinct, high-tin group. (○), Shang ceremonial vessels (34 points); (+), Early Chou ceremonial vessels (31 points); (△), Middle and Late Chou ceremonial vessels (31 points); (*), Chin and Han ceremonial vessels (5 points); (#), "Later" ceremonial vessels (19 points); ($), mirrors (27 points); (<), Ming Knife Coins (8 points); (>), belt hooks up to Han (147 points); (:), later belt hooks (6 points).

There are only three vessels whose tin content lies over 20%. This is rather an interesting fact in itself. Vessels lie generally between 10–20% tin along the pure-tin axis or off towards the 40% pure-lead corner of the plot, and this confirms Ziebold's earlier plot. The belt hooks range upwards to 100% copper. They form a distribution which runs down along the lead axis with a bulge into the high-leaded vessel area. The distributions of belt hooks and vessels look different, although there is a large area of interpenetration.

Probably the most interesting aspect of this plot is the distinct compositional difference between the mirrors and all other bronzes. Mirrors form a distinct and separate compositional group, centering around 25% tin and 5% lead. Using this plot, we have chosen a 71% copper–26%

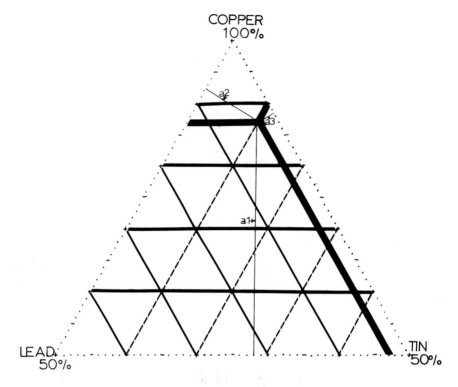

Figure 3. *A representation of the position of an 87% copper – 10% tin – 3% lead alloy using the ternary diagram.*

This is actually the top one fourth of a full ternary diagram of the Cu–Sn–Pb system and includes compositions with 50–100% copper. The heavy lines intersect at the 87-10-3 composition. Contours of equal Cu composition are shown as solid horizontal lines, Sn as dotted lines with positive slope, and Pb as light solid lines with negative slope. If the heavy intersecting lines are extended parallel to these contours, they intersect the sides of the triangle at 87% towards the Cu vertex, 10% towards the Sn vertex, and 3% towards the Pb vertex. Altitude lengths can also be used to determine the composition: a1 = Cu%; a2 = Sn%; a3 = Pb%; a1 + a2 + a3 = 100%. For further discussion of the properties and applications of the ternary diagram, see Refs. 12 and 13 and the early work of Willard Gibbs, after whom this is often called the "Gibbs Triangle."

tin–3% lead mixture as the nominal mixture to use for Chinese mirror test reproductions.

Metallurgical Implications

Before examining other plots in more detail, it should be obvious that the compositional differences shown on this diagram have strong metallurgical implications.

Let us look at a simplified plot which indicates some of the properties of the ternary plotting method. Figure 3 is a hand-drawn plot with

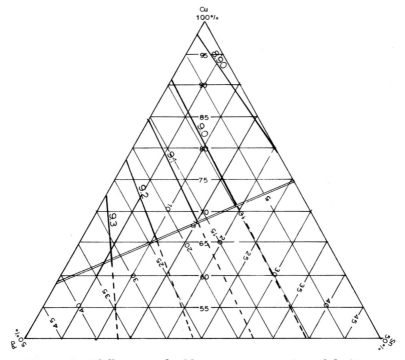

Figure 4. Chill-cast, tin–lead bronzes: contours of equal density

lines to indicate 10% compositional difference and heavy lines to show the location of one particular composition on the plot. The horizontal lines show compositions with fixed amounts of copper. Thus, the upper horizontal line shows compositions with 90% copper ranging from 10% lead on the left to 10% tin on the right. The lowest horizontal line shows compositions with 60% copper and 40% lead–40% tin and, in the middle, 20% tin and 20% lead. This is the intersection of two diagonal lines with the horizontal line. The diagonal dotted line which goes from the upper right to lower left of the plot shows, since it is the second diagonal line from the left-hand side, that the mixture has 20% tin. The diagonal line from upper left to lower right shows that the mixture has 20% lead. A mixture of 10% tin and 3% lead is plotted in the upper portion of the diagram with heavy lines. Since 10% tin plus 3% lead equals 13%, the copper must equal 87%, and the horizontal line is 13 divisions down from the upper vertex of the diagram where copper is 100%. Altitudes (labeled here 21, 22, 23) can also be used to derive the composition.

As a further illustration, it is intuitively obvious that as lead is added to the alloys, the density will increase. Figure 4 shows contours of equal density superimposed on the ternary composition diagram. As

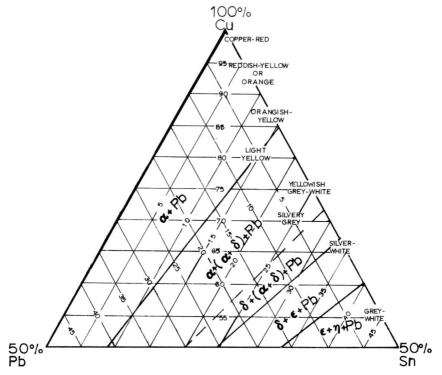

Figure 5.   Chill-cast, tin–lead bronzes: colors and equilibrium phases at room temperature

the lead content increases (i.e., as the composition gets closer to the lower left corner of the triangle), the density also increases.

Other individual properties of copper–tin–lead alloys can, of course, be drawn out on the ternary plot. Figure 5 shows colors of these alloys. One unfortunate problem is that most published data has been obtained on pure copper–tin binary alloys. Not as much data exists on the ternary alloys; usually the right-hand axis is well defined, and the rest of the diagram is somewhat hazy. The colors on the right-hand side are from Dono (*14*). The two colors in the middle of the diagram are supplied from bronzes cast for the Freer Gallery of Art by Rob Pond of Windsor Metalcrystals. Bronze colors range from the salmon-red of copper through reddish yellow, orange–yellow, and yellowish grey–white into a silver–white and a greyish white. The lead alloy tends to be a less shiny grey–white, but we have no definitions of their colors in a graded series yet. Of course, these colors are as seen on polished, metallographically prepared surfaces. For interest the equilibrium phases at room temperature have also been indicated on Figure 5 (*15*).

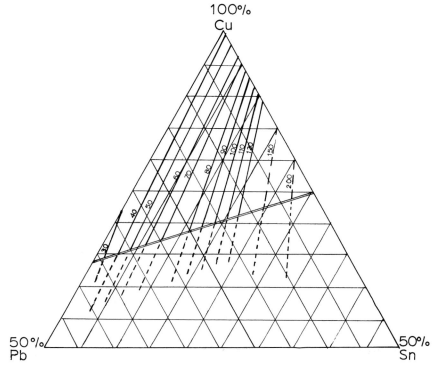

Figure 6.   Chill-cast, tin–lead bronzes: Brinnell hardness

Mechanical properties can also be graphed out on the diagram and can be correlated interestingly with the alloys of the various bronzes. Figures 6, 7, and 8 show Brinnell hardness numbers, tensile strength (tons/sq in.), and elongation (%). The properties of the alloys at the centroids of the various plots below are shown in Table II. Alloys in the mirror range are hard and brittle. Alloys in the vessel range are less hard but are still harder than an annealed carbon steel and are less brittle, for as the copper increases, the alloys get more ductile.

Comparing Figure 7 (tensile strength) with the Figures 10–18, the vessels and particularly some of the weapons lie in the area of higher tensile strength whereas many of the belt hooks lie in an area which has lower tensile strength. The data on Figures 4–8 are derived from studies of chill-cast, tin–lead bronzes (*15*). Our alloys are not for the most part truly chill cast, but they are also not cooled at an equilibrium rate as can be seen by the frequent appearance of cored α (alpha) dendrites that are evident in almost every metallographic section taken from ancient Chinese bronzes.

The melting points and liquidus lines can also be graphed out on the diagram (Figure 9; data from Ref. *15*). It would be interesting to

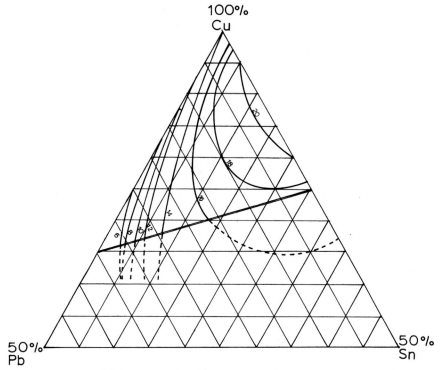

Figure 7.   Chill-cast, tin–lead bronzes: tensile strength (tons/sq in.)

have an extension of this diagram into the 40–50% tin + lead region and even further (*see* Ref. *16*).

Figure 9 incorporates a great deal of information and needs some explanation. As the center portion, the figure uses an adaptation of the diagram of liquidus and separation surfaces of the copper–tin–lead system (*15*) drawn in our standard representation method and scale. On the left is the portion of the copper–lead equilibrium diagram covering 0–50% lead down to 600°C, drawn backwards; on the right is the upper left portion of the copper–tin equilibrium diagram.

The actual ternary diagram would be a solid triangular prism, with the copper–tin equilibrium diagram on the right face, the copper–lead diagram on the left face, and the lead–tin diagram on the face towards the bottom of the page, in this instance. In other words, we are looking at the top and two side walls of this solid prism unfolded here. The base of the prism would show the equilibrium phases at room temperature as in Figure 5.

The liquidus surface would run from the liquidus on the copper–tin diagram to that on the copper–lead and tin–lead diagrams. In Figure 9,

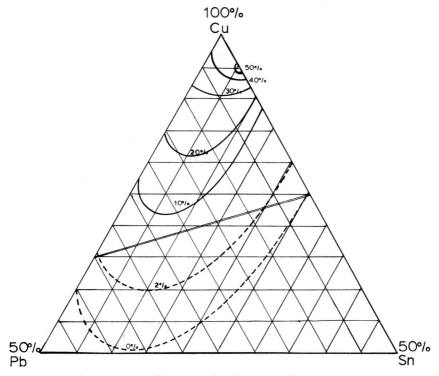

Figure 8. Chill-cast, tin–lead bronzes: elongation (%)

the liquidus is shown as contours (the highest at 1080°C, running from about 98% copper–2% tin to about 96% copper–4% lead at the top of the triangle). The intersections of the liquidus contours on the triangle with those on the side walls are shown by large black diamonds.

The contours which intersect with the liquidus contours define the separation surface where two immiscible liquids separate in the melt. This can be seen to be 953°C on the pure-lead side, falling to 930°, 900°, etc. as one increases the tin.

The dark line from A to B shows the limit of miscibility. This gives the maximum lead/tin ratio with which an alloy free from segregation can be produced using ordinary casting methods. To cast an alloy free from segregation is simple at temperatures and compositions above this line. For instance, an alloy of about 71% copper, 21% tin, and 8% lead will begin to solidify at 810°C. Since the separation surface in this region is also at 810°C, the alloy will not segregate before freezing.

Below this line, however, the temperature must be raised sufficiently to maintain the emulsion of two liquids; this is shown by the curved contours of the separation surfaces below the A–B line. These alloys also must be cooled rapidly so that solidification can take place before segregation.

When this diagram is superimposed on the diagram of ancient Chinese bronzes, one sees the tendency of the compositions to come down from the high-tin area and cluster along the limit of miscibility. The

Table II. Centroids

No. of Analyses	Object[a]	Figure Reference
4	Early Shang vessels, excavated	16
34	Shang vessels, FGA analyses	3
11	Shang vessels, FGA (TLD)	13
3	Shang–Chou Transition vessels, FGA	13
17	Shang bronzes, excavated	16
31	Early Chou vessels, FGA	3
19	Early Chou vessels, FGA (TLD)	13
18	Early Chou bronzes, excavated	16
15	Middle Chou vessels, FGA	3
5	Middle Chou vessels, FGA (TLD)	13
16	Late Chou vessels, FGA	3
4	Late Chou vessels, FGA (TLD)	13
26	Late Chou bronzes, excavated	16
3	Ch'in bronzes, excavated	16
16	Han bronzes, excavated	16
5	Han vessels, FGA	3
19	"Recent" vessels, FGA	3
151	Belt-hooks, FGA	3
36	Weapons excavated	17
3	Shang weapons, FGA	15
12	Early Chou weapons, FGA	15
5	Late Chou weapons, FGA	15
3	Mirrors, excavated	17
24	Mirrors, Chou & Han, FGA	11
3	Mirrors, T'ang, FGA	11
56	Mirrors, Riederer analyses	11
8	Ming-Tao coinage, FGA	11

[a] FGA, Freer Gallery of Art analyses; TLD, thermoluminescence-dated.
[b] Colors as shown on Figure 5.

founders put in a maximum amount of lead in several cases. In some cases the ancient Chinese even went over the maximum amount of lead, below the lower limit of miscibility. This seems to be particularly true in terms of late Chou coinage where lead ranges up to 50% and even above. These alloys must have been solidified rapidly from elevated temperatures as discussed below.

of Distributions

Centroid of Distribution				Hard-ness (BHN)[c]	Elon-gation (%)[c]	Tensile Strength (tons/in.²)[c]
Cu	Sn	Pb	Color[b]			
80.1	6.1	13.8	⎫	70	19	16
79.2	13.7	7.2	⎪	115	9	19
80.2	14.5	5.3	⎬ orangish yellow	125	8	18.5
73.1	13.5	13.4	⎪	105	8	17
88.6	10.9	0.5	⎭	120	12	20.5
79.3	14.4	6.2	light yellow	125	8	18.5
81.0	14.5	4.4	light yellow	135	8	19
83.2	12.1	4.7	orangish yellow	115	9	19
77.9	9.9	12.1	⎫ orangish yellow to	85	12	16.5
75.0	10.1	14.9	⎭ light yellow	80	11	17
76.1	9.9	14.0	⎫ light yellow to	80	11	16.5
74.2	11.8	14.0	⎬ yellowish grey–	115	9.5	16.5
74.7	16.9	8.4	⎭ white	135	4	18
75.2	14.8	10.0	light yellow	115	8	17.5
90.4	8.5	1.2	orange	90	29	20
79.5	8.6	12.0	orange	73	17	16
77.8	9.3	12.8	orange	75	15	16
83.5	4.5	12.0	copper–red to orange	50	22	15
83.3	12.9	3.8	orangish yellow	110	10	19
91.9	7.7	0.4	orange	90	32	21
86.7	12.2	1.1	orangish yellow	130	10	21
76.0	18.7	5.3	yellowish grey– white	150	3	18
73.9	20.6	5.5	⎫	170	2	17.5
70.4	25.4	4.2	⎪ yellowish grey–	200	0	17
70.3	24.7	4.9	⎬ white	200	0	17
73.6	20.6	5.8	⎭	170	2	17.5
54.0	1.6	44.4	orange–yellow?	20?	0(?)	< 6

[c] Physical properties extrapolated from Figures 6–8.

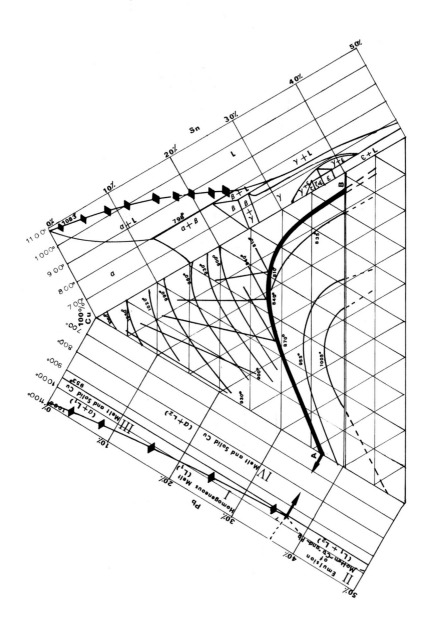

Ternary Representations of Groups of Bronzes

Figure 10 shows our analyses of Chinese mirrors and coinage. It illustrates the vast compositional differences between them and the extremely high lead content of the coinage. Ziebold has attempted to have objects cast with this high lead composition. His foundryman said that it was impossible to do so, as the copper and lead segregated into two separate liquid phases in the crucible and could not be poured together!

In Figure 11 analyses of mirrors made in Berlin by Josef Riederer (18) have been added to the analyses in Figure 10. Riederer's data and the Freer data seem to have different centroids although they fall in generally the same area on the diagram (*see also* Table II). Riederer's data seem to reflect a slightly higher copper composition in general, but this may result from the analytical method. The more interesting part of this diagram is, however, the group of five outliers in Riederer's data which fall over toward the left-hand side of the diagram. These are five mirrors of anomalous composition. By exercising the labeling option of our program it would be easy to pick out exactly which items these are, as the object numbers of the outliers can be printed on the plot.

Figure 12 shows all of the ceremonial vessel analyses from the Freer Gallery bronze book plus two analyses from a bronze Hsien. The 20% tin + lead line emphasizes that the idea of a constant alloy with 80% copper and 20% tin + lead is quite false. The analyses really tend towards the separation line. On this plot we have included zinc values over 1%. (The zinc values are displaced upwards a half-line space and to the right one full line space from the points to which they refer.) The vessels with zinc fall, for the most part, in the area of fairly high lead content, although one falls up in the area of high copper content. None of them fall along the high-tin line.

Figure 13 shows ceremonial vessels in the Freer Gallery of Art which have been thermoluminescence dated. This figure shows quite clearly the dependence of composition on date. The picture is not unambiguous, but the Shang bronzes tend to fall with the Early Chou bronzes in the high-tin area. The later Chou bronzes contain more lead. The Han bronze falls off towards higher copper percentages, and the recent bronzes have a very high lead content indeed. The Hsien mentioned above is also on this diagram and has what we considered an anomalously

Figure 9. (opposite) Liquidus and separation surfaces of the copper–tin–lead system after Ref. 13. The appropriate portions of the phase equilibrium diagrams for Cu–Sn and Cu–Pb appear on the sides of the triangle. (♦), Intersection of the liquidus contours with the liquidus of the binary phase diagrams. One should really picture this as a solid triangular prism viewed from the top; the sides of the prism would show the binary phase equilibrium diagrams.

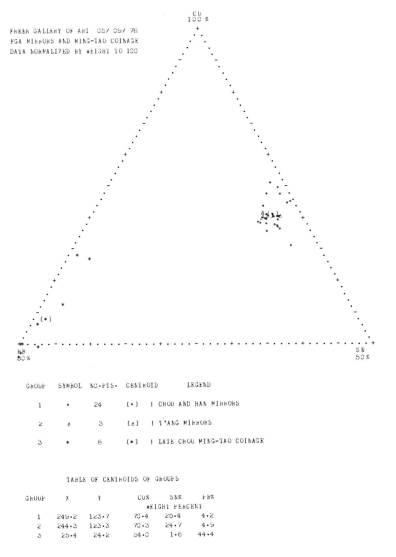

*Figure 10.   Freer Gallery of Art mirrors and Ming-Tao coinage;
data normalized by weight to 100%*

high lead content for a supposedly An-yang Shang bronze. The data
listing for this plot is reproduced as Appendix V.

If we take the 120 ceremonial vessels from the Freer Gallery bronze
book and plot them with an indicator of the recent bronzes, the supposed
date, we come out with Figure 14. There seem to be more recent bronzes
in the high-lead areas, but recent bronzes also fall in the high-copper
and high-tin areas. This suggests that recent imitations can have almost

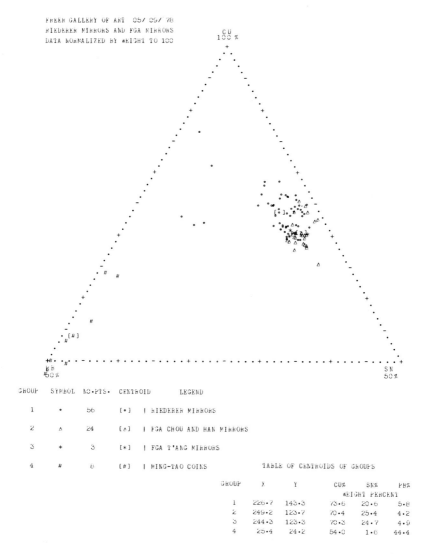

Figure 11.   Riederer and Freer Gallery of Art mirrors; data normalized by weight to 100%

any composition, possibly even a composition close to that of a genuine piece. Figure 15 shows a plot of Early Chinese weapons analyzed at the Freer Gallery. The weapons generally seem to fall in along the high-tin line, tending up to about 5% lead. The later Chou swords tend to have a higher tin content, and there are two Late Chou weapons which have a very high tin + lead content and which fall down in an area of generally lower tensile strength. These weapons were probably made for

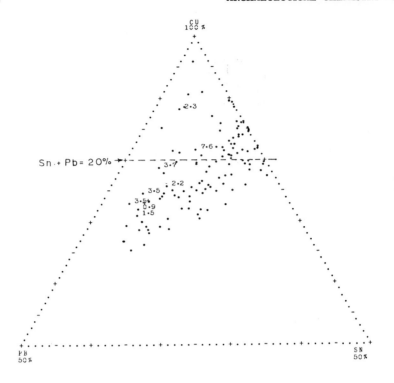

Figure 12. All ceremonial vessels plus Hsien with zinc; data normalized by weight to 100%. (·), All ceremonial vessels (120 points); (+), Hsien (2 points). Where Zn > 0%, the amount appears on the plot.

a ceremonial or display function. One weapon contains 100% copper. This is an early Shang weapon, made in the pre-An-yang phase. Other weapons of this composition have been published by Dono (14). In the pre-An-yang phase the metallurgy was not as well understood as it was in the An-yang phase when the other two Shang weapons on the plot were made.

All of the above charts were made from analyses of Chinese bronzes which are unprovenanced or at least not archaeologically excavated. To see how they compare with excavated material we have included Figures 16 and 17, which show analyses of excavated material as published in the

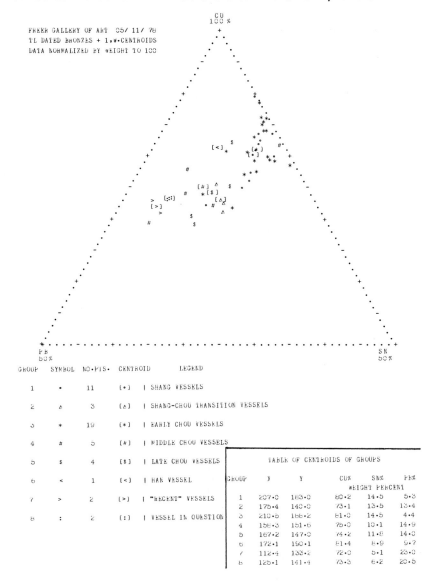

Figure 13.   Thermoluminescence-dated bronzes and Hsien, with centroids; data normalized by weight to 100%

composite table in Ref. 4, supplemented by four analyses from a recent publication on excavated pre-An-yang bronzes (*17*). The same pattern is evident in this data and in the listing of centroids (Table II).

Shang bronzes tend to be high-tin, although there are some outliers. The early Shang material falls in a very wide range, one vessel being

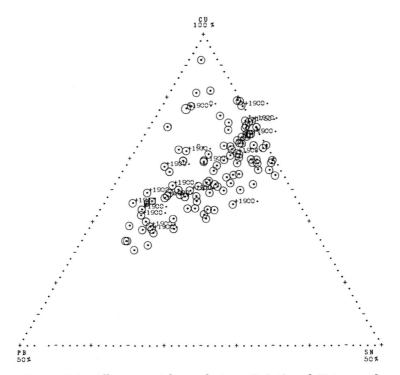

Figure 14. All ceremonial vessels from Ref. 7 and Hsien, with dates; data normalized by weight to 100%. (⊙), Ceremonial vessel; (□), Hsien.

fairly high copper, one being moderately high in both tin and lead, and two being very high in lead. Again, this indicates (on very little data) that metallurgy was developing in the pre-An-yang phase, and, with the move to An-yang, Chinese metallurgy reached its developed state. The Shang and Early Chou bronzes on this plot overlap. Late Chou bronzes (which include some weapons here) fall in the high-lead area with some outliers, and the Ch'in and Han bronzes fall in a wide range including very high copper ones and very high lead ones. The typological separation on Figure 18 shows the same things we have seen before in terms of distinct compositions of mirrors, weapons being somewhat higher in copper than vessels, etc.

One of the long-standing problems in the study of Chinese metallurgy is the interpretation of formulas given in the *K'ao Kung Chi* section of the *Ch'ou Li* concerning bronze compositions. It would be nice if these ternary diagrams could help us with the problem (compare, Refs. 4, and 14). The specified compositions from the *Ch'ou Li* are indicated on the right-hand axis of Figure 18 and are repeated in Table III. The *Ch'ou Li*

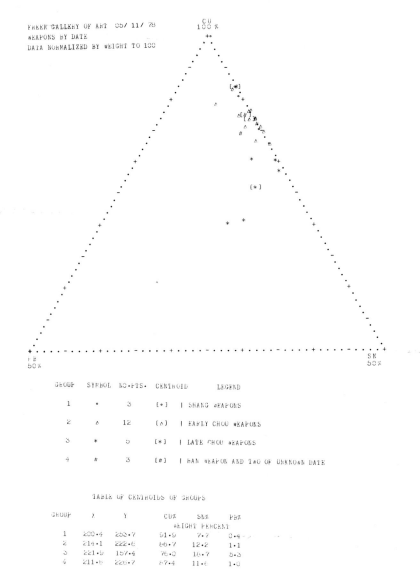

Figure 15.   Weapons by date; data normalized by weight to 100%

only mentions copper and tin although the translation of these Chinese terms could be open to question (4).

We have grouped the bronzes according to the groupings in the *Ch'ou Li* and have plotted these groups in Figure 18. The centroids of the various groups are listed as "actual" in Table III. Group I, vessels and bells, does agree reasonably well with the *Ch'ou Li* formula with

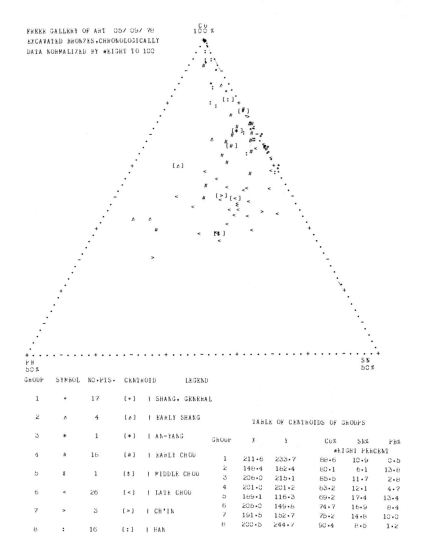

Figure 16.   Excavated bronzes, chronologically; data normalized by weight to 100%

some added lead, as does group II. In fact, if one took the formula from the *Ch'ou Li* and simply added lead to it, one would get bronzes falling in the areas of the group I and II centroids.

The actual centroids for groups III, IV, and V, however, do not look significantly different than those for group II; if anything, groups III and IV have less rather than more tin. The *Ch'ou Li* formulas do not hold for these groups. Group V contains more lead and a bit more tin than

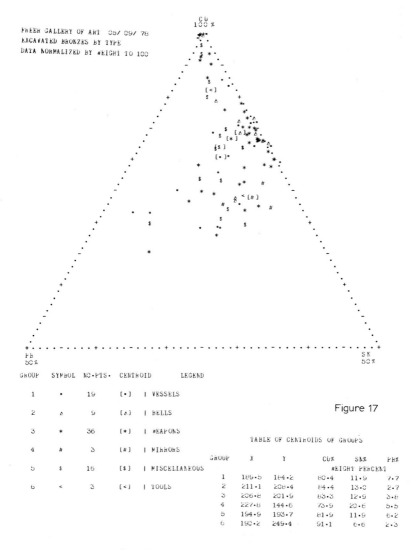

Figure 17. Excavated bronzes by type; data normalized by weight to 100%

groups III and IV but less tin than group II. The mirrors (group VI) do contain significantly more tin, but not nearly as much as is specified in the *Ch'ou Li*. The formulas do not, as a whole, stand up to our inspection.

However, the *Ch'ou Li* formulas do have some basis with regards to the properties of the various alloys. Vessels and bells do not need great hardness; they should be durable and, in the case of bells, somewhat

Table III.   Chou-Li Formulas

Type		Cu	Sn	Pb
Vessels and bells	formula	85.71	14.29	0
	actual	78.7	12.4	8.9
Axes	formula	83.33	16.67	0
	actual	82.1	15.9	2.0
Halberds and spears	formula	80.0	20.0	0
	actual	84.9	12.2	2.9
Swords and knives	formula	75.0	25.0	0
	actual	84.4	13.2	2.4
Erasing knives and arrows	formula	71.43	28.57	0
	actual	82.6	12.3	5.1
Mirrors	formula	50.0	50.0	0
	actual	70.8	24.9	4.4

[a] *See* Figure 18.

tough. Color also does not seem to be an important critrion. The *Ch'ou Li* formula would be satisfactory even with the added lead. The next four groups, all weapons, are required to hold an edge. In fact, as one proceeds down Table III the *Ch'ou Li* formulations get harder, more brittle, and whiter. Mirrors do not need to hold an edge, but it is helpful if they can have a hard surface to stand repeated polishings and use; color is probably the most important criterion here. While the *Ch'ou Li* formulas may not reflect accurately ancient Chinese alloying practice, their ranking of objects by use in terms of increasing tin does make metallurgical sense. One hopes that this question will be taken up in detail by Joseph Needham and his co-workers in the volume of "Science and Civilization in China" dealing with non-ferrous metallurgy.

Conclusions

Unfortunately, it is not possible to go into detail on individual analyses presented graphically in this paper. As shown in Appendix V, even the data for one of the more modest plots take up a great deal of space. While the data on ceremonial vessels has been published quite completely (7), data of many of the objects have not. We can, however, draw some conclusions from these ternary representations.

(1) After an initial development phase, the Chinese bronze founder knew exactly what he was doing in terms of alloy composition. The

and Corresponding Centroids [a]

Color	Hardness (BHN)	Elongation (%)	Tensile Strength (tons/in.²)
orangish yellow	150	7	21
light yellow	105	9	18
orangish yellow to yellowish	200	2	20
gray–white	150	5	20
light yellow			
yellowish gray–white	300	0	17
orangish yellow	120	10	20
silver–white	> 300	0 (?)	16
orangish yellow	120	9	20
silver–white	> 300	0 (?)	15?
light yellow	110	10	19
gray–white	> 300	0 (?)	12?
silvery gray	200	0	17

casters became more sophisticated as time went on, and in the late Chou and Han periods several individual formulas were in vogue for different types of objects. While the formulas in the *Ch'ou Li* reflect this situation, they do not seem to be an accurate indication of actual practice.

(2) The control of alloying was quite good, as can be seen from the close grouping of mirrors on the ternary plots.

(3) The picture of ancient Chinese technology which we have derived from unprovenanced objects in Western collections agrees with that from objects excavated scientifically in China.

(4) Except for the pre-An-yang vessels, early vessels seem to have lower lead than later vessels. The break seems to occur after the early Chou period (*see also* Table II). This might coincide with the demise of the An-yang bronze foundries, which persisted for some time after the Chou conquest. After this, more lead comes into the vessel composition, and the compositions become more variable.

We have seen how the simple idea of ternary plotting of the compositions of Chinese bronzes has given us new insights into the consistency and complexity of Chinese metallurgical practice in ancient times. With the increased interest in Chinese archaeology and the continuing flow of useful data from the People's Republic of China along with the emerging picture of regional Chinese bronze styles and such analytical methods as lead isotope ratio analyses, these ternary diagrams may prove even more useful in the future than they have already.

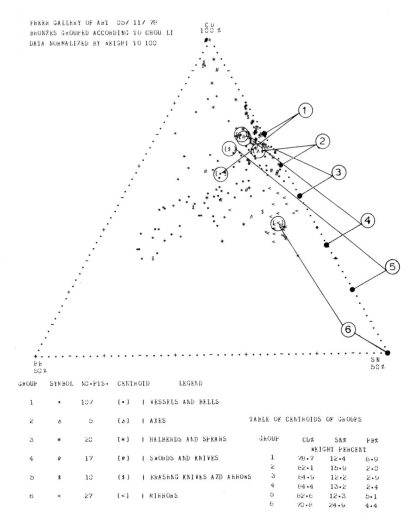

Figure 18.   Bronzes grouped according to Chou Li; data normalized by weight to 100%

Appendix I

The computer program for drawing these plots is tied to a modified teletype ASR 33 which has had Selsyn motors and electronics attached to turn it into a point plotter. The unit was sold by the Typagraph Corp. and is their model T-3. This works in a time-sharing mode with a commercial time-sharing computer company (Dialcom, Inc., Silver Spring, MD.).

The program (TGPL78) and its associated subroutine (AX12) written in FORTRAN, are reproduced as Appendixes II and III. A computer-drawn flow chart of the program is shown in Appendix IV. In the computer memory we have a large file of our Chinese bronze analyses both for major and trace elements. Some additional data as to the bronze type, date, provenance, whether it has been dated by thermoluminescence, and what the lead isotope ratios are if measured are also in this file. For the 150 Chinese belt hooks which we have analyzed, we also include three lines of shape and typological data.

The only thing which is used in the TGPL78 program is the first line on each object in these data files. The first line includes the accession number (or analysis number), a classification number which includes type and date, the supposed date of the object expressed in years (minus being equal to B.C.), five columns of compositional figures (copper, tin, lead, zinc, and iron), the analytical total, and the length or height of the object in centimeters. Sometimes the last figure is omitted in the file. The format for these lines is given in line 810 of the program.

Operation of the program is simple from the Typagraph. It requests the file which the operator wishes to plot. Before running the plot, the operator has to insert delimiters in the file to change the plot characters on the Typagraph and to terminate the program (*see* lines 320 and 330). After the user has typed in the name of the file which he wishes to plot, the program asks for options controlling the plot. These include a label to be inserted at the top of the plot; whether the centroids will be computed, plotted, or listed; whether normalization is to be as to weight percentage, volume percentage, or atomic percentage; and whether the individual points are to be labeled.

Possible labels are the accession number, the date, and the zinc content. After the options have been chosen, the computer plots the individual points for the objects and then draws the axes. Symbols are printed out underneath in order, and the user can type a legend for each symbol. After the list of symbols has been labeled, the program terminates.

The plotting of axes is done by subroutine AX12, Appendix III. The data which was plotted can also be printed out after the plot simply by typing LIST DATA. Appendix V shows the DATA file for the plot in Figure 13. The DATA file consists of the accession number; the original copper, tin, and lead contents; and the total of these three added together. This is followed by the normalized copper, tin, and lead content and the plotting points which are fed into the Typagraph. These plotting points correspond to x and y coordinates in 0.002-in. increments from the origin (50% lead).

Appendix II. TGPL78 Ternary Plotting Program

```
100 ♦ TRIANGULAR PLOTTING PROGRAM : H+B+C=100% BY WEIGHT
110 ♦ W.T.CHASE, FREER GALLERY OF ART, FEBRUARY, 1976; REVISED 3/3/78
120 ♦ PROGRAM FOR NO MORE THAN 20 GROUPS, AND NO MORE THAN 200 ITEMS PER GROUP
130 ♦ GROUPS IN STATEMENT 10, ITEMS IN STATEMENT FOLLOWING (IF YOU WANT
140 ♦ TO CHANGE THEM.)
145 ♦ ------------------------------------------------------- ♦
150 ♦ ♦♦♦♦♦♦----SPECIFICATION STATEMENTS---------------♦♦♦♦♦♦ ♦
160   DIMENSION  PLC(20), COUNT(20), AC(1), IDUM (2500), IBCD(30)
170 + ,IFILE(4),LPL(20),LPT(20),IRSC(72),INOR,NR1(8),NR2(8),NR3(8),
180 + LPOT(3),XCTR(20),YCTR(20),ICK(2),ICENT(1)
190   COMMON IDUM
200   INTEGER AC,CHAR,PLC,COUNT
210   DOUBLE PRECISION AC
220   DATA  (PLC)/1H&,1H\,1H#,1H♦,1H@,1H$,1H<,1H>,1H:,1H!,
230 + 1HA,1HB,1HC,1HD,1HE,1HF,1HG,1HH,1HI,1HJ,1HK/ ,(ICENT)/0/,
240 + (INOR(KK),KK=1,8)/2HDA,2HTA,2H N,2HOR,2HMA,2HLI,2HZE,2HD /,
250 + (NR1(KL),KL=1,8)/2HBY,2H W,2HEI,2HGH,2HT ,2HTD,2H 1,2H00/,
260 + (NR2(KM),KM=1,8)/2HBY,2H V,2HOL,2HUM,2HE ,2HTD,2H 1,2H00/,
270 + (NR3(KO),KO=1,8)/2HTD,2H A,2HTO,2HMI,2HC ,2HPE,2HRC,2HNT/,
280 + (LPT(KN),KN=1,6)/2HFR,2HEE,2HR ,2HGA,2HLL,2HER/,
290 + (LPT(KN),KN=7,11)/2HY ,2HOF,2H A,2HRT,2H  /,
300 + (LPT(KN),KN=14,16)/2H/ ,2H/ ,2H/ /,(ICK(KP),KP=1,2)/1H[,1H]/
305 ♦ ------------------------------------------------------- ♦
310 ♦ ♦♦♦♦♦♦ ---------SETTING PLOT LABELS---------------♦♦♦♦♦♦ ♦
320 ♦ NOTE : GROUPS DIVIDED BY D 'DATE' ;   DATE >2000. = GROUP DIVIDER;
330 ♦ DATE > 3000. = STOP PROGRAM
340   CALL DATE(IM,ID,IY); LPT(13)=IM; LPT(15)=ID; LPT(17)=IY
350   WRITE(9,105); 105 FORMAT('FILE TO BE NORMALIZED AND PLOTTED')
360   READ(9,110)IFILE; 110 FORMAT(4A2)
370   CALL DEFINE(1,IFILE)
380   CALL DEFINE(2,5HDATA,)
390   IWV=1HW ; DO 630 NO=1,8 ; 630 INOR(NO+8)=NR1(NO)
400   WRITE(9,120); 120 FORMAT('OPTIONS? (TYPE 0 FOR PLOT, 1 FOR OPTIONS ) ')
410   READ(9,130)IOP; 130 FORMAT(I2)
420   IF (IOP.EQ.0) GO TO 140
430   WRITE(9,150); 150 FORMAT('LABEL')
440   READ(9,160)LPL; 160 FORMAT(20A2)
450   WRITE(9,151); 151 FORMAT ('WITH CENTROIDS ? (0=NO; 1=YES)')
470   READ(9,152) ICENT(1); 152 FORMAT(I2)
480   WRITE(9,170); 170 FORMAT('NORMALIZATION TO WEIGHT(W),')
490   WRITE(9,175); 175 FORMAT('VOLUME(V), OR ATOMIC PERCENT(A) ?')
500   READ(9,180)IWV; 180 FORMAT(A1)
510   IF (IWV.EQ.1HV) GO TO 600
520   IF (IWV.EQ.1HA) GO TO 625
530   GO TO 650
540 600 DO 640 NO=1,8; 640 INOR(NO+8)=NR2(NO);  GO TO 650
550 625 DO 645 NO=1,8; 645 INOR(NO+8)=NR3(NO)
560 650 CONTINUE
570   WRITE(9,190); 190 FORMAT('POINTS UNLABELLED(0), OR LABELLED BY')
580   WRITE(9,191); 191 FORMAT('ACCESSION NUMBER(1)')
590   WRITE(9,192); 192 FORMAT('DATE(2)')
600   WRITE(9,193); 193 FORMAT('ZINC(3)')
610   WRITE(9,194); 194 FORMAT('OTHER(4) ?')
620   READ(9,196)LPOI; 196 FORMAT (I2)
630   IF(LPOI.EQ.4) WRITE(9,195)
640 195 FORMAT('OTHER LABELS NOT NOW CODED')
650   WRITE(2,657)LPT; 657 FORMAT(17A2)
660   WRITE(2,655)LPL; 655 FORMAT(20A2)
670   WRITE(2,660)INOR; 660 FORMAT(16A2)
680   WRITE(2,656); 656 FORMAT(/,/)
685 ♦ ------------------------------------------------------- ♦
690 ♦ ♦♦♦♦♦♦---INITIALIZATION OF PLOT AND TOP LABELS-----♦♦♦♦♦♦ ♦
690 ♦ INITIALIZATION OF PLOTTING (SETS POINT OF ORIGIN)
700 140 CALL TGPLOT(7.5,0.0)/♦INITIALIZATION OF PLOT(SETS ORIGIN)♦/
710   CALL TGORIG (-0.50,0.0)
720   IF (LPL(1).EQ.1H .AND.LPL(2).EQ.1H .AND.LPL(3).EQ.1H ) GO TO 10
730   CALL TGLABL(0.0,7.4,LPT,34,0)
740   CALL TGLABL (0.0,7.2,LPL,40,0)
750   CALL TGLABL(0.0,7.0,INOR,32,0)
760 10 DO 20 I=1,20
770   XCTR(I)=0; YCTR(I)=0
780   XC=0; YC=0
790   DO 50 I2=1,200
800   READ(1,11)AC,CL,DA,CU,SN,PB,ZN,FE,TOT,CM
```

```
810  11 FORMAT (6X,3A2,1X,F5.1,1X,F6.0,3(1X,F5.1),2(1X,F6.2),
820  + 1X,F6.1,1X,F5.1)
830  IF (DA.GT.1999.) GO TO 100
835  * ------------------------------------------------------- *
840  * ++++++-------NORMALIZATION TO 100% ---------++++++ *
850  TN=CU+PB+SN ;  FACTR = 100.0/TN
860  CUN = CU+FACTR ;  PBN = PB+FACTR ;  SNN = SN+FACTR
870  IF (IWV.EQ.1HW) GO TO 215
880  IF (IWV.EQ.1HA) GO TO 205
900   CUV=CUN/8.96;  SNV=SNN/7.30;  PBV=PBN/11.68
910  VTOT=CUV+SNV+PBV;  VOLF=100.0/VTOT
920  CUN=VOLF+CUV;  SNN=VOLF+SNV;  PBN=VOLF+PBV
930  GO TO 215 /+ ATOMIC PERCENT NORMALIZATION BELOW+/
940  205 SIGP=(CUN/63.54)+(SNN/118.69)+(PBN/207.19)
950  CUA=(100.0+CUN)/(SIGP+63.54) ;  SNA=(100.0+SNN)/(SIGP+118.69)
960  PBA=(100.0+PBN)/(SIGP+207.19)
970  CUN=CUA;  SNN=SNA;  PBN=PBA
980  * LATER,INCLUDE FACTORS AND PROVISION FOR ZN
990  * SET BASELINE;  THIS PROGRAM HAS BASELINE SET AT 50% PB-SN
1000 215 BLF = 50.0
1005 * ------------------------------------------------------- *
1010 * ++++++---------TRANSFORM TO COORDINATES-----++++++ *
1020 YT=((CUN-BLF)+(303.1/(100.-BLF)))
1030 XT=175.+((SNN-PBN)+(175./(100.-BLF)))
1040 WRITE (2,12) AC,CU,SN,PB,TOT,CUN,SNN,PBN,XT,YT
1050 12 FORMAT (3A2,2X,3(2X,F5.1),2X,F6.1,3(2X,F5.1),3X,F6.1,2X,F6.1)
1060 IY = YT+.5 ;   IX = XT+.5 ;   ISYM = PLC(I)
1070 XC=XC+XT; YC=YC+YT
1080 CALL TGMOVE (IX,IY,ISYM)
1090 COUNT(I)=COUNT(I)+1
1100 XCT=XC/COUNT(I) ; YCT=YC/COUNT(I)
1105 * ------------------------------------------------------- *
1110 * ++++++---------LABELLING OF POINTS-------------++++++ *
1120 X=(IX+.02)+.05 ; Y=(IY+.02)+.05
1130 IF (LPOI.EQ.0) GO TO 50
1140 IF (LPOI.EQ.4) GO TO 50
1150 CALL RESCAN (1)
1160 IF (LPOI.EQ.3) GO TO 240
1170 IF (LPOI.EQ.2) GO TO 230
1180 READ (1,730)LPOT ; 730 FORMAT (6X,3A2,60X) ; GO TO 780
1190 230 READ (1,731)LPOT ; 731 FORMAT (19X,3A2,45X) ; GO TO 780
1200 240 READ (1,732)LPOT ; 732 FORMAT (45X,3A2,18X) ; GO TO 780
1210 780 CALL TGLABL(X,Y,LPOT,6,0)
1220 50 CONTINUE
1230 100 WRITE (2,13); 13 FORMAT (/)
1235 * ------------------------------------------------------- *
1237 * ++++++--------------PLOTTING CENTROIDS---------++++++ *
1240 XCTR(I)=XCT; YCTR(I)=YCT
1250 IX = XCTR(I)+.5; IY=YCTR(I)+.5
1255 IF (ICENT(1).EQ.0) GO TO 19
1270 CALL TGMOVE(IX-4,IY,ICK(1)); CALL TGMOVE (IX,IY,ISYM)
1280 CALL TGMOVE(IX+4,IY,ICK(2))
1325 19 IF (DA.EQ.3000.) GO TO 333
1330 20 CONTINUE
1340 333 CALL AX12(BLF)
1345 * -------------------------------------------------------
1347 * ++++++---DRAWING AXES AND WRITING TABLES------++++++ *
1350 IF(ICENT(1).EQ.1) GO TO 756
1360 WRITE(9,700)
1370 700 FORMAT(/,'SYMBOL  NO.PTS.   LEGEND',/)
1380 DO 720 I7=1,I
1390 WRITE(9,707)PLC(I7),COUNT(I7)
1400 707 FORMAT (2X,A1,3X,I5,3X)
1410 READ(9,710)IANS; 710 FORMAT(A1)
1420 720 CONTINUE
1430 GO TO 761
1440 756 WRITE(9,757)
1450 757 FORMAT(/,'GROUP    SYMBOL  NO.PTS.  CENTROID       LEGEND',/)
1460 DO 759 I7=1,I
1470 WRITE(9,758) I7,PLC(I7),COUNT(I7),PLC(I7)
1480 758 FORMAT (2X,I2,6X,A1,3X,I5,6X,1H(,A1,1H),3X)
1490  READ (9,760)IANS; 760 FORMAT (A1)
1500 759 CONTINUE
1510 WRITE (9,754)
1520 754 FORMAT (/,/,10X,'TABLE OF CENTROIDS OF GROUPS',/,
1530 + /,'GROUP    X      Y',9X,'CU%    SN%    PB%')
1532 IF(IWV.EQ.1HV) GO TO 601
1533 IF(IWV.EQ.1HA) GO TO 602
```

```
1534 WRITE(9,765); 765 FORMAT(30X,'WEIGHT PERCENT'); GO TO 603
1536 601 WRITE(9,766); 766 FORMAT(30X,'VOLUME PERCENT'); GO TO 603
1538 602 WRITE(9,767); 767 FORMAT(30X,'ATOMIC FRACTION'); GO TO 603
1540 603 AE=(303.1/(100.-BLF)); BE=(175./(100.-BLF))
1550 DO 764 I7=1,I
1560 CUN=(YCTR(I7)/AE)+BLF
1570 SNN=(XCTR(I7)-175.)/(2*BE)+50.-(CUN/2)
1580 PBN=100.-(SNN+CUN)
1590 WRITE (9,755) I7,XCTR(I7),YCTR(I7),CUN,SNN,PBN
1600 755 FORMAT(2X,I2,3X,F6.1,2X,F6.1,2X,3(3X,F5.1))
1610 764 CONTINUE
1620
1630 761 STOP
1640 END
```

Appendix III.   Subroutine AX12

```
?LIST AX12

 100 SUBROUTINE AX12(BLF)
 245  DIMENSION IDUM(2500),ISYM(10)
 250 COMMON IDUM
 700 DATA ISYM /1H+,1H.,1H.,1H.,1H.,1H-,1H.,1H.,1H.,1H./
 800 IX = 0
 900 IY = 0
1000 DO 100 I1=1,5
1010 DO 100 I2=1,10
1100 CALL TGMOVE(IX,IY,ISYM(I2))
1200 100 IX =IX+7
1205 CALL TGMOVE(IX,IY,ISYM(1))
1300 CALL TGMOVE(335,-7,1HS)
1400 CALL TGMOVE (340,-7,1HN)
1500 CALL TGMOVE(335,-14,1H5)
1525 CALL TGMOVE(340,-14,1H0)
1530 CALL TGMOVE(345,-14,1H%)
1540 IX=0 ;IY=0
1600 X = IX
1700 DO 200 I1=1,5
1710 DO 200 I2=1,10
1800 CALL TGMOVE (IX,IY,ISYM(I2))
1900 X=X+3.5
2000 Y=SQRT (3.0)*X
2100 IX=X+.5; IY=Y+.5
2200 200 CONTINUE
2300 CALL TGMOVE (172,317,1HC)
2400 CALL TGMOVE (177,317,1HU)
2500 CALL TGMOVE (167,312,1H1)
2505 CALL TGMOVE (172,312,1H0)
2510 CALL TGMOVE (177,312,1H0)
2515 CALL TGMOVE (184,312,1H%)
2600 DO 300 I1=1,5
2610 DO 300 I2=1,10
2700 CALL TGMOVE (IX,IY,ISYM(I2))
2800 X = X+3.5; Y = (SQRT(3.0)*350.0)-(SQRT(3.0)*X)
2900 IX=X+.5 ; IY=Y+.5
3000 300 CONTINUE
3100 CALL TGMOVE (0,-7,1HP)
3200 CALL TGMOVE (5,-7,1HB)
3300 CALL TGMOVE (0,-14,1H5)
3310 CALL TGMOVE (5,-14,1H0)
3321 CALL TGMOVE (10,-14,1H%)
3400 CALL TGEND
3500 RETURN
3600 END
```

Appendix IV. Computer-Produced Flow Diagram for TGPL78

```
♦ TRIANGULAR PLOTTING PROGRAM : A+B+C=100% BY WEIGHT
♦ W.T.CHASE, FREER GALLERY OF ART, FEBRUARY, 1976: REVISED 3/3/78
♦ PROGRAM FOR NO MORE THAN 20 GROUPS, AND NO MORE THAN 200 ITEMS P
♦ GROUPS IN STATEMENT 10, ITEMS IN STATEMENT FOLLOWING (IF YOU WAN
♦ TO CHANGE THEM.)
♦ ------------------------------------------------------- ♦
♦ ♦♦♦♦♦♦----SPECIFICATION STATEMENTS----------------♦♦♦♦♦ ♦
DIMENSION  PLC(20), COUNT(20), AC(1), IDUM (2500), IBCD(30)
 ,IFILE(4),LPL(20),LPT(20),IRSC(72),INOR(20),NR1(8),NR2(8),NR3(8)
 LPOT(3),XCTR(20),YCTR(20),ICK(2),ICENT(1)
COMMON IDUM
INTEGER AC,CHAR,PLC,COUNT
DOUBLE PRECISION AC
```

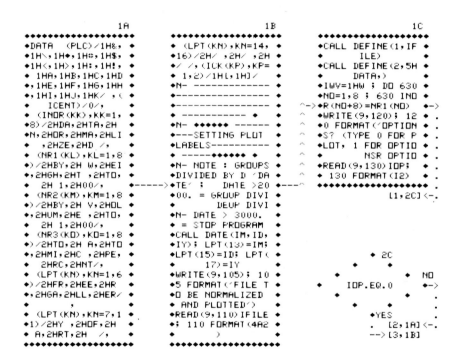

```
             1A
♦♦♦♦♦♦♦♦♦♦♦♦♦♦♦♦♦♦♦♦♦
♦DATA   (PLC)/1H&, ♦
♦1H\,1H♦,1H♯,1H$, ♦
♦1H<,1H>,1H:,1H!, ♦
♦ 1HA,1HB,1HC,1HD ♦
♦,1HE,1HF,1HG,1HH ♦
♦,1HI,1HJ,1HK/ ,( ♦
♦   ICENT)/0/,    ♦
♦ (INOR(KK),KK=1, ♦
♦8)/2HDA,2HTA,2H  ♦
♦N,2HOR,2HMA,2HLI ♦
♦ ,2HZE,2HD /,    ♦
♦ (NR1(KL),KL=1,8 ♦
♦)/2HBY,2H W,2HEI ♦
♦,2HGH,2HT ,2HTO, ♦
♦ 2H 1,2H00/,     ♦
♦ (NR2(KM),KM=1,8 ♦
♦)/2HBY,2H V,2HOL ♦
♦,2HUM,2HE ,2HTO, ♦
♦ 2H 1,2H00/,     ♦
♦ (NR3(KO),KO=1,8 ♦
♦)/2HTO,2H A,2HTO ♦
♦,2HMI,2HC ,2HPE, ♦
♦ 2HRC,2HNT/,     ♦
♦ (LPT(KN),KN=1,6 ♦
♦)/2HEE,2HEE,2HR  ♦
♦,2HGA,2HLL,2HER/ ♦
♦          ,      ♦
♦ (LPT(KN),KN=7,1 ♦
♦1)/2HY ,2HOF,2H  ♦
♦ A,2HRT,2H  /,   ♦
♦♦♦♦♦♦♦♦♦♦♦♦♦♦♦♦♦♦♦♦♦
```

```
             1B
♦♦♦♦♦♦♦♦♦♦♦♦♦♦♦♦♦♦♦♦♦
♦ (LPT(KN),KN=14, ♦
♦16)/2H/ ,2H/ ,2H ♦
♦/ /,(ICK(KP),KP= ♦
♦ 1,2)/1HL,1HJ/   ♦
♦N- --------------♦
♦-----------------♦
♦ ----------- ♦  ♦
♦N- ♦♦♦♦♦♦ ------ ♦
♦---SETTING PLOT  ♦
♦LABELS---------- ♦
♦ -----♦♦♦♦♦♦ ♦  ♦
♦N- NOTE : GROUPS ♦
♦DIVIDED BY D /DA ♦
♦TE' ;   DATE >20 ♦
♦00. = GROUP DIVI ♦
♦        DEUP DIVI♦
♦N- DATE > 3000.  ♦
♦ = STOP PROGRAM  ♦
♦CALL DATE(IM,ID, ♦
♦IY); LPT(13)=IM; ♦
♦LPT(15)=ID; LPT( ♦
♦      17)=IY     ♦
♦WRITE(9,105); 10 ♦
♦5 FORMAT('FILE T ♦
♦O BE NORMALIZED  ♦
♦ AND PLOTTED')   ♦
♦READ(9,110)IFILE ♦
♦; 110 FORMAT(4A2 ♦
♦        )        ♦
♦♦♦♦♦♦♦♦♦♦♦♦♦♦♦♦♦♦♦♦♦
```

```
             1C
♦♦♦♦♦♦♦♦♦♦♦♦♦♦♦♦♦♦♦♦♦
♦CALL DEFINE(1,IF ♦
♦      ILE)       ♦
♦CALL DEFINE(2,5H ♦
♦      DATA,)     ♦
♦IWV=1HW ; DO 630 ♦
♦NO=1,8 ; 630 INO ♦
^->♦R(NO+8)=NR1(NO)  ♦->
^  ♦WRITE(9,120); 12 ♦ .
^  ♦O FORMAT('OPTION ♦ .
^  ♦S? (TYPE 0 FOR P ♦ .
^  ♦LOT, 1 FOR OPTIO ♦ .
^  ♦       NSR OPTIO ♦ .
^  ♦READ(9,130)IOP;  ♦ .
^  ♦ 130 FORMAT(I2)  ♦ .
^  ♦♦♦♦♦♦♦♦♦♦♦♦♦♦♦♦♦♦♦♦♦
^         [1,2C]<-.
```

```
              ♦ 2C
        ♦            ♦ NO
     ♦    IOP.EQ.0   ♦->
        ♦            ♦ .
          ♦YES       .
        . [2,1A]<-.
        -->[3,1B]
```

```
.
-->[2,3C]
```

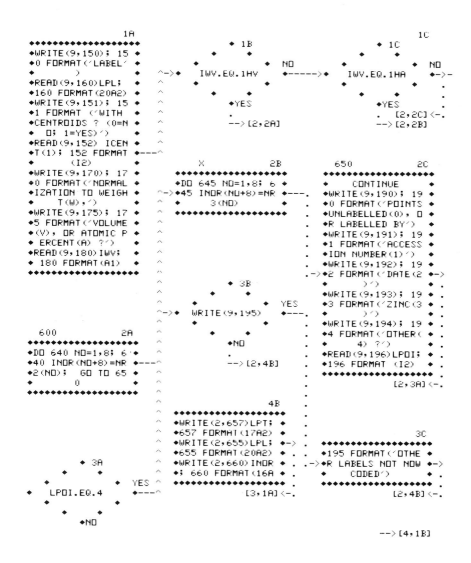

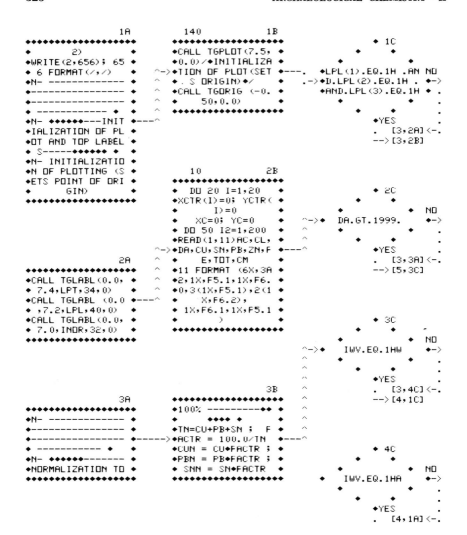

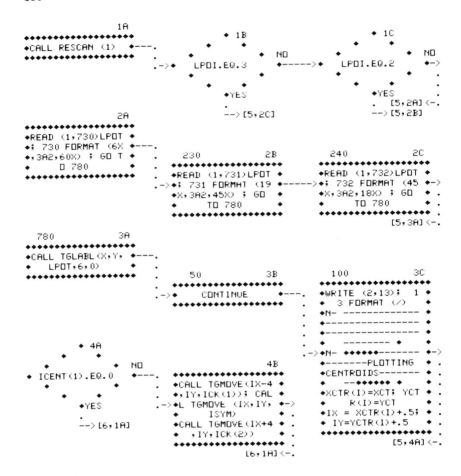

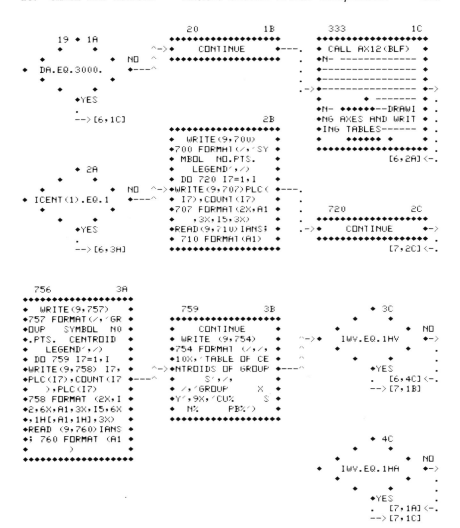

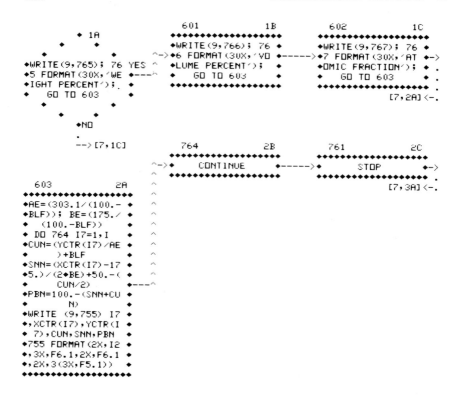

```
                 ◆ 1A
          ◆          ◆
       ◆        ◆        ◆
  ◆WRITE(9,765); 76 YES ^
  ◆5 FORMAT(30X,'WE ◆---^
  ◆IGHT PERCENT'); .
  ◆    GO TO 603        ◆
       ◆          ◆
          ◆NO
           .
          --> [7,1C]
```

```
          601            1B
     ◆◆◆◆◆◆◆◆◆◆◆◆◆◆◆◆◆◆◆◆
     ◆WRITE(9,766); 76 ◆
  ^->◆6 FORMAT(30X,'VO ◆----->
     ◆LUME PERCENT');  ◆
     ◆   GO TO 603     ◆
     ◆◆◆◆◆◆◆◆◆◆◆◆◆◆◆◆◆◆◆◆
```

```
          602            1C
     ◆◆◆◆◆◆◆◆◆◆◆◆◆◆◆◆◆◆◆◆
     ◆WRITE(9,767); 76 ◆
     ◆7 FORMAT(30X,'AT ◆->
     ◆OMIC FRACTION'); ◆ .
     ◆   GO TO 603     ◆ .
     ◆◆◆◆◆◆◆◆◆◆◆◆◆◆◆◆◆◆◆◆ .
                      [7,2A] <-.
```

```
          603            2A
  ◆◆◆◆◆◆◆◆◆◆◆◆◆◆◆◆◆◆◆◆
  ◆AE=(303.1/(100.- ◆      ^
  ◆BLF)); BE=(175./ ◆      ^
  ◆  (100.-BLF))    ◆      ^
  ◆ DO 764 I7=1,I   ◆      ^
  ◆CUN=(YCTR(I7)/AE ◆      ^
  ◆    )+BLF        ◆      ^
  ◆SNN=(XCTR(I7)-17 ◆      ^
  ◆5.)/(2◆BE)+50.-( ◆      ^
  ◆    CUN/2)       ◆----^
  ◆PBN=100.-(SNN+CU ◆
  ◆    N)           ◆
  ◆WRITE (9,755) I7 ◆
  ◆,XCTR(I7),YCTR(I ◆
  ◆ 7),CUN,SNN,PBN  ◆
  ◆755 FORMAT(2X,I2 ◆
  ◆,3X,F6.1,2X,F6.1 ◆
  ◆,2X,3(3X,F5.1))  ◆
  ◆◆◆◆◆◆◆◆◆◆◆◆◆◆◆◆◆◆◆◆
```

```
          764            2B
     ◆◆◆◆◆◆◆◆◆◆◆◆◆◆◆◆◆◆◆◆
  ^->◆    CONTINUE     ◆----->
     ◆◆◆◆◆◆◆◆◆◆◆◆◆◆◆◆◆◆◆◆
```

```
          761            2C
     ◆◆◆◆◆◆◆◆◆◆◆◆◆◆◆◆◆◆◆◆
  ◆◆       STOP       ◆->
     ◆◆◆◆◆◆◆◆◆◆◆◆◆◆◆◆◆◆◆◆ .
                      [7,3A] <-.
```

```
                  3A
     ◆◆◆◆◆◆◆◆◆◆◆◆◆◆◆◆◆◆◆◆
     ◆       END        ◆
     ◆◆◆◆◆◆◆◆◆◆◆◆◆◆◆◆◆◆◆◆
```

Appendix V. DATA *File for Figure 13*

23·1	80·0	14·3	4·7	99·0	80·8	14·4	4·7	208·9	186·8
35·12	75·1	15·6	9·5	100·2	75·0	15·6	9·5	196·3	151·2
39·53	70·1	11·6	15·6	97·3	72·0	11·9	16·0	160·6	133·6
40·11	82·8	14·6	3·7	101·1	81·9	14·4	3·7	212·7	193·4
42·14	74·4	15·7	7·8	97·6	76·0	16·0	8·0	203·2	157·6
46·31	86·0	9·7	0·4	96·1	89·5	10·1	0·4	208·9	239·4
47·11	79·6	13·8	3·1	96·5	82·5	14·3	3·2	213·8	196·9
51·19	77·9	17·2	2·4	97·5	79·9	17·6	2·5	228·1	181·2
53·83	76·3	15·3	6·9	98·5	77·5	15·5	7·0	204·8	166·5
54·15	82·5	15·4	0·9	98·8	83·5	15·6	0·9	226·4	203·1
60·18	82·9	13·6	2·8	99·3	83·5	13·7	2·8	213·1	203·0
30·26	67·4	14·0	13·4	94·8	71·1	14·8	14·1	177·2	127·9
30·26	70·4	13·7	12·8	94·8	72·7	14·1	13·2	178·3	137·3
61·33	73·4	11·3	12·5	97·2	75·5	11·6	12·9	170·7	154·7
11·38	76·0	14·8	7·9	98·7	77·0	15·0	8·0	199·5	163·7
11·53	82·6	13·2	1·5	97·3	84·9	13·6	1·5	217·1	211·5
11·54	85·6	11·0	0·2	96·8	88·4	11·4	0·2	214·0	233·0
15·102	77·2	16·2	3·2	96·6	79·9	16·8	3·3	222·1	181·4
16·480	78·2	17·1	3·2	98·5	79·4	17·4	3·2	224·4	178·2
17·193	78·5	10·3	8·5	97·3	80·7	10·6	8·7	181·5	186·0
24·14	82·0	13·8	1·8	97·6	84·0	14·1	1·8	218·0	206·2
30·54	77·9	20·3	0·9	99·1	78·6	20·5	0·9	243·5	173·4
31·10	70·3	10·0	14·7	95·0	74·0	10·5	15·5	157·7	145·5
33·2	71·1	15·7	12·3	99·1	71·7	15·8	12·4	187·0	131·8
35·21	82·4	12·2	1·0	96·3	86·2	12·8	1·0	216·0	219·4
35·22	84·4	13·3	1·1	99·5	85·4	13·5	1·1	218·2	214·7
38·20	82·0	13·2	3·6	98·8	83·0	13·4	3·6	209·0	200·0
46·4	76·0	16·0	6·0	98·0	77·6	16·3	6·1	210·7	167·0
47·12	74·7	12·6	5·4	92·7	80·6	13·6	5·8	202·2	185·4
50·7	77·7	14·9	5·5	98·1	79·2	15·2	5·6	208·5	177·0
60·19	81·2	14·0	1·4	96·6	84·1	14·5	1·4	220·7	206·5
68·28	83·8	13·2	0·4	97·7	86·0	13·6	0·4	221·0	218·5
68·29	78·0	17·8	2·7	98·8	79·2	18·1	2·7	228·7	176·9
07·33	66·1	5·1	24·1	95·3	69·4	5·4	25·3	105·2	117·4
09·259	71·9	7·9	17·3	97·1	74·0	8·1	17·8	141·1	145·8
11·60	76·5	6·4	15·3	98·2	77·9	6·5	15·6	143·3	169·1
61·31	79·7	17·1	0·7	97·5	81·7	17·5	0·7	233·9	192·4
39·5	70·8	12·8	14·7	98·3	72·0	13·0	15·0	168·2	133·5
40·23	68·5	11·7	18·9	99·4	69·1	11·8	19·1	149·6	115·9
48·24	68·6	8·9	6·0	83·8	82·2	10·7	7·2	187·2	194·9
57·22	68·8	10·5	18·3	97·6	70·5	10·8	18·7	147·0	124·2
61·30	74·1	13·6	10·8	7098·5	75·2	13·8	11·0	184·9	152·9

61·3	79·4	8·7	9·5	98·1	81·4	8·9	9·7	172·1	190·1
09·254	68·9	3·8	21·8	98·0	72·9	4·0	23·1	108·3	138·9
11·49	68·4	5·9	22·0	97·8	71·0	6·1	22·8	116·5	127·5
V63·75	71·4	5·9	20·3	98·3	73·2	6·0	20·8	123·4	140·4
V63·75	71·5	6·2	19·6	97·9	73·5	6·4	20·1	126·8	142·4

Literature Cited

1. "Historical Relics Unearthed in New China," Foreign Language Press, Peking, 1972.
2. Loehr, M., "Relics of Ancient China from the Collection of Paul Singer," p. 58–61, The Asia Society, New York, 1965.
3. Barnard, N., "The First Radiocarbon Dates from China," Monographs of Far Eastern History 8, Australian National University, Canberra, 1972.
4. Barnard, N., Tamotsu, S., "Metallurgical Remains of Ancient China," p. 20, Nichosha, Tokyo, 1975.
5. Rudolf, R. C., "Preliminary Notes on Sung Archaeology," *J. Asian Stud.* (1963) **22**(2), 169–177.
6. Barnard, N., "The Incidence of Forgeries Among Archaic Chinese Bronzes —Some Preliminary Notes," *Monumenta Serica* (1968) **27**, 91–168.
7. Gettens, R. J., "The Freer Chinese Bronzes, Volume II, Technical Studies," *Smithsonian Institutional Oriental Studies* (1969) **7**, 45.
8. Chia-pao, W., "Ku ch'i-wu yen-chiu chuan-k'an," Li Chi, Ed., Academia Sinica, Nan-kang, 1964–1973.
9. Barnard, N., "Bronze Casting and Alloys of Ancient China," *Monumenta Serica* (1961) **14**.
10. Needham, J., "Science and Civilization in China," Vol. 5, No. 2, Cambridge University, 1974.
11. Chase, W. T., "Examinations of Art Objects in the Freer Gallery Laboratory," *Ars Orientalis* (1973) **9**, Chart 1.
12. Prince, A., "Alloy Phase Equilibria," Elsevier, Amsterdam, 1966.
13. Rhines, F. N., "Phase Diagrams in Metallurgy: Their Development and Application," McGraw-Hill, New York, 1956.
14. Dono, T., "The Chemical Investigation of the Ancient Metallic Culture," Asakura, Tokyo, 1967.
15. Hanson, D., Pell-Walpole, W. T., "Chill-cast Tin Bronzes," Edward Arnold & Co., London, 1951.
16. Hoffmann, W., "Lead and Lead Alloys; Properties and Technology," p. 152, Springer-Verlag, New York, 1970.
17. Hupei Provincial Museum, "The Shang Dynasty Bronzes of the Erh Li Kang. Unearthed at the Ancient City of P'an Lung in Huang P'i County, Hupei Province," *Wen-Wu* (1976) **2**, 37.
18. Riederer, J., "Metallanalyser Chinesischer Spiegel," *Berl. Beitrag. Archaeom.* (1977) **2**, 6–18.

RECEIVED September 19, 1977.

Prehistoric Copper Artifacts in the Eastern United States

SHARON I. GOAD and JOHN NOAKES

Departments of Anthropology and Geochronology, University of Georgia, Athens, GA 30602

Spectrographic and activation analysis are used to identify trace elements in native copper ores and artifacts based on data from sources throughout the U. S. and on several artifacts from Copena sites. The data were grouped using the statistical techniques of principal components and cluster analysis. Spectrographic data were grouped into discrete geographic clusters while the activation data formed one undifferentiated cluster. The spectrographic analysis suggested that the copper used to fabricate the Copena artifacts was from Great Lakes deposits, and the mechanisms of exchange responsible for the movement and dispersal of these artifacts from their source to Copena is described. Future applications of activation analysis to native copper may result in more discrete cluster groups and the differentiation of specific copper quarries.

The use of semi-quantitative and quantitative trace element analyses to identify copper artifacts and their geologic sources has been used with varying degrees of success by archaeologists and by chemical and physical scientists. The techniques used include x-ray fluorescence, optical emission spectroscopy, and, more recently, mass spectrometry and neutron activation analysis (1–6). These investigations have attempted to survey the trace element impurities of copper ores and to associate these elements with the impurities of copper artifacts. The present study uses two techniques—optical emission spectroscopy and neutron activation analysis—for the trace element analysis of native copper ores and artifacts.

The native copper ores analyzed are from provenienced deposits in the eastern and southwestern United States. The artifacts analyzed are

0-8412-0397-0/78/33-171-**335**$05.00/1

from Middle Woodland Copena (A.D. 200–500) mound sites in the Tennessee River valley of northern Alabama. These artifacts were manufactured from thin sheets of cold hammered native copper and include reels (x-shaped breastplates), single rivet earspools, and tubular copper beads.

Sample ores were surface collected from copper deposits throughout the eastern U.S. These samples represent small copper nuggets weighing approximately 200 mg and contained no matrix material or crystalline impurities. Artifact samples were removed from the sides and back of the artifact. In artifacts manufactured from multiple sheets, portions of each sheet were included in the sample.

Optical Emission Spectroscopy

Twenty native copper ores and 24 Copena artifacts were used in this analysis. The specific technique described uses a combination of optical and computer microphotometer analyses which gives a higher degree of accuracy and sensitivity than that obtained using standard optical procedures (7).

Table I.   Optical Emission Spectrographic Techniques (8)

Electrodes: cathode, ASTM type C-6, 2 in. long; anode, 1/4 in. diam. thin-walled graphite (Ultra no. 3170)

Electrode charge: 15 mg sample + 15 mg GeO + 30 mg graphite (type UCP-2/200 mesh)

Spectrograph: 3.4 m Ebert design (Mark III)

Power source: 325 V open circuit d-c arc, resistance controlled

Excitation: 15 amp d-c arc set with empty graphite electrodes

Arc gap: 4 mm, maintained throughout burn

Exposure: 20 sec at 5 amp followed by 130 sec at 15 amp, continuous burn

Atmosphere: 70% argon + 30% oxygen; 6.6 L/min flow rate with top of Helz jet nozzle 2 mm below top of electrode

Wavelength range: 2300–4700 A; first order

Grating: 600 groove/mm; 5 A/mm reciprocal linear dispersion

Slit: 25 m wide × 2 mm high

Filter: 47% transmission neutral-density filter

Illumination: arc image focused on collimator by 450 mm focal-length lens at slit.

Mask of collimator: 1.8 cm

Emulsion: Kodak III-0 (102- by 508-mm plates)

Processing: Kodak D-19 developed, 31/4 min at 20°C; Kodak indicator stop bath, 30 sec; Kodak fixer, 8 min; wash, 20–30 min at 20°C; and dry with warm air for 20 min.

Sample Preparation. Approximately 200 mg of copper were cut from each sample artifact and ore using sterilized scissors. Portions of the sample were later removed for analysis with a carbide-tipped drill. The samples were cleaned in a reagent-grade $0.5N$ HCl bath ($1N$) and rinsed with distilled water. The samples were then dissolved in HNO_3 (reagent grade, mixed with triple distilled water to form a $1N$ solution). Germanium dioxide and graphite were added in the ratio of 1:1:2 (1 part sample: 1 part GeO: 2 parts carbon). This solution was evaporated to dryness and powdered to insure homogeneity. Forty-five mg of this powder was loaded into a carbon electrode and arced. An argon–oxygen atmosphere was introduced during arcing to clear the cyanogen-band region, making many additional analytical lines available for analysis and helping to obtain a set of constant conditions by controlling burn temperature. Arcing and plate developing specifications are listed in Table I. Cadmium reference lines (2748.58 A and 4415.704 A) were superimposed before each burn for plate calibration and line finding (8).

Analysis. After development the plate is placed on a Moore no. 3 measuring machine (438 mm) and is scanned by a microphotometer (9). As each sample is scanned, the trace element spectra are recorded on magnetic tape, along with the wavelength intensity of each line, the background readings, the concentration determination, and the spectral interference allowance. This analysis determines 68 elements with a maximum of 500 spectral lines. The reduction of the cyanogen bands and the computerized scanning enable the recording of very weak spectral lines at predetermined wavelengths, greatly increasing the sensitivity and precision of this technique (Table II).

After the plate is scanned and the transmittance data are recorded on a magnetic computer tape, the tape is evaluated using an IBM 360 computer. The transmittances are converted to intensities, backgrounds are subtracted, interferences are recognized, and one final answer is chosen from several preliminary answers for each element. Concentrations are obtained from analytical curves prepared prior to sample run and stored in computer memory.

Only the wavelength intensities were used in this analysis since the concentrations generated by the program are semiquantitative in that the concentration is an average of the spectra scanned for each element. Copper standards were prepared, arced, and scanned along with the copper samples. Calibration curves were drawn for each analytical standard. The wavelength intensities of the sample spectra were compared with these curves and the quantitative trace element composition derived.

Neutron Activation Analysis

This technique was selected for its precision and sensitivity and because it is nondestructive. Twenty copper ores and 20 artifacts identi-

Table II. Approximate Visual Lower Limits of Determination for the Elements Analyzed by the Optical Emission Spectrographic Technique[a]

Element	Analytical Line	Concentration Limit (%)
Ag	3382.891	0.000010
As	2860.452	0.0100
Bi	2897.975	0.0010
Co	3260.818	0.000464
Cr	3421.211	0.0001
Fe	2941.342	0.001
Mg	2779.834	0.00215
Mn	3070.266	0.0001
Ni	2943.914	0.0005
P	2554.930	0.0464
Pb	2614.178	0.0010
Sb	2877.915	0.00681
Ti	2641.099	0.0002
V	3198.012	0.000147
Zn	3302.578	0.00147

[a] Some combinations of elements affect the limits of determination. In favorable materials, values lower than above may be detected; in unfavorable materials these limits of determination may not be attained.

cal to those ores and to artifacts used in the spectrographic analysis were analyzed.

Sample Preparation. Copper samples weighing ~ 100 mg were cut from larger samples using a carbide-tipped drill. These samples were cleaned in reagent-grade HCl (1N) and rinsed in triple distilled water to remove surface contaminants. Sample material was placed in polyethylene vials and thermally sealed. Four USGS standards—numbers AGV-1, BRC-1, DTS-1 and W-1—were weighed and placed in similar vials to be used as absolute standards.

Irradiation. Samples were irradiated for 7 hr in the Georgia Institute of Technology research reactor at a flux of approximately 10^{13} neutrons/cm^2/sec. Samples were cooled for 7 days to allow the copper (^{64}Cu) to decay from the samples. Prior to gamma ray analysis, samples were transferred to new polyethylene vials to eliminate any analytical error from trace metal contaminants in the original irradiated polyethylene vials.

Analysis. The analytical system used for gamma-ray measurements consisted of a lithium drifted germanium (GeLi) crystal detector, a 4096 multi-channel analyzer, a PDP 11 computer, and a cassette magnetic tape storage. The germanium detector crystal has a volume of 55 cm^3 with FWHM resolution of 2.3 keV at 1.33 MeV. The computer was used to analyze the gamma ray spectra, to identify the radio isotopes, and to calculate the concentration (Table III).

Results and Discussion

Fifteen trace elements were identified by the spectrographic analysis. Of these, five elements (Mg, Ni, Co, Bi, and Fe) vary similarly from sample to sample. These elements were used as variables in the subsequent statistical analysis.

Principal Components Analysis. Several multivariate statistical techniques were used to group the optical emission data. The first technique used, principal components analysis, is a data transformation technique in which each of the variables measured is a factor of variability. During principal components anlysis the data are transformed to describe the total variance with the same number of factors as variables such that the first factor accounts for as much of the total variance as possible, the second factor accounts for as much remaining variance as possible, while being uncorrelated with the first, etc. (*11*). Normally the first two or three factors account for most of the total variance, with the remaining variance being accounted for by a number of small factors which are generally discounted from further consideration.

The program used in this study was a modified version of Dixon's BMD08M factor analysis with varimax rotation (*12*). The principal components analysis was conducted using covariance matrices. Five factors were created from the data set. Examination of the individual proportion of the total variance contributed by each of the factors demonstrated that 96.3% of the total variance could be accounted for by the first three factors. These three factors were used in the following cluster analysis.

Cluster Analysis. The general technique used for grouping ore and artifact samples is cluster analysis. The computational method used was Ward's method (*13*).

Table III. Detection Limits for Neutron Activation Analysis

Energy	Element	LLD^a at 95% (ppm except as noted)
103	Sm	0.01
122	Eu	0.014
136	Se	0.22
160	Ca (Sc-47)	660
279	Hg	0.46
312	Th (Pa-233)	0.13
320	Cr	1.2
412	Au	0.078 ppb
559	As	0.015
603	Sb	0.11
658	Ag	0.059
937	Ag	0.12
1099	Fe	69
1120	Sc	0.008
1173	Co	0.14
1597	La	0.027

^{a} As defined in Ref. *10*.

All cluster analyses attempt to group variables into clusters so that the elements within a cluster have a high degree of natural association among themselves, although at the same time the clusters are relatively distinct from one another (*14*). Ward's method is an hierarchical agglomerative clustering technique in which clustering proceeds by progressive fusion beginning with the individual case (a copper sample) and ending with the total population. This error sum-of-squares objective function attempts to find at each stage of fusion those two cases whose merger gives the minimum increase in the total within group error sum of squares (*15*).

The results of this cluster analysis is shown in Figure 1. Three broad clusters are evident from this figure: Cluster I, the southwestern ores and those from Virginia; Cluster II, the southeastern ore sources; and Cluster III, ore sources from Michigan and Wisconsin and the Copena copper artifacts.

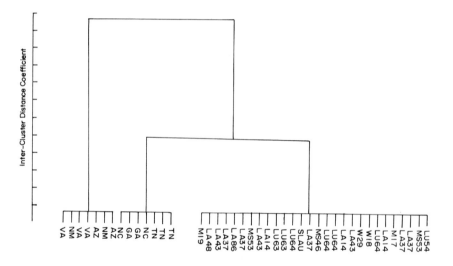

Figure 1.   Dendrogram of complete-linkages copper sources and artifacts

Fifteen trace elements were identified by neutron activation analysis: Hg, Ag, Sm, Eu, Se, Ca, Th, Cr, Au, As, Sb, Fe, Sc, Co, and La. A problem encountered in this analysis is the interference in gamma-ray spectra caused by the high silver content of the native ore samples. Because of this effect, some elements in some samples may have gone undetected, and other elements may have been analyzed only semi-quantitatively.

A principal components factor analysis was run using these 15 elements. Fifteen factors were created with seven factors accounting for 93.4% of the total variance. These factors were used as input data to a

Ward's method cluster analysis. The resultant dendrogram grouped all data into one undifferentiated cluster.

Subsequent principal components and cluster analyses were performed eliminating variables appearing in only one or two cases or using only those variables such as silver, iron, or mercury which seemed to co-vary consistently between cases. Regardless of these transformations, discrete clusters were not produced.

Chemical Implications of the Results. The results of this analysis indicate the usefulness of optical emission spectroscopy as a technique for analyzing native copper. This technique produced ore and artifact groupings consistent with geological and archaeological expectations. However, Rapp, using a larger number of samples, failed to obtain satisfactory ore groupings using optical emission spectroscopy and has since used activation analysis in his studies of native copper (*16*).

Although the neutron activation analysis produced indefinitive results, we suggest, based on our preliminary analysis, that several factors must be addressed in depth before the success or failure of this technique can be assessed in relation to that of optical emission spectroscopy.

First, the method of sample preparation must be examined and a method of sampling and sample preparation developed. Second, copper standards for use in neutron activation analysis need to be prepared and tested. This will eliminate many of the problems encountered using USGS or other standards. Third, the relationship between certain elements and certain deposits or regions needs to be clarified. This can be accomplished by analyzing a large number of samples from each deposit and by increasing overall sample size. Initial investigations of Great Lakes copper sources by Rapp (*16*) indicate that this approach may be successful. Finally, the question of the masking effect of silver needs to be resolved either by developing computer and statistical techniques that take this effect into account, by the use of stripping techniques, or by the computer analysis of complex spectra.

Archaeological Implications of the Results. Past analyses of copper artifacts have used either unprovenienced artifacts or artifacts representing divergent geographical and temporal spans. Although the analysis of copper ores has been more systematic, little attention, outside of physical sampling, has been given to the selection of artifacts to be analyzed. Many analyses have been rendered less valuable as a result of this lack of foresight in artifact selection.

The use of chemical and physical analyses in archaeology has come about in part as a response to the study of the exchange and economy of prehistoric societies. The primary result of these analyses was the establishment of probable source areas. In applying the results of such an analysis to archaeological problem solving, the establishment of source areas must be considered only a beginning. Problem formulation must go beyond the simple question of "where does this artifact come from?" Entire sets of artifacts from specifically defined groups or localities are necessary if internal or external exchange and economic mechanisms are to be delineated.

Copena Exchange. The archaeological intent of this study was not only to identify general native copper sources but to delineate the mechanisms of exchange by which copper ores or artifacts were brought into and distributed within the Copena area. The implicit hypothesis was the verification or rejection of a northern source area for copper artifacts in one section of the southeastern U.S. The results of the optical emission spectrographic analysis (Figure 1) indicated that Copena copper artifacts were manufactured from ores quarried in the Great Lakes area.

Briefly, once the probable source areas are established, exchange routes by which the materials moved can be formulated, and the possible inter- and intra-regional exchange mechanisms can be established. By tabulating site size and the number of artifacts per site, both exotic and local, it was possible to identify two large transactional centers that may have acted as regional exchange centers for the Copena area (Figure 2). These two centers, the Wright and Roden Mounds at the eastern and western periphery of the Copena area contained large quantities of copper and other exotic goods, as well as large numbers of local materials, such as greenstone (Table IV).

It is postulated that these centers acted as pooling centers for goods entering the Copena area and that they redistributed a portion of these goods to sites within the area. In addition to the two large transaction centers, four other sites are designated as local or intermediate transac-

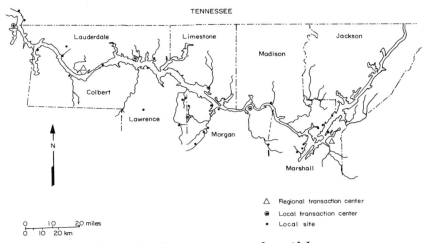

Figure 2.   Copena sites in northern Alabama

Table IV. Copper Artifacts Used in This Analysis

Sample No.	Site No.	Artifact Type	Museum No.[a]
1	Lu54	bead	A16
2	Lu63	reel	A36
3	La14	earspool	A51
4	La14	reel	A37
5	Lu64	earspool	29-1A
6	La37	reel	A36
7	La43	earspool	A31
8	Lu64	reel	A35
9	Slau	earspool	17/233H
10	La43	reel	A13
11	La37	bead	A21
12	Ms53	bead	17/3115H
13	Lu86	bead	A55
14	Ms53	earspool	17/223H
15	La37	bead	A24
16	La43	bead	A20
17	La14	reel	A13
18	Lu64	earspool	A12
19	La37	reel	A29
20	La48	bead	A52
21	La37	bead	A23
22	Lu64	earspool	A22
23	Lu63	earspool	A2
24	Ms46	earspool	17/311H

[a] Artifacts with museum numbers A are from the collection of the University of Alabama and the Moundville State Park. Artifacts with museum numbers H are from the collection of the Heye Museum of the American Indian.

tional centers using the criteria listed above. Evenly spaced approximately every 29 miles along the Tennessee River, these sites received a portion of goods from the regional transaction centers, distributing part of these items to small local sites in their vicinity. In a similar manner locally manufactured goods were collected by local centers and were sent to the regional centers for exchange outside of the Copena area.

The differential distribution of copper artifacts within Copena suggests that copper artifacts were manufactured or received as finished goods by one of the two regional transaction centers depending on the artifact's style. The distribution of copper beads, reels, and earspools is shown in Figure 3. From this graph the distribution of earspools shows a west→east distributional pattern. Large numbers of earspools are found at the western regional transaction center, Wright, with a decreasing frequency in eastern sites. In a similar manner, reels and beads reflect an east→west distribution from the Roden regional transaction center decreasing in frequency in western sites.

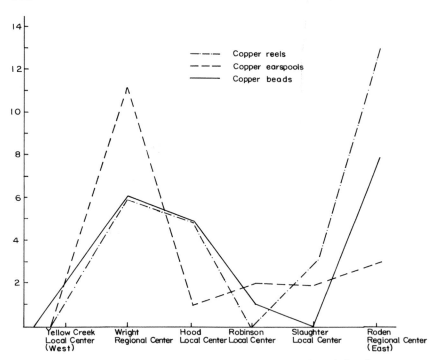

Figure 3. Distribution of copper artifacts at regional and local transaction centers

Two historic Indian trails, the Natchez Trace and the Great South crossed the Tennessee River near the two regional centers (Figure 4). The Wright mounds are at the point where the Natchez Trace branched southward to Mississippi, and the Roden mounds are near the point where the Great South turned southeasterly to Georgia and Florida. It is suggested that Copena copper artifacts or ores moved southward from the Great Lakes along or near these two historic trails and entered the Copena area at the two regional transaction centers.

Further validation of this hypothesis is suggested by the cluster analysis (Figure 1). This analysis suggests a relationship between artifact style and source area, i.e., beads and reels from Michigan sources. Further analysis is required to establish or reject this relationship.

Summary

This chapter has presented the preliminary analysis of native copper ores and artifacts from the eastern U.S. Although the number of samples analyzed to date is too small to obtain a statistically significant distribution of the trace element impurities of native copper ores, the results of the spectrographic and neutron activation analyses indicate that it is possible to differentiate native copper source areas and to assign artifacts to these broad regional areas. Before the technique of neutron activation analysis can be evaluated fully for the analysis of native copper, four areas need to be addressed: sample preparation; preparation of copper standards; identification of regionally specific trace elements; and masking effect of silver.

The number of artifacts analyzed, although small, is a representative sample of copper from a specific regional and temporal prehistoric population. By limiting the areal distribution of the artifacts chosen for analy-

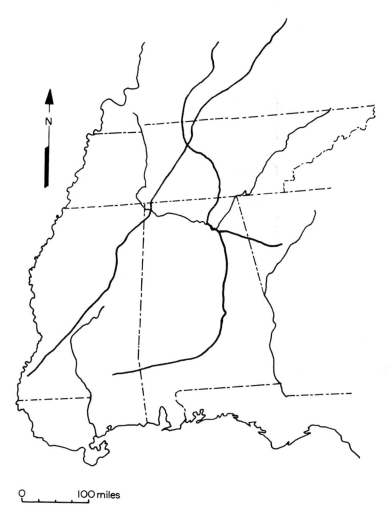

N

0 _____ 100 miles

Indian Trails of the Southeast

Figure 4. Historic Indian trails (17)

sis, the data furnished by the chemical analysis are useful not only in identifying probable source areas but also in partially validating archaeological hypotheses concerning the mechanisms of both inter- and intraregional exchange during the Middle Woodland period in northern Alabama.

Acknowledgments

We wish to thank Jim Spaulding for his assistance in all phases of the activation analysis, and Chung Ho Lee, who prepared the figures. The Smithsonian Institution, the Moundville State Museum, and the Heye Foundation provided samples for this analysis. The optical emission spectrographic analysis was conducted at the USGS spectrographic labs, Reston, VA. Funds for this research were provided by Goad's National Science Foundation Grant No. BNS76-23059.

Literature Cited

1. Balmuth, M. S., Tylecote, R. F., *J. Field Archaeol.* (1976) 3, 195.
2. Bastian, T., "Lake Superior Copper and the Indians," J. B. Griffin, Ed., *Anthropological Papers*, No. 17, p. 150, Museum of Anthropology, University of Michigan, 1961.
3. Fields, P. R. et al., "Science and Archaeology," R. H. Brill, Ed., Chap. 9, M.I.T., Cambridge, 1971.
4. Freeman, A. M. et al., *Science* (1966) 152, 1504.
5. Gilmore, G. R., *Modern Trends* (1976) 2, 1187.
6. Olsen, E. J., *American Antiquity* (1962) 28, 234.
7. Helz, A. W., *J. Res. U.S. Geol. Surv.* (1973) 1(4), 475.
8. Dorrzapf, A. F., Jr., *J. Res. U.S. Geol. Surv.* (1973) 1(5), 559.
9. Walthall, F. G., *J. Res. U.S. Geol. Surv.* (1974) 2(1), 61.
10. Pasternack, B. S., Harley, N. H., "Detection Limits for Radionuclides in the Analysis of Multi-Component Gamma Ray Spectrometer Data," *Nucl. Instrum. Methods* (1971) 91, 533.
11. Daultrey, S., *Concepts and Techniques in Modern Geography No. 8* (1972) 3.
12. Dixon, W. J., *Biomedical Computer Programs* (1973), 255.
13. Anderberg, M. R., "Cluster Analysis for Application," Academic, New York, 1973.
14. Ibid., 1973, xi.
15. Ibid., 1973, 142–145.
16. Rapp, G., personal communication.
17. Meyer, W. E., "Indian Trails of the Southeast," Bureau of American Ethnology, Smithsonian Institution, 42nd Annual Report, 1928.

RECEIVED October 6, 1977.

Chemical Compositions of Copper-Based Roman Coins. Augustan Quadrantes, ca. 9–4 B.C.

GILES F. CARTER

Department of Chemistry, Eastern Michigan University, Ypsilanti, MI 48197

Concentrations of the trace elements, Fe, Ni, Ag, Sn, Sb, and Pb, in 245 Augustan quadrantes, which were analyzed by x-ray fluorescence, correlate with the year of manufacture; the range of composition varies for different issues struck within a given year. Trends in weight, size, and thickness as well as in concentration of Fe, Ag, Sb, and Sn, suggest that the ca.-8 B.C. quadrantes are older than the ca.-9 B.C. quadrantes. Four out of eight pairs of coins produced from the same pair of obverse and reverse dies have the same composition within experimental error, whereas only four out of 33 pairs of single die-linked coins (having either an obverse or reverse die link) had essentially the same compositions.

The existence of literally millions of ancient coins enables us to study much that has not been recorded in man's history. However, numismatic investigations must be coupled with scientific studies to answer many of the questions raised by the existence of these coins. Previous analyses of Roman coins have given much information:

(1) Debasement of Roman silver and brass coinage occurred gradually sometimes and rapidly at other times (*1, 2*);

(2) The coins produced in the major mints of Rome and Lugdunum (Lyons) were essentially pure copper or brass and were quite different from coins which were made in less important mints and which contained relatively high concentrations of impurities such as tin, lead, and zinc;

(3) Antioch was a major mint of Augustus and Tiberius, as indicated by the fact that the copper-based coins were made of pure bronze (copper–tin alloy);

0-8412-0397-0/78/33-176-**347**$07.50/1
© 1978 American Chemical Society

(4) Roman copper coins may frequently be dated rather exactly from their compositions;

(5) Unique sources of ore were used for coins in 23 B.C. and in A.D. 22–30 when Roman coins contained comparatively high concentrations of nickel (3);

(6) The trace metal contents of so-called "imitation" coins of Caligula and Claudius, some of which may have been minted in France and England for the Roman Army, differ significantly from those of coins minted in Rome (4);

(7) Brass coins of Nero minted in Rome appear to have a significantly lower content of zinc than those minted in Lugdunum at the same time (4, 5).

Much other information has been obtained by analyzing Roman coins (1–8).

This chapter covers the analyses of 180 Augustan quadrantes which have not previously been reported. Analyses of 65 quadrantes published elsewhere (7, 9) are included for completeness and because changes have been made in some of the compositions as a result of recalculation based on new standards.

Augustan quadrantes are small, essentially pure copper coins possibly minted in the four years, 9, 8, 5, and 4 B.C. (10) (Figure 1). However, the exact years during which Augustan quadrantes were produced remain open to question. The quadrantes analyzed in this work were studied in detail by C. E. King (11). Since quadrantes were not struck prior to 9 B.C. and since they were not made again until about 40 years later during the reign of Caligula, they are a very compact and homogeneous group of coins. Quadrantes were likely made primarily for votives and small change. Most of the coins in this study were dredged from the Tiber River near Rome; presumably many of them were tossed into the river to appease a god or to make a wish. In another well known instance, thousands of Roman coins were found near the remains of a Roman bridge, indicating that they were deliberately thrown into the river (12).

The objectives of this work are to present the compositions of 245 Roman copper coins minted from about 9 to 4 B.C.; to determine whether chemical compositions may be correlated closely with the date of manufacture (year); to determine whether different issues of coins within a given year have the same range of compositions, i.e., whether the issues were struck concurrently, separately, or only partly concurrently; and to determine whether coins having either obverse or reverse die links (pairs of coins struck from the same die) have the same composition within experimental error.

Experimental Method

The analytical method of x-ray fluorescence has been described in several texts (13, 14, 15). Since the fluorescent x-rays are usually not

Figure 1a. Augustan quadrans (2.5×). R-1021 "5 B.C.";
obverse (left), reverse (right).

Figure 1b. Augustan quadrans (2.5×). R-1011 "5 B.C.";
obverse (left), reverse (right).

Figure 1c. Augustan quadrans (2.5×). R-972 "4 B.C.";
reverse (left), obverse (right).

very energetic, they penetrate only about 5 μ of a metal such as copper.
Therefore, coins must be prepared carefully for analysis by electrolytic
(cathodic) reduction of surface oxides followed by abrasion with chemically pure aluminum oxide powder carried in a stream of compressed air.

This removes about 10 μ of metal from the surface and exposes metal that is representative of the average composition of the coin. Fairly pure copper coins, such as Augustan quadrantes, are essentially homogeneous and can give very good analyses using x-ray fluorescence (XRF).

Fourteen standard copper and brass alloys, the compositions of which have been certified by the National Bureau of Standards, have been used to calculate the concentrations of various elements in the coins (NBS C-1100. 1101, C-1102, 1106, C-1109, C-1111, C-1112, C-1115, 1116, C-1120, 63C, 62D, 157A, and 158A). All standards were prepared metallographically, ending with a diamond polish to obtain a surface representative of the interior of the standard. Excellent calibration curves were obtained with very good precision; for both standards and coins, the calculated standard deviation is about 0.003% for Fe, 0.004% for Ni, 0.005% for Ag, 0.002% for Sn, 0.004% for Sb, and 0.003% for Pb.

During XRF determinations for various elements, the coins were always oriented identically to minimize variations in absorption caused by uneven surfaces (16). Also it was necessary to normalize the concentrations to 100% for each coin. The roughness of the surface causes greater absorption of the fluorescing x-rays compared with polished, flat specimens. Therefore, nearly all determinations for coins are somewhat lower than they should be because of the surface roughness; the rougher the surface, the greater the absorption. The quadrantes in this study had normalization factors of 0.98–1.10, with the average factor being about 1.03. The normalization factor is the number that when multiplied times the concentrations of various elements causes the sum of all concentrations to be $100 \pm 0.1\%$. Further details of the experimental method may be found in Refs. 3 and 17.

Physical Measurements of Coins

Since each Roman coin was hand struck from a hot, cast blank, the weight, size, and shape vary from coin to coin. In order to identify uniquely a specific coin it is necessary to give its catalog number, weight, maximum thickness, and size (maximum and minimum diameters). It is also useful to record other physical data such as the relative orientation of the obverse and reverse dies and the density. Unless a coin subsequently is physically damaged, the above measurements are sufficient to identify uniquely each ancient coin. It is extremely improbable that two coins having the same catalog number would have the same physical measurements within experimental error. Identification of each coin undergoing analysis is necessary if present analyses are to be useful in the future because one must be able to avoid repeating the analysis of a given coin unknowingly.

Table I summarizes physical measurements for the Augustan quadrantes. The average weight of quadrantes made in 9 and 8 B.C. is about 2.3 g. However, these coins have been worn in circulation, corroded to varying degrees, and cleaned; hence their weights are probably 0.2–0.5 g lower than when they were struck. The average weights of quadrantes made in 5 and 4 B.C. are slightly higher, about 2.5 g. It is surprising that the average weights actually increase with elapsed time since the reverse is the normal pattern for coinage. The average weights are 2.30 g for the "8 B.C." quadrantes (now believed to be the oldest, *see* the next paragraph), 2.32 g for the "9-B.C." coins, 2.37 g for the "5 B.C." group, and 2.47 g for the "4 B.C." quadrantes.

The 8 B.C. quadrantes are significantly thinner and larger than the 9 B.C. coins (Table II). The 5 B.C. coins are thicker than the 9 B.C. coins, and the 4 B.C. quadrantes are the thickest of all: average maximum thicknesses are 0.181 cm for "8 B.C," 0.201 cm for "9 B.C.," 0.204 for "5 B.C.," and 0.232 cm for "4 B.C." The dates tentatively assigned to these coins by Mattingly (*10*) are likely incorrect. Several factors, to be discussed later, indicate that the "8 B.C." coins in reality were likely minted before the "9 B.C." coins. If this is true, then the coins continually become thicker, heavier, and smaller from the beginning to the end of the series: 9 B.C. coins (Table II). The 5 B.C. coins are thicker than the 9 B.C. 1.61 cm for "5 B.C.," and 1.52 cm for "4 B.C." coins. Almost all of the 8 B.C. coins are thinner and larger than almost all of the 4 B.C. coins, particularly the Catullus issue, BMC 270 (British Museum Catalog No. 270; *see* Reference *10*). The 4 B.C. coins ars also more elliptical than the other coins; the average ratio of maximum diameter/minimum diameter for "4 B.C." is 1.09, compared with 1.08 for "5 B.C.," 1.06 for "9 B.C.," and 1.07 for "8 B.C."

The densities of the coins are very close to the density of pure copper, 8.92 g/cm^3. These measurements show that the coins are nonporous, and even the coins that apparently have lower densities may not be porous. Occasionally a coin that has a rough surface traps air at the surface, causing the weight of the coin in water to be less than it really is and thus decreasing the apparent density. The results in Table I also show that the densities of coins may be measured with high precision; the average density is 8.92 ± 0.03 g/cm^3 (standard deviation = 0.03 g).

Since Roman coins were struck by hand using two separate dies, the relative orientation of the obverse (heads) and reverse (tails) dies is variable. If the obverse is normally oriented with the bust facing left or right, the obverse is oriented at "12:00" (the major vertical axis points in the direction of the hour hand at 12:00). If the coin is rotated 180° around the 12:00 axis, the major vertical axis of the reverse may now point in any direction, and the relative orientation is given as the direc-

Table I. Physical Measurements of Augustan Quadrantes

Coin No.	Wt. (g)	Max. Thickness (cm)	Max. Diam. (cm)	Min. Diam. (cm)	Max. Diam./Min. Diam.	Dens. (g/cm³)	Die Orientation	BMC No.
			PTR Quadrantes, "8 B.C."					
R-794	2.0953	0.182	1.662	1.575	1.055	8.90	7:00	204
R-817	2.0349	0.162	1.763	1.695	1.040	8.93	12:30	204
R-728	2.6935	0.161	1.78	1.70	1.05	8.98	—	205
R-792	2.4113	0.167	1.661	1.599	1.039	8.92	7:00	205
R-813	2.1722	0.166	1.760	1.665	1.057	8.94	1:00	205
R-999	2.5087	0.191	1.751	1.683	1.040	8.92	7:00	205
R-793	2.2309	0.180	1.682	1.534	1.096	8.92	10:00	207
R-809	1.7926	0.183	1.745	1.505	1.159	8.93	6:00	207
R-821	2.5929	0.199	1.732	1.590	1.089	8.92	4:00	207
R-995	2.4037	0.187	1.701	1.624	1.047	8.92	7:00	207
R-996	2.4790	0.225	1.640	1.512	1.085	8.94	1:00	207
R-997	1.7595	0.139	1.867	1.688	1.106	8.98	3:30	207
R-998	2.7213	0.217	1.589	1.539	1.032	8.92	4:30	207
			LSA Quadrantes, "9 B.C."					
R-797	2.3104	0.199	1.669	1.545	1.080	8.92	10:30	200
R-812	2.8982	0.178	1.791	1.712	1.046	8.95	4:00	200
R-824	2.0295	0.163	1.655	1.624	1.019	8.89	3:30	200
R-833	2.0634	0.217	1.671	1.613	1.036	8.92	7:00	200
R-842	2.4046	0.211	1.598	1.533	1.042	8.91	7:30	200
R-851	2.2167	0.218	1.48	1.41	1.05	8.98	—	200
R-857	2.2794	0.167	1.688	1.655	1.020	8.90	10:30	200
R-863	2.4889	0.208	1.585	1.526	1.039	8.93	4:00	200
R-869	2.1957	0.201	1.583	1.521	1.041	8.91	10:00	200
R-875	2.8495	0.253	1.568	1.438	1.090	8.91	5:00	200
R-893	2.4049	0.177	1.772	1.648	1.075	8.93	3:00	200
R-897	2.3536	0.184	1.703	1.600	1.064	8.92	4:30	200
R-901	2.2728	0.207	1.533	1.494	1.026	8.90	1:00	200
R-905	2.0307	0.181	1.541	1.494	1.031	8.94	11:00	200
R-909	2.5827	0.199	1.600	1.559	1.026	8.92	1:00	200
R-913	1.9130	0.192	1.564	1.509	1.036	9.07	5:00	200
R-917	2.7210	0.223	1.726	1.617	1.067	8.95	2:00	200
R-921	2.5421	0.197	1.707	1.665	1.025	8.95	4:00	200
R-925	2.1655	0.177	1.639	1.540	1.064	8.94	1:30	200
R-928	2.2722	0.208	1.584	1.524	1.039	8.94	7:00	200
R-930	2.2685	0.181	1.707	1.612	1.059	8.92	10:00	200
R-932	1.9796	0.178	1.581	1.522	1.039	8.91	6:30	200
R-934	2.3939	0.208	1.646	1.571	1.048	8.93	1:00	200
R-936	2.0129	0.197	1.509	1.414	1.067	8.92	1:30	200
R-938	2.2357	0.207	1.562	1.465	1.066	8.91	10:30	200
R-940	2.4932	0.231	1.506	1.465	1.028	8.92	4:30	200
R-942	2.2091	0.224	1.513	1.364	1.109	8.92	4:00	200
R-944	2.5506	0.223	1.642	1.569	1.047	8.92	10:30	200

Table I. Continued

Coin No.	Wt. (g)	Max. Thickness (cm)	Max. Diam. (cm)	Min. Diam. (cm)	Max. Diam. / Min. Diam.	Dens. (g/cm³)	Die Orientation	BMC No.
R-946	2.5443	0.222	1.668	1.395	1.196	8.94	10:30	200
R-948	1.8344	0.181	1.595	1.542	1.034	8.93	10:00	200
R-804	2.4428	0.180	1.679	1.608	1.044	8.86	1:00	201
R-808	1.8300	0.170	1.57	1.44	1.09	9.01	—	201
R-820	2.7310	0.245	1.549	1.455	1.065	8.94	5:00	201
R-830	1.6530	0.135	1.561	1.510	1.034	8.91	1:30	201
R-839	2.1753	0.183	1.652	1.546	1.069	8.89	9:00	201
R-848	2.1531	0.195	1.552	1.490	1.042	8.91	1:00	201
R-887	1.7474	0.164	1.621	1.544	1.050	8.90	1:00	201
R-894	2.8515	0.263	1.530	1.489	1.028	8.93	7:00	201
R-898	2.1084	0.188	1.685	1.555	1.084	8.94	10:00	201
R-902	2.7275	0.237	1.625	1.521	1.068	8.94	2:00	201
R-906	1.9645	0.196	1.754	1.640	1.070	8.96	1:30	201
R-910	2.0368	0.179	1.644	1.556	1.057	8.95	1:00	201
R-914	2.0730	0.171	1.587	1.504	1.055	8.94	7:00	201
R-918	2.4503	0.203	1.652	1.595	1.036	8.94	1:00	201
R-922	2.6183	0.199	1.704	1.594	1.069	8.90	1:00	201
R-926	3.0724	0.226	1.649	1.584	1.041	8.91	7:00	201
R-929	2.4879	0.211	1.622	1.565	1.036	8.93	5:00	201
R-931	3.0155	0.249	1.647	1.536	1.072	8.91	1:00	201
R-933	2.3730	0.204	1.779	1.534	1.160	8.90	1:00	201
R-935	1.6234	0.179	1.622	1.499	1.082	8.89	10:30	201
R-937	2.0795	0.168	1.659	1.529	1.085	8.92	4:30	201
R-939	2.6274	0.204	1.676	1.578	1.062	8.91	1:30	201
R-941	2.5719	0.224	1.633	1.573	1.038	8.94	7:00	201
R-943	2.5716	0.224	1.610	1.549	1.039	8.93	4:00	201
R-945	2.4800	0.165	1.788	1.618	1.105	8.93	7:30	201
R-947	1.7920	0.178	1.562	1.513	1.032	8.93	12:30	201
R-949	2.2593	0.204	1.633	1.591	1.026	8.92	1:00	201
R-950	2.6015	0.209	1.644	1.549	1.061	8.92	7:00	201
R-799	1.4327	0.162	1.572	1.452	1.083	8.92	7:00	202
R-816	1.1718	0.187	1.443	1.323	1.091	9.00	1:30	202
R-827	2.5473	0.230	1.575	1.492	1.056	8.91	7:00	202
R-836	1.7981	0.193	1.512	1.458	1.037	8.88	1:30	202
R-845	2.0688	0.216	1.525	1.473	1.035	8.91	6:30	202
R-854	2.3910	0.211	1.573	1.478	1.064	8.93	7:00	202
R-860	2.0814	0.214	1.612	1.463	1.102	8.91	12:30	202
R-866	1.9175	0.217	1.520	1.422	1.069	8.93	11:00	202
R-872	2.3366	0.212	1.761	1.562	1.127	8.90	8:00	202
R-878	1.7131	0.209	1.550	1.419	1.092	8.92	2:00	202
R-881	2.2287	0.219	1.683	1.569	1.073	8.88	7:00	202
R-882	2.2976	0.210	1.557	1.464	1.064	8.92	1:00	202
R-884	2.9513	0.231	1.629	1.580	1.031	8.92	7:00	202
R-895	2.5755	0.226	1.713	1.455	1.177	8.93	10:30	202
R-899	2.4356	0.229	1.593	1.534	1.038	8.93	10:00	202

Table I. Continued

Coin No.	Wt. (g)	Max. Thickness (cm)	Max. Diam. (cm)	Min. Diam. (cm)	Max. Diam. / Min. Diam.	Dens. (g/cm³)	Die Orientation	BMC No.
R-903	3.0562	0.280	1.552	1.433	1.083	8.91	2:00	202
R-907	1.9930	0.180	1.657	1.558	1.064	8.94	7:00	202
R-911	2.2393	0.214	1.606	1.573	1.021	8.94	10:00	202
R-915	2.7523	0.272	1.500	1.421	1.056	8.93	7:00	202
R-919	2.3348	0.233	1.711	1.569	1.091	8.93	1:00	202
R-923	2.5356	0.196	1.724	1.640	1.051	8.93	7:00	202
R-927	2.8249	0.210	1.730	1.667	1.038	8.92	4:00	202

AGMS Quadrantes, "5 B.C."

Coin No.	Wt. (g)	Max. Thickness (cm)	Max. Diam. (cm)	Min. Diam. (cm)	Max. Diam. / Min. Diam.	Dens. (g/cm³)	Die Orientation	BMC No.
R-873	2.3106	0.206	1.606	1.518	1.058	8.91	6:00	243 var.?
R-798	2.6251	0.217	1.699	1.532	1.109	8.92	7:00	246
R-822	1.6600	0.154	1.599	1.530	1.045	8.86	6:30	247
R-1002	2.6547	0.220	1.579	1.489	1.060	8.94	7:00	247
R-1009	1.5913	0.166	1.535	1.442	1.064	8.94	12:00	250
R-1010	2.2576	0.195	1.686	1.550	1.088	8.94	6:00	250
R-1011	2.6770	0.188	1.786	1.627	1.098	8.93	6:30	250
R-805	2.5744	0.224	1.578	1.416	1.114	8.94	4:00	252
R-828	2.6033	0.197	1.566	1.501	1.043	8.91	10:30	252
R-849	2.1270	0.165	1.557	1.514	1.028	8.91	3:30	252
R-1003	2.2585	0.192	1.790	1.587	1.128	8.94	10:30	252
R-1004	2.8007	0.225	1.633	1.516	1.077	8.92	12:30	252
R-1005	2.1063	0.185	1.576	1.395	1.130	8.93	1:30	252
R-1006	2.6317	0.225	1.527	1.458	1.047	8.92	4:00	252
R-1007	2.0783	0.199	1.564	1.466	1.067	8.94	4:30	252
R-1008	2.4370	0.202	1.612	1.491	1.081	8.95	10:30	252
R-810	2.3324	0.195	1.646	1.546	1.065	8.95	10:30	253
R-814	1.6509	0.197	1.475	1.384	1.066	8.97	11:00	253
R-858	1.4490	0.163	1.415	1.385	1.022	8.92	10:30	253
R-876	2.4001	0.202	1.575	1.473	1.069	8.92	3:00	253
R-1013	2.8863	0.235	1.561	1.481	1.054	8.95	6:30	253
R-1014	1.8552	0.168	1.595	1.475	1.081	8.89	12:30	253
R-1015	2.8584	0.255	1.534	1.444	1.062	8.95	1:00	253
R-1016	2.5583	0.231	1.466	1.394	1.052	8.96	3:30	253
R-806	3.2337	0.276	1.497	1.347	1.111	8.93	—	256
R-861	1.7276	0.178	1.592	1.442	1.104	8.90	9:30	256
R-1012	2.2725	0.184	1.569	1.528	1.027	8.91	9:00	256
R-864	2.4092	0.203	1.745	1.518	1.150	8.86	6:30	258
R-879	1.9268	0.178	1.493	1.448	1.031	8.92	12:30	258
R-1017	2.8877	0.247	1.561	1.459	1.070	8.94	7:00	258
R-1018	2.5248	0.215	1.543	1.464	1.054	8.95	12:30	258
R-1024	1.9546	0.181	1.680	1.462	1.149	8.95	3:00	258 or 264
R-1029	2.1036	0.216	1.655	1.386	1.194	8.96	4:00	258
R-831	2.3170	0.213	1.512	1.450	1.043	8.91	8:00	259
R-834	2.0239	0.148	1.787	1.640	1.090	8.90	12:00	259

Table I. Continued

Coin No.	Wt. (g)	Max. Thick-ness (cm)	Max. Diam. (cm)	Min. Diam. (cm)	Max. $\frac{Diam.}{Min. Diam.}$	Dens. (g/cm³)	Die Ori-enta-tion	BMC No.
R-885	2.9636	0.274	1.653	1.302	1.270	8.91	4:00	259
R-1000	2.4847	0.208	1.610	1.528	1.054	8.90	6:00	259
R-1020	2.3075	0.172	1.755	1.695	1.035	8.95	7:00	259
R-1021	2.9737	0.177	1.790	1.693	1.057	8.94	12:00	259
R-840	2.0704	0.204	1.724	1.445	1.193	8.93	4:00	260
R-843	2.3182	0.192	1.660	1.611	1.030	8.92	8:00	260
R-852	2.0922	0.215	1.597	1.521	1.050	8.90	7:00	260
R-1022	2.5365	0.202	1.602	1.523	1.052	8.93	9:30	260?
R-1025	2.5181	0.202	1.657	1.484	1.117	8.93	12:00	261
R-80	3.1269	0.266	1.573	1.424	1.105	8.89	7:00	261 var.
R-801	2.4986	0.214	1.612	1.436	1.123	8.94	8:00	261 var.
R-870	2.1436	0.147	1.836	1.648	1.114	8.91	11:30	261 var.
R-1026	3.5137	0.302	1.666	1.391	1.198	8.95	6:00	261 var.
R-818	2.3634	0.200	1.618	1.520	1.064	8.83	1:00	262
R-846	1.8900	0.171	1.604	1.527	1.050	8.91	12:30	262
R-855	2.1897	0.181	1.610	1.516	1.062	8.92	12:00	262
R-1027	2.8605	0.262	1.530	1.461	1.047	8.94	4:30	262
R-1028	2.6747	0.188	1.696	1.553	1.092	8.93	12:30	262
R-825	2.6207	0.207	1.520	1.498	1.015	8.91	12:30	264
R-837	2.7508	0.240	1.752	1.448	1.210	8.74	4:30	264
R-1001	2.226	0.222	1.448	1.416	1.023	8.92	—	264
R-1030	2.7413	0.225	1.537	1.501	1.024	9.96	7:00	264
R-1031	3.2909	0.235	1.631	1.577	1.034	8.94	4:30	264
R-867	2.3364	0.207	1.522	1.402	1.086	8.90	5:30	C.351ᵃor261
R-795	1.9259	0.166	1.609	1.550	1.038	8.96	12:30	C.374ᵃ
R-791	2.3109	0.164	1.667	1.620	1.029	8.92	7:00	C.375ᵃ
R-796	1.5959	0.133	1.705	1.630	1.046	8.95	10:00	C.424ᵃ
R-889	1.9733	0.169	1.557	1.441	1.080	8.96	3:30	C.424?ᵃ
R-1019	3.2914	0.319	1.551	1.368	1.134	8.91	1:00	C.535ᵃ
R-892	1.6732	0.189	1.502	1.382	1.087	8.92	9:30	?
R-1023	1.5607	0.167	1.413	1.371	1.031	8.95	10:30	—

BBCC Quadrantes, "4 B.C."

Coin No.	Wt. (g)	Max. Thick-ness (cm)	Max. Diam. (cm)	Min. Diam. (cm)	Max. $\frac{Diam.}{Min. Diam.}$	Dens. (g/cm³)	Die Ori-enta-tion	BMC No.
R-748	2.4746	0.206	1.61	1.54	1.05	8.93	—	265
R-802	2.1407	0.221	1.591	1.392	1.143	8.93	1:00	265
R-819	2.5312	0.239	1.491	1.399	1.066	8.94	1:30	265
R-832	1.7277	0.158	1.642	1.519	1.081	8.91	1:30	265
R-844	2.1952	0.245	1.398	1.346	1.039	8.91	7:00	265
R-853	2.3496	0.221	1.483	1.388	1.068	8.91	1:30	265
R-982	3.0038	0.231	1.711	1.565	1.093	8.91	1:00	265
R-983	2.2028	0.211	1.614	1.375	1.173	8.92	7:30	265
R-984	3.4833	0.263	1.603	1.534	1.045	8.80	1:00	265
R-985	2.5085	0.193	1.658	1.438	1.153	8.93	10:00	265
R-986	2.3548	0.191	1.641	1.573	1.043	8.92	7:00	265

Table I.   Continued

Coin No.	Wt. (g)	Max. Thickness (cm)	Max. Diam. (cm)	Min. Diam. (cm)	Max. Diam. / Min. Diam.	Dens. (g/cm³)	Die Orientation	BMC No.
R-987	2.0707	0.192	1.691	1.487	1.137	8.93	12:30	265
R-988	2.2959	0.230	1.444	1.351	1.069	8.94	12:30	265
R-989	2.3269	0.189	1.562	1.477	1.058	8.94	12:30	265
R-990	2.2247	0.187	1.560	1.510	1.033	8.93	6:30	265
R-991	2.0991	0.196	1.574	1.461	1.077	8.91	1:00	265
R-992	2.2951	0.246	1.444	1.397	1.034	8.90	7:00	265
R-993	2.8970	0.339	1.331	1.220	1.091	8.94	10:00	265
R-994	1.4000	0.163	1.473	1.394	1.057	8.93	10:00	265
R-727	3.1543	0.276	1.50	1.44	1.04	8.94	—	267
R-800	2.4266	0.211	1.667	1.400	1.191	8.93	6:00	267
R-811	2.6356	0.245	1.613	1.268	1.272	8.93	7:00	267
R-826	2.8858	0.290	1.418	1.311	1.082	8.92	6:30	267
R-838	2.2160	0.212	1.514	1.424	1.063	8.90	6:00	267
R-967	3.3912	0.332	1.407	1.364	1.032	8.91	9:30	267
R-968	2.7811	0.245	1.514	1.383	1.095	8.91	10:00	267
R-969	2.6576	0.263	1.504	1.245	1.208	8.91	7:00	267
R-970	2.6185	0.281	1.436	1.330	1.080	8.95	1:30	267
R-971	2.5173	0.210	1.682	1.396	1.205	8.94	12:00	267
R-972	3.7389	0.292	1.599	1.499	1.067	8.94	9:30	267
R-973	2.0755	0.221	1.541	1.456	1.058	8.92	8:00	267
R-974	3.4941	0.276	1.520	1.477	1.029	8.94	10:00	267
R-803	2.5164	0.230	1.446	1.405	1.029	8.91	9:30	269
R-807	2.0646	0.237	1.396	1.353	1.032	8.94	1:00	269
R-823	2.4407	0.208	1.543	1.498	1.030	8.90	10:00	269
R-835	2.3053	0.220	1.548	1.465	1.057	8.90	10:00	269
R-847	2.9905	0.255	1.687	1.487	1.134	8.93	3:30	269
R-856	1.1635	0.133	1.757	1.461	1.203	8.96	3:30	269
R-862	2.1847	0.210	1.540	1.477	1.043	8.89	3:30	269
R-868	1.7905	0.184	1.535	1.412	1.087	8.92	1:00	269
R-951	3.2166	0.264	1.633	1.536	1.063	8.94	10:00	269
R-952	2.9225	0.223	1.715	1.471	1.166	8.95	1:00	269
R-953	2.8016	0.242	1.521	1.456	1.045	8.91	10:00	269
R-954	2.9637	0.267	1.503	1.445	1.040	8.92	1:00	269
R-955	2.2325	0.201	1.505	1.417	1.062	8.95	1:00	269
R-956	2.5316	0.246	1.495	1.352	1.106	8.94	7:00	269
R-957	2.5668	0.226	1.583	1.432	1.105	8.87	7:00	269
R-958	2.4985	0.240	1.584	1.508	1.050	8.87	1:00	269
R-959	2.8986	0.269	1.486	1.397	1.064	8.92	6:00	269
R-960	1.5318	0.160	1.603	1.450	1.106	8.95	10:00	269
R-961	2.1095	0.205	1.818	1.433	1.269	8.91	12:30	269
R-962	2.8476	0.260	1.489	1.400	1.064	8.94	12:00	269
R-963	2.6483	0.261	1.503	1.461	1.029	8.92	7:00	269
R-964	3.1131	0.250	1.722	1.425	1.208	8.87	8:00	269
R-965	2.4368	0.218	1.507	1.483	1.016	8.94	12:30	269

Table I. Continued

Coin No.	Wt. (g)	Max. Thick- ness (cm)	Max. Diam. (cm)	Min. Diam. (cm)	Max. Diam. Min. Diam.	Dens. (g/ cm³)	Die Ori- enta- tion	BMC No.
R-966	2.5591	0.215	1.598	1.514	1.055	8.93	11:30	269
R-815	2.6567	0.261	1.474	1.315	1.121	8.92	8:00	270
R-829	2.7689	0.248	1.467	1.330	1.103	8.92	1:00	270
R-841	3.0434	0.262	1.712	1.277	1.341	8.89	1:30	270
R-850	2.4205	0.245	1.530	1.362	1.123	8.92	7:00	270
R-859	2.4896	0.261	1.407	1.351	1.041	8.89	6:30	270
R-865	2.3651	0.208	1.50	1.45	1.03	9.00	—	270
R-871	2.3397	0.264	1.511	1.237	1.222	8.93	1:00	270
R-874	2.2234	0.246	1.355	1.298	1.044	8.91	7:30	270
R-877	2.8595	0.238	1.466	1.433	1.023	8.92	4:00	270
R-880	2.3312	0.224	1.515	1.395	1.086	8.89	5:00	270
R-883	2.5179	0.244	1.399	1.333	1.050	8.92	12:00	270
R-886	2.3941	0.240	1.497	1.319	1.135	8.90	1:00	270
R-888	1.2763	0.165	1.425	1.213	1.175	8.93	5:00	270
R-890	1.9364	0.215	1.400	1.326	1.056	8.93	10:00	270
R-891	0.8853	0.132	1.480	1.373	1.078	9.00	12:30	270
R-896	1.9260	0.189	1.565	1.535	1.020	8.88	12:30	270
R-900	2.5224	0.231	1.697	1.383	1.227	8.93	9:00	270
R-904	2.0942	0.232	1.405	1.367	1.028	8.95	4:00	270
R-908	2.7244	0.295	1.367	1.253	1.091	8.94	2:00	270
R-912	2.7751	0.252	1.435	1.324	1.084	8.93	6:30	270
R-916	1.8616	0.180	1.504	1.443	1.042	8.96	7:30	270
R-920	2.0724	0.221	1.492	1.365	1.093	8.93	1:00	270
R-924	3.0493	0.288	1.456	1.314	1.108	8.92	8:00	270
R-975	2.4182	0.221	1.495	1.370	1.091	8.91	1:00	270
R-976	2.2465	0.234	1.477	1.316	1.122	8.94	7:00	270
R-977	2.5377	0.252	1.493	1.396	1.069	8.94	3:30	270
R-978	2.5509	0.241	1.421	1.324	1.073	8.93	10:00	270
R-979	3.3063	0.263	1.506	1.446	1.041	8.94	7:00	270
R-980	1.7501	0.166	1.584	1.304	1.215	8.97	3:30	270
R-981	2.4233	0.279	1.338	1.262	1.060	8.91	1:30	270

[a] C: Cohen classification number.

tion of the reverse axis in terms of the hour hand at a given time. In Table I the relative orientations for all the coins are given in this manner. Only one orientation is presented for each coin since the obverse die is always oriented at 12:00. If one actually chooses the wrong side as the obverse, the orientation is still the same since the description of the die orientation is independent of which side is chosen as the obverse.

In the Augustan quadrantes there are four highly preferred orientations: approximately 1:00, 4:00, 7:00, and 10:00 (Figure 2). The orientations half way in between, namely 2:30, 5:30 etc., are almost totally

Table II.   Average Weight, Size, Shape, and Density of Augustan Quadrantes

Coin No.	Wt. (g)	Max. Thick- ness (cm)	Max. Diam. (cm)	Min. Diam. (cm)	Max. Diam. Min. Diam.	Dens. (g/ cm³)	Die Ori- enta- tion	BMC No.
PTR Quadrantes, "8 B.C."								
Mean	2.36	0.175	1.724	1.649	1.045	8.927	5:00	205
S^a	0.17	0.014	0.055	0.044	0.010	0.012	3:30	
Mean	2.28	0.190	1.708	1.570	1.088	8.933	5:10	207
S	0.38	0.028	0.088	0.067	0.041	0.022	2:50	
Mean	2.30	0.181	1.718	1.608	1.069	8.932	4:50	all
S	0.32	0.024	0.072	0.073	0.037	0.024	3:00	all
LSA Quadrantes, "9 B.C."								
Mean	2.34	0.200	1.624	1.543	1.054	8.928	5:50	200
S	0.27	0.021	0.077	0.083	0.035	0.031	3:30	200
Mean	2.34	0.199	1.641	1.549	1.060	8.920	4:00	201
S	0.40	0.030	0.065	0.044	0.028	0.022	3:15	201
Mean	2.26	0.216	1.604	1.500	1.070	8.922	5:40	202
S	0.47	0.026	0.087	0.082	0.036	0.024	3:25	202
Mean	2.32	0.201	1.625	1.533	1.060	8.924	5:10	all
S	0.37	0.035	0.076	0.074	0.033	0.026	3:30	all
AGMS Quadrantes, "5 B.C."								
Mean	2.38	0.199	1.603	1.491	1.075	8.928	5:40	252
S	0.28	0.020	0.082	0.055	0.038	0.015	4:10	252
Mean	2.36	0.214	1.529	1.437	1.065	8.938	5:50	253
S	0.61	0.038	0.072	0.063	0.024	0.025	4:25	253
Mean	2.30	0.207	1.613	1.456	1.108	8.930	3:35	258
S	0.38	0.026	0.096	0.042	0.065	0.037	2:50	258
Mean	2.51	0.199	1.684	1.551	1.092	8.918	4:10	259
S	0.38	0.044	0.112	0.156	0.089	0.021	3:30	259
Mean	2.40	0.200	1.612	1.515	1.063	8.906	1:20	262
S	0.39	0.036	0.059	0.034	0.018	0.044	1:45	262
Mean	2.72	0.226	1.578	1.488	1.061	8.994	4:10	264
S	0.38	0.013	0.117	0.061	0.083	0.088	2:40	264
Mean	2.37	0.204	1.606	1.488	1.080	8.923	5:10	all
S	0.46	0.036	0.093	0.083	0.051	0.034	3:35	all
BBCC Quadrantes, "4 B.C."								
Mean	2.35	0.217	1.55	1.44	1.077	8.916	4:30	265
S	0.47	0.043	0.11	0.09	0.041	0.033	3:50	265
Mean	2.82	0.261	1.52	1.38	1.108	8.925	6:50	267
S	0.52	0.038	0.08	0.08	0.081	0.016	3:20	267
Mean	2.53	0.228	1.57	1.45	1.083	8.917	5:20	269
S	0.43	0.030	0.09	0.04	0.064	0.027	4:10	269
Mean	2.47	0.232	1.48	1.35	1.104	8.923	4:35	270
S	0.38	0.037	0.08	0.07	0.074	0.022	3:15	270
Mean	2.47	0.232	1.52	1.40	1.091	8.923	5:00	all
S	0.49	0.038	0.10	0.08	0.066	0.027	3:40	all

a S: standard deviation.

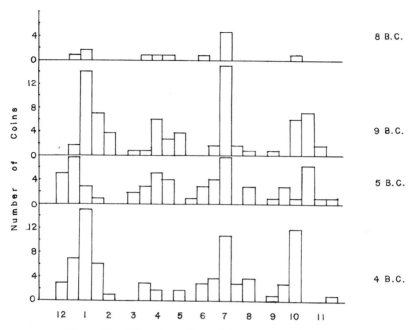

Figure 2.   Die orientations of Augustan quadrantes

absent. Probably the four major orientations result from two factors: from use of a die with an oblong (elliptical) or square shank so that one would hold it, either by hand or with tongs, in one of only two or four orientations and from left-handed and right-handed employees. The first factor would cause an elliptical-shank die to produce relative orientations differing by 180° (90° for a square shank). The second factor would cause a die with an elliptical shank to produce relative orientations differing by about 90° (3 hr) if they were used first by a left-handed person and then by a right-handed employee. This assumes that the fixed, anvil die is clamped in position with a 12:00 orientation. Laing (*18*) comments that 640 coins produced between 975 and 1017 A.D. in England all lie in one of four axes, corresponding with the four sides of the die.

Seventy-nine quadrantes have a die orientation of 1:00 ± 1 hr; 70 have an orientation of 7:00 ± 1 hr, but 39 die orientations are 4:00 ± 1 hr and 46 are 10:00 ± 1 hr. No coins had die orientations of 2:30 or 8:30, with only one coin at 5:30 and two coins at 11:30 (*see* Figure 2).

Chemical Compositions of 245 Augustan Quadrantes

Augustan quadrantes were produced during four years, ca. 9, 8, 5, and 4 B.C. One other denomination of essentially pure copper, the "as," was produced sporadically by Augustus. Previous analysis of a few Augustan asses showed that the composition of the asses, which were produced from approximately 13 to 6 B.C. (again the dates are not known precisely), differed significantly from that of the quadrantes with respect to nickel, silver, and antimony (7). The asses, with only one exception, contain 0.04–0.10% nickel, compared with less than 0.015% for most quadrantes. The asses usually have higher silver contents than the quadrantes, 0.04–0.10% compared with 0.03–0.06% for most quadrantes. In addition the asses contain significantly less antimony than the quadrantes, 0.01–0.03% compared with 0.03–0.15% for most quadrantes. Because of the consistent major differences in trace metal contents between asses and quadrantes, I conclude that all asses were likely produced before the quadrantes, instead of quadrantes in ca. 9 B.C. and 8 B.C., asses in 7 and 6 B.C., and more quadrantes in 5 and 4 B.C. as suggested by Mattingly (10). No asses were produced again until about 15 A.D., and quadrantes were not subsequently produced until the reign of Caligula, beginning in 37 A.D.

A second problem of chronology is the relative dates of the quadrantes themselves. It has not been possible to date the coins exactly according to historical records or legends on the coins. Mattingly has tentatively placed the quadrantes of Pulcher, Taurus, and Regulus (PTR) during 8 B.C., and the coins of Lamia, Silius, and Annius (LSA) during 9 B.C. Evidence in this chapter indicates that the order of these coins should be reversed: the PTR quadrantes appear to be the first quadrantes made, and the LSA quadrantes seem to be more recent. There are four reasons for this conclusion. First, when the PTR quadrantes are placed first, the quadrantes gradually become smaller, thicker, and heavier throughout the four years of issue. Second, the PTR quadrantes are less common than the other three groups, possibly as a result of a mid-year decision to make quadrantes and then expenditure of time for tooling up for production; this would account for the fact that fewer PTR ("8 B.C.") coins exist compared with the other issues. Third, the range of composition of various trace elements is relatively small for the PTR quadrantes, again most likely indicating a shorter production period for these quadrantes compared with the other three years. Fourth, the compositions of the coins also indicate that the PTR coins are probably the oldest because of several consistent trends throughout the four-year period.

Chemical Compositions of 13 PTR Quadrantes ("ca. 8 B.C.")

Pulcher, Taurus, and Regulus were probably the first three moneyers commissioned to produce Augustan quadrantes. Table III contains the

Table III.   Compositions of PTR Augustan Quadrantes, "ca. 8 B.C."[a]

Coin No.	wt % Fe	wt % Ni	wt % Cu	wt % Ag	wt % Sn	wt % Sb	wt % Pb	BMC No.
R-794	0.056	0.004	99.6	0.068	0.011	0.123	0.18	204
R-817	0.106	0.004	99.6	0.063	0.011	0.138	0.18	204
R-728	0.06	0.025	99.6	0.042	0.02	0.13	0.04	205
R-792	0.079	0.010	99.6	0.063	0.015	0.121	0.17	205
R-813	0.087	0.008	99.6	0.062	0.013	0.111	0.16	205
R-999	0.068	0.015	99.6	0.064	0.011	0.120	0.11	205
R-793	0.131	0.004	99.6	0.064	0.013	0.113	0.16	207
R-809	0.145	0.011	99.6	0.054	0.013	0.124	0.11	207
R-821	0.122	0.012	99.5	0.069	0.013	0.121	0.15	207
R-995	0.046	0.010	99.6	0.062	0.006	0.110	0.16	207
R-996	0.088	0.010	99.5	0.047	0.014	0.157	0.16	207
R-997	0.089	0.011	99.4	0.052	0.016	0.179	0.21	207
R-998	0.222	0.015	99.4	0.032	0.013	0.218	0.11	207

[a] PTR: Pulcher, Taurus, and Regulus, moneyers.

compositions of 13 PTR quadrantes. Table IV contains the average compositions and standard deviations of the determinations for each element. Compared with the other quadrantes, these coins contain consistently high antimony contents, $0.14 \pm 0.04\%$. In fact, none of the coins contains less than 0.110%. Since the precision for antimony determinations is $s = 0.004\%$, these percentages are significantly higher than in most other quadrantes, and it is important that they are consistently high. Tin concentrations tend to correlate with antimony, so it is not surprising that the PTR quadrantes also contain significantly higher amounts of tin on the average than contained by other groups of quadrantes, except for the Catullus quadrantes (BMC 270) of 4 B.C. The mean iron contents of the PTR coins is lower than that of LSA quadrantes, and the PTR coins are also somewhat higher in nickel than LSA coins. None of the 13 PTR quadrantes has a highly unusual composition or even an unusual concentration for any given element.

Chemical Compositions of 80 LSA Quadrantes ("ca. 9 B.C.")

Table V contains the compositions of 80 LSA quadrantes. Lamia, Silius, and Annius were the three moneyers who produced these quadrantes, and their names are on all the coins, tentatively ascribed to 9 B.C. by Mattingly (10). The LSA quadrantes comprise three issues of coins, that is, three different types of coins having separate catalog numbers. The same three types comprised the PTR quadrantes, thereby closely linking these two years.

Thirty BMC-200 coins were analyzed as well as 28 BMC-201 coins and 22 BMC-202 coins. Therefore sufficient analyses have been per-

Table IV. Average Concentrations of Augustan Quadrantes

Coin No.	wt % Fe	wt % Ni	wt % Cu	wt % Ag	wt % Sn	wt % Sb	wt % Pb	BMC No.
			PTR Quadrantes, "8 B.C."					
Mean	0.100	0.011	99.55	0.055	0.013	0.136	0.15	all 8 B.C.
S^a	0.047	0.0055	0.1	0.012	0.003	0.031	0.03	all 8 B.C.
Mean	0.120	0.010	99.5	0.054	0.013	0.146	0.15	207
S	0.056	0.003	0.1	0.012	0.003	0.041	0.03	207
			LSA Quadrantes, "9 B.C."					
Mean	0.129	0.006	99.6	0.050	0.011	0.085	0.15	200
S	0.111	0.004	0.2	0.014	0.007	0.056	0.08	200
Mean	0.111	0.006	99.6	0.053	0.012	0.091	0.13	201
S	0.052	0.004	0.2	0.016	0.004	0.041	0.08	201
Mean	0.117	0.004	99.6	0.048	0.009	0.076	0.13	202
S	0.065	0.004	0.1	0.014	0.006	0.042	0.07	202
Mean	0.120	0.006	99.6	0.051	0.011	0.085	0.14	all 9 B.C.
S	0.081	0.004	0.2	0.015	0.006	0.047	0.07	all 9 B.C.
			AGMS Quadrantes, "5 B.C."					
Mean	0.172	0.011	99.5	0.051	0.008	0.077	0.16	252
S	0.089	0.006	0.1	0.020	0.003	0.037	0.05	252
Mean	0.119	0.008	99.6	0.045	0.004	0.050	0.12	253
S	0.036	0.003	0.1	0.011	0.003	0.025	0.04	253
Mean	0.097	0.009	99.7	0.056	0.007	0.047	0.12	258
S	0.043	0.004	0.1	0.018	0.006	0.010	0.04	258
Mean	0.153	0.008	99.5	0.045	0.009	0.084	0.22	259
S	0.177	0.002	0.2	0.008	0.004	0.023	0.03	259
Mean	0.093	0.010	99.7	0.050	0.005	0.044	0.11	260
S	0.015	0.004	0.04	0.015	0.002	0.020	0.06	260
Mean	0.149	0.008	99.6	0.036	0.011	0.091	0.14	262
S	0.123	0.003	0.2	0.008	0.005	0.046	0.05	262
Mean	0.106	0.007	99.6	0.047	0.009	0.061	0.15	264
S	0.057	0.003	0.1	0.024	0.002	0.012	0.06	264
Mean	0.152	0.009	99.6	0.046	0.008	0.067	0.15	all 5 B.C.
S	0.104	0.004	0.2	0.015	0.004	0.031	0.06	all 5 B.C.
			BBCC Quadrantes, "4 B.C."					
Mean	0.174	0.010	99.5	0.044	0.007	0.069	0.15	265
S	0.109	0.005	0.1	0.018	0.004	0.035	0.07	265
Mean	0.172	0.011	99.5	0.030	0.009	0.082	0.16	267
S	0.084	0.005	0.2	0.017	0.004	0.032	0.06	267
Mean	0.156	0.009	99.6	0.037	0.008	0.072	0.15	269
S	0.100	0.003	0.2	0.014	0.004	0.033	0.07	269
Mean	0.251	0.010	99.4	0.037	0.032	0.105	0.20	270
S	0.176	0.005	0.3	0.011	0.070	0.048	0.08	270
Mean	0.194	0.010	99.5	0.038	0.016	0.084	0.17	all 4 B.C.
S	0.137	0.005	0.2	0.015	0.043	0.042	0.07	all 4 B.C.

[a] S: standard deviation.

Table V. Compositions of LSA Augustan Quadrantes, "ca. 9 B.C."[a]

Coin No.	wt % Fe	wt % Ni	wt % Cu	wt % Ag	wt % Sn	wt % Sb	wt % Pb	BMC No.
R-797	0.074	0.009	99.6	0.045	0.016	0.12	0.15	200
R-812	0.072	0.008	99.6	0.058	0.007	0.094	0.14	200
R-824	0.039	0.006	99.6	0.049	0.009	0.097	0.12	200
R-833	0.094	0.002	99.8	0.036	0.005	0.034	0.08	200
R-842	0.087	0.004	99.5	0.028	0.010	0.057	0.24	200
R-851	0.071	0.011	99.5	0.052	0.012	0.248	0.10	200
R-857	0.060	0.004	99.6	0.044	0.007	0.086	0.15	200
R-863	0.130	N	99.5	0.062	0.007	0.057	0.23	200
R-869	0.082	N	99.6	0.067	0.004	0.051	0.24	200
R-875	0.100	0.001	99.6	0.033	0.011	0.055	0.18	200
R-893	0.304	0.008	99.4	0.052	0.016	0.116	0.07	200
R-897	0.075	0.016	99.6	0.07	0.018	0.116	0.10	200
R-901	0.081	0.007	99.7	0.032	0.010	0.046	0.14	200
R-905	0.187	0.009	99.5	0.056	0.013	0.096	0.16	200
R-909	0.063	0.003	99.7	0.064	0.011	0.057	0.06	200
R-913	0.085	0.005	99.7	0.036	0.007	0.045	0.10	200
R-917	0.588	0.009	99.1	0.030	0.008	0.055	0.16	200
R-921	0.238	0.008	99.5	0.063	0.016	0.115	0.11	200
R-925	0.069	0.004	99.6	0.068	0.012	0.116	0.09	200
R-928	0.074	N	99.6	0.053	0.009	0.040	0.26	200
R-930	0.085	0.012	99.4	0.079	0.017	0.172	0.21	200
R-932	0.090	0.010	99.7	0.037	0.006	0.061	0.13	200
R-934	0.156	0.007	99.7	0.035	0.004	0.036	0.07	200
R-936	0.220	0.006	99.1	0.067	0.030	0.237	0.39	200
R-938	0.068	0.008	99.9	0.035	0.006	0.032	N	200
R-940	0.098	N	99.6	0.065	0.007	0.053	0.18	200
R-942	0.055	0.004	99.6	0.051	0.004	0.040	0.24	200
R-944	0.300	0.002	99.4	0.047	0.012	0.063	0.21	200
R-946	0.139	N	99.7	0.037	0.006	0.033	0.11	200
R-948	0.095	0.013	99.6	0.052	0.031	0.129	0.10	200
R-804	0.053	0.005	99.7	0.063	0.014	0.16	0.16	201
R-808	0.132	0.010	99.5	0.045	0.017	0.124	0.16	201
R-820	0.150	0.014	99.6	0.045	0.011	0.057	0.19	201
R-830	0.118	0.002	99.4	0.074	0.010	0.065	0.35	201
R-839	0.160	0.009	99.6	0.052	0.011	0.091	0.15	201
R-848	0.076	0.009	99.6	0.030	0.011	0.069	0.19	201
R-887	0.038	0.001	99.0	0.076	0.009	0.128	0.11	201
R-894	0.072	N	99.6	0.064	0.013	0.061	0.23	201
R-898	0.051	0.008	99.8	0.052	0.012	0.103	N	201
R-902	0.123	0.005	99.7	0.030	0.007	0.049	0.14	201
R-906	0.106	0.008	99.6	0.052	0.016	0.126	0.10	201
R-910	0.276	0.009	99.4	0.031	0.012	0.058	0.24	201
R-914	0.117	0.014	99.5	0.043	0.015	0.157	0.12	201
R-918	0.170	N	99.4	0.074	0.009	0.077	0.23	201
R-922	0.042	0.005	99.7	0.078	0.011	0.111	0.09	201
R-926	0.089	0.002	99.6	0.039	0.012	0.083	0.14	201
R-929	0.065	0.006	99.8	0.034	0.009	0.073	0.07	201
R-931	0.098	0.005	99.7	0.056	0.010	0.063	0.10	201

Table V. Continued

Coin No.	wt % Fe	wt % Ni	wt % Cu	wt % Ag	wt % Sn	wt % Sb	wt % Pb	BMC No.
R-933	0.148	0.003	99.4	0.083	0.007	0.210	0.12	201
R-935	0.113	0.004	99.7	0.058	0.010	0.074	0.04	201
R-937	0.111	0.007	99.6	0.053	0.010	0.084	0.09	201
R-939	0.124	N	99.5	0.064	0.009	0.048	0.23	201
R-941	0.111	0.009	99.7	0.051	0.010	0.087	0.04	201
R-943	0.208	0.008	99.6	0.043	0.010	0.052	0.08	201
R-945	0.107	0.011	99.6	0.061	0.011	0.116	0.11	201
R-947	0.060	0.006	99.7	0.036	0.017	0.056	0.13	201
R-949	0.124	0.008	99.6	0.071	0.024	0.128	0.05	201
R-950	0.076	0.010	99.8	0.037	0.018	0.030	0.06	201
R-799	0.092	0.013	99.6	0.028	0.009	0.042	0.14	202
R-816	0.118	N	99.5	0.076	0.001	0.035	0.24	202
R-827	0.246	0.003	99.4	0.068	0.005	0.029	0.16	202
R-836	0.066	0.006	99.7	0.036	0.008	0.071	0.15	202
R-845	0.079	0.001	99.7	0.047	0.003	0.041	0.15	202
R-854	0.068	0.008	99.7	0.044	0.008	0.085	0.14	202
R-860	0.167	0.003	99.5	0.034	0.007	0.052	0.25	202
R-866	0.108	0.003	99.6	0.030	0.006	0.061	0.19	202
R-872	0.177	0.002	99.5	0.054	0.008	0.122	0.12	202
R-878	0.077	N	99.7	0.029	0.008	0.057	0.15	202
R-881	0.028	0.006	99.8	0.028	0.001	0.040	0.16	202
R-882	0.083	0.002	99.6	0.056	0.015	0.125	0.15	202
R-884	0.037	0.009	99.8	0.031	0.005	0.038	0.16	202
R-895	0.164	0.006	99.7	0.050	0.006	0.109	N	202
R-899	0.116	0.013	99.7	0.052	0.007	0.047	0.02	202
R-903	0.159	0.002	99.6	0.058	0.014	0.130	0.03	202
R-907	0.249	0.008	99.5	0.057	0.010	0.052	0.17	202
R-911	0.135	N	99.6	0.044	0.017	0.099	0.07	202
R-915	0.095	N	99.5	0.052	0.015	0.070	0.25	202
R-919	0.034	N	99.7	0.047	0.016	0.074	0.08	202
R-923	0.214	0.006	99.6	0.059	0.018	0.106	0.04	202
R-927	0.067	0.006	99.5	0.068	0.021	0.197	0.11	202

[a] LSA: Lamia, Silius, and Annius, moneyers.

formed to delineate the range of composition for each of six trace elements: Fe, Ni, Ag, Sn, Sb, and Pb in each of the three issues. Previous analyses have shown no zinc, cobalt, or arsenic detectable by the XRD-6 spectrometer, but calcium and sulfur are present in very low and apparently uncorrelatable concentrations (9). Titanium and manganese usually are not detectable in these quadrantes by this XRF spectrometer.

The three issues of LSA quadrantes have significantly different averages for several trace elements (*see* Table IV). For instance 30 BMC-200 coins have an average tin content of 0.011% compared with 0.012% for 28 BMC-201 coins and 0.009% for 22 BMC-202 coins. More significantly the BMC-201 quadrantes have a mean antimony concentration of 0.091% compared with an average of only 0.076% for the BMC-202 coins. Since the standard deviation for antimony determinations is $s = 0.004\%$, there is indeed a substantial difference between the two issues of coins. It is concluded that the three issues of coins were not produced during exactly the same periods. Undoubtedly there was overlap—times when two or even all three issues were being made—but there were periods when no more than two issues were being made.

As shown in the previous section, the LSA quadrantes are significantly different on the average from the PTR coins in trace metal contents of iron, nickel, and antimony (*see* Table IV). In the next section the differences in composition between the LSA quadrantes and the AGMS quadrantes (ca. 5 B.C.) are discussed. The quadrantes produced during each of the four years had significantly different average concentrations for three or more trace elements.

Several of the LSA quadrantes have relatively high iron contents, > 0.300%. Two coins ave > 0.030% tin, three coins contain > 0.200% antimony, and two coins contain > 0.30% lead.

Chemical Compositions of 66 AGMS Quadrantes ("ca. 5 B.C.")

In 5 B.C. (tentative date of Mattingly) there were four moneyers. Since three of the four names are moderately long, there simply was not enough room to place all four names on the obverse of the coin, such as was done for both the PTR and LSA quadrantes. Instead two names were placed on the obverse and two names on the reverse. Since presumably all four men shared the position of moneyers equally, the names were rotated so that there was no preferred order in which one name was always the most prominent. The four names, Apronius, Galus, Messalla, and Sisenna (AGMS), may be arranged in 24 permutations.

In the AGMS quadrantes there is still one more variable; the altar appearing on the reverse of the coin appears in two styles: a bowl-shaped altar and a pedestal-shaped altar (Figure 1). The bowl-shaped altar was used exclusively in the earlier PTR and LSA coins. Therefore in 5 B.C. there were 48 different types of quardantes possible. In this chapter 19 different types are represented among the 66 AGMS quadrantes, whereas only slightly more than half the total possible permutations are represented in the vast collections of the British Museum and the Bibliothèque Nationale. Presumably many of the permutations were not produced at all because of time limitations or perhaps a deficiency in scheduling. However, each of the four names does appear first on a reasonable proportion of the coins.

Since 25–30 varieties of coins were struck during 5 B.C., it is probable that no single issue was produced for more than three or four weeks.

The length of production for a single issue undoubtedly depended on how many dies were produced for the given type and on how long the dies lasted. A rough estimate is that the dies lasted about one to two weeks on the average (*19, 20*). Because of the relatively large number of issues of quadrantes in 5 B.C. causing short production runs, the analyses should cluster unless there is as much day-to-day variation in composition as month-to-month variation.

Table VI contains analyses of 66 AGMS quadrantes, "ca. 5 B.C." In comparing the AGMS quadrantes with the PTR and LSA, one sees from Table IV that the average iron content increases from PTR (0.100%) and LSA (0.120%) to AGMS (0.148%); presumably the oldest coins have the lowest iron. This trend continues through the 4 B.C. quadrantes, as discussed in the next section. The average tin and antimony contents decrease significantly with time (PTR: 0.136% Sb, 0.013% Sn; LSA: 0.085% Sb, 0.011% Sn; AGMS: 0.067% Sb, 0.008% Sn). These concentration trends give credence to the hypothesis that the PTR quadrantes are older than the LSA quadrantes, contrary to Mattingly's designation.

Table IV shows marked differences in the average trace metal contents for various issues of the AGMS quadrantes. Three issus of coins have relatively high average contents of antimony, tin, iron, and lead:

Table VI.   Compositions of AGMS Augustan Quadrantes, ca. 5 B.C.[a]

Coin No.	wt % Fe	wt % Ni	wt % Cu	wt % Ag	wt % Sn	wt % Sb	wt % Pb	BMC No.
R-873	0.227	0.006	99.4	0.044	0.007	0.077	0.14	243 var.?
R-798	0.082	0.009	99.6	0.047	0.004	0.059	0.29	246
R-822	0.059	0.007	99.7	0.053	0.006	0.056	0.14	247
R-1002	0.060	0.015	99.6	0.019	0.012	0.124	0.19	247
R-1009	0.179	0.019	99.4	0.043	0.011	0.101	0.22	250
R-1010	0.180	0.020	99.6	0.017	0.006	0.079	0.09	250
R-1011	0.364	0.016	99.3	0.030	0.004	0.113	0.22	250
R-805	0.094	0.003	99.6	0.069	0.006	0.074	0.14	252
R-828	0.392	0.009	99.3	0.059	0.009	0.086	0.14	252
R-849	0.152	0.007	99.4	0.060	0.011	0.149	0.21	252
R-1003	0.132	0.013	99.6	0.041	0.002	0.049	0.18	252
R-1004	0.130	0.004	99.6	0.048	0.009	0.092	0.11	252
R-1005	0.181	0.015	99.6	0.038	0.006	0.038	0.13	252
R-1006	0.161	0.021	99.4	0.023	0.012	0.110	0.25	252
R-1007	0.107	0.012	99.6	0.038	0.006	0.059	0.15	252
R-1008	0.197	0.015	99.6	0.088	0.008	0.039	0.09	252
R-810	0.083	0.004	99.7	0.056	0.004	0.041	0.07	253
R-814	0.166	0.009	99.6	0.057	0.005	0.043	0.10	253
R-858	0.082	0.008	99.7	0.051	0.004	0.043	0.10	253
R-876	0.150	0.007	99.7	0.053	0.003	0.030	0.07	253
R-1013	0.077	0.004	99.7	0.029	0.005	0.043	0.11	253

Table VI. Continued

Coin No.	wt % Fe	wt % Ni	wt % Cu	wt % Ag	wt % Sn	wt % Sb	wt % Pb	BMC No.
R-1014	0.133	0.006	99.5	0.034	0.010	0.111	0.20	253
R-1015	0.110	0.010	99.7	0.035	0.002	0.042	0.11	253
R-1016	0.15	0.012	99.6	0.046	0.002	0.050	0.16	253
R-806	0.389	0.002	99.8	0.053	0.005	0.030	N	256
R-861	0.068	0.008	99.6	0.039	0.009	0.082	0.20	256
R-1012	0.091	0.009	99.6	0.025	0.010	0.090	0.22	256
R-864	0.041	0.014	99.7	0.048	0.012	0.056	0.10	258
R-879	0.153	0.003	99.7	0.065	0.002	0.031	0.08	258
R-1017	0.104	0.007	99.6	0.083	0.005	0.046	0.12	258
R-1018	0.080	0.009	99.7	0.046	0.004	0.049	0.10	258
R-1024	0.066	0.012	99.7	0.039	0.003	0.043	0.14	258 or 264
R-1029	0.136	0.007	99.6	—	0.016	0.057	0.19	258
R-831	0.484	0.009	99.1	0.049	0.013	0.115	0.21	259
R-834	0.072	0.005	99.5	0.053	0.012	0.103	0.22	259
R-885	0.221	0.006	99.4	0.051	0.011	0.063	0.24	259
R-1000	0.019	0.011	99.7	0.036	0.008	0.091	0.16	259
R-1020	0.040	0.009	99.6	0.044	0.003	0.069	0.26	259
R-1021	0.081	0.009	99.6	0.035	0.006	0.061	0.22	259
R-840	0.093	0.007	99.7	0.067	0.004	0.072	0.20	260
R-843	0.106	0.009	99.7	0.047	0.008	0.038	0.07	260
R-852	0.101	0.014	99.7	0.054	0.003	0.029	0.09	260
R-1022	0.071	0.013	99.7	0.031	0.005	0.057	0.09	260?
R-1025	0.358	0.010	99.3	0.032	0.012	0.083	0.18	261
R-80	0.086	0.009	99.7	0.055	0.007	0.049	0.10	261 var.
R-801	0.196	0.005	99.8	0.069	0.005	0.025	0.09	261 var.
R-870	0.099	0.013	99.7	0.035	0.015	0.052	0.11	261 var.
R-1026	0.152	0.011	99.6	0.049	0.002	0.039	0.19	261 var.
R-818	0.035	0.003	99.8	0.046	0.009	0.038	0.08	262
R-846	0.082	0.010	99.6	0.029	0.016	0.125	0.16	262
R-855	0.141	0.009	99.4	0.031	0.015	0.141	0.21	262
R-1027	0.356	0.008	99.3	0.038	0.014	0.102	0.14	262
R-1028	0.131	0.010	99.7	—	0.003	0.047	0.13	262
R-825	0.091	0.009	99.6	0.049	0.012	0.064	0.14	264
R-837	0.032	0.002	99.7	0.070	0.007	0.046	0.11	264
R-1001	0.181	0.011	99.5	0.022	0.009	0.078	0.25	264
R-1030	0.083	0.007	99.8	—	0.007	0.057	0.09	264
R-1031	0.142	0.007	99.6	—	0.008	0.060	0.15	264
R-867	0.272	0.013	99.4	0.045	0.016	0.062	0.15	261 or C.351[b]
R-795	0.171	0.006	99.8	0.051	0.005	0.031	0.12	C.374[b]
R-791	0.205	0.007	99.7	0.050	0.010	0.073	0.15	C.375[b]
R-796	0.185	0.002	99.7	0.055	0.007	0.060	0.09	C.424[b]
R-889	0.142	0.014	99.5	0.033	0.010	0.096	0.21	C.424?[b]
R-1019	0.497	0.001	99.3	0.042	0.009	0.036	0.08	C.535[b]
R-892	0.229	0.010	99.3	0.068	0.010	0.148	0.24	?
R-1023	0.127	0.008	99.7	0.038	0.004	0.048	0.05	—

[a] ACMS: Apronius, Galus, Messalla, and Sisenna, moneyers.
[b] C: Cohen classification number.

BMC 252 (0.077% Sb, 0.008% Sn, 0.172% Fe, 0.16% Pb); BMC-259 (0.084% Sb, 0.009% Sn, 0.153% Fe, 0.22% Pb); and BMC 262 (0.091% Sb, 0.011% Sn, 0.149% Fe, 0.14% Pb). Since tin contents correlate somewhat with antimony, tin contents should be higher on the average when antimony contents are high. Three issues have on the average relatively low antimony, tin, iron, and lead contents: BMC 253 (0.050% Sb, 0.004% Sn, 0.119% Fe, 0.12% Pb); BMC 258 (0.047% Sb, 0.007% Sn, 0.097% Fe, 0.12% Pb); and BMC 260 (0.044% Sb, 0.005% Sn, 0.093% Fe, 0.11% Pb). BMC 264 are somewhat intermediate, with 0.061% Sb, 0.009% Sn, 0.106% Fe, 0.15% Pb.

Probably BMC 252, 259, and 262 were made either early or late in the year, and BMC 253, 258, and 260 were made correspondingly late or early. BMC 250 probably belongs to the BMC 252, 259, and 262 grouping (high antimony) and BMC 261-variety belong to the BMC 253, 258, 260 grouping (low antimony). Changes in trace metal contents from day to day or year to year depend on the variation in trace metal content of the copper ore and the methods and conditions used for reducing the ore and refining the copper.

Six coins contain over 0.300% iron. Since the iron contents appear highly variable in every issue, it is likely that iron in the form of an oxide, silicate, or sulfide was a constituent of the slag. Also iron was undoubtedly present in unrefined copper so that iron should be highly variable from one batch of copper to the next. Even within one batch of copper it is probable that the iron content might vary appreciably as a function of time, if, as believed, the slag contained iron. Furthermore, the liquid copper may have been exposed to iron in the form of ladles or stirrers. No coins have unusually high contents of Ni, Ag, Sn, Sb, or Pb.

Chemical Compositions of 86 BBCC Quadrantes ("ca. 4 B.C.")

During 4 B.C., which is the tentative date assigned to the last group of Augustan quadrantes by Mattingly, the moneyers were Bassus, Blandus, Capella, and Catullus (BBCC). Instead of using the complicated system of having permutations of all four names on each coin, as in 5 B.C., each coin in 4 B.C. bears only one name. Therefore only four issues were produced in 4 B.C.

The average iron concentration increased in each of the four years during which quadrantes were made, assuming the PTR quadrantes to be the oldest—PTR: 0.100%; LSA: 0.120%, AGMS: 0.152%; BBCC: 0.194%. The silver concentrations apparently descreased with time— PTR: 0.055%; LSA: 0.051%; AGMS: 0.046%; BBCC: 0.038%. Within experimental error, lead remains essentially constant. Antimony and tin decrease for the first three years but increase during the fourth year of production, primaily because of the unusual issue of Catullus, BMC 270 (Table VII). The compositional trends tend to confirm assignment of PTR quadrantes as the oldest, rather than the LSA coins.

The average compositions of three series, BMC 265, 267, and 269, are close to the average compositions for the high-antimony group of coins of 5 B.C. The fourth series in 4 B.C., BMC 270, Catullus, has by far the highest average concentrations of any series of quadrantes in iron, tin, and lead. The average compositions of BMC 265 and BMC 269 are very close to each other for all elements.

Five of 19 Bassus quadrantes, BMC 265, contain > 0.300% iron. The concentrations of the remainder of the trace elements are quite normal in these 19 coins. The Capella coins, BMC 267, are quite ordinary with only one high iron content and one high lead content, 0.30% lead. The Blandus quadrantes, BMC 269, are also normal with only a few of the 24 coins having high iron or lead. No coin from these three series has relatively high tin (> 0.200%) or high antimony (> 0.200%).

The Catullus series is remarkable compared with all the other quadrantes; four of the 30 coins contain more than 0.400% iron. Coins R-841 and R-978 contain 0.286 and 0.291% tin respectively, more than 10 times the amount of tin of any other quadrans, and six other Catullus quadrantes contain > 0.020% tin; two of the coins contain > 0.200% antimony, and four Catullus coins have > 0.030% lead.

Obviously the Catullus series is quite different in trace metal content compared with other series. The few coins that contain very high tin perhaps had some bronze objects remelted along with copper from copper ore. Because of their high purity, most of the quadrantes were undoubtedly made solely from copper obtained from ore and not through remelting of older coins or other metallic objects. Very likely the Catullus series, BMC 270, was carried over for a period beyond the production of the other three series in 4 B.C. At least the Catullus coins must have been struck for an appreciable period by themselves since the other series do not have coins representative of the unusually high antimony, tin, and iron contents of BMC 270. It is reasonable that this period was at the end of the year rather than at the beginning since 5 B.C. coins, the AGMS quadrantes, also do not have representative coins having very high tin contents. It is also reasonable to believe that production of quadrantes was phased out during the period in which only the Catullus quadrantes were being made.

Compositions of Die-Linked Coins

Roman coins were individually hand struck from heated blanks using two hand-tooled dies, a fixed anvil die for the obverse, and a second,

Table VII. Compositions of BBCC Augustan Quadrantes, ca. 4 B.C.[a]

Coin No.	wt % Fe	wt % Ni	wt % Cu	wt % Ag	wt % Sn	wt % Sb	wt % Pb	BMC No.
R-748	0.14	0.025	99.7	0.042	0.005	0.035	N	265
R-802	0.305	0.002	99.7	0.073	0.008	0.035	0.16	265
R-819	0.131	N	99.6	0.070	0.005	0.028	0.12	265
R-832	0.322	0.001	99.4	0.073	0.004	0.033	0.2	265
R-844	0.198	0.010	99.6	0.053	0.011	0.069	0.08	265
R-853	0.071	0.008	99.7	0.070	0.007	0.044	0.10	265
R-982	0.110	0.008	99.6	0.041	0.003	0.040	0.16	265
R-983	0.098	0.006	99.8	0.029	0.004	0.047	0.05	265
R-984	0.061	0.012	99.6	0.023	0.009	0.114	0.14	265
R-985	0.307	0.015	99.4	0.029	0.007	0.067	0.18	265
R-986	0.332	0.012	99.4	0.054	0.012	0.055	0.10	265
R-987	0.109	0.008	99.5	0.023	0.007	0.089	0.28	265
R-988	0.341	0.010	99.4	0.020	0.005	0.068	0.17	265
R-989	0.100	0.010	99.6	0.041	0.001	0.050	0.18	265
R-990	0.313	0.012	99.3	0.036	0.018	0.132	0.19	265
R-991	0.046	0.013	99.7	0.049	0.004	0.038	0.14	265
R-992	0.036	0.015	99.6	0.053	0.008	0.090	0.18	265
R-993	0.149	0.014	99.5	0.024	0.010	0.097	0.20	265
R-994	0.145	0.017	99.4	0.028	0.011	0.145	0.24	265
R-727	0.09	0.02	99.7	0.01	0.01	0.08	0.04	267
R-800	0.172	0.006	99.8	0.059	0.008	0.040	0.13	267
R-811	0.252	0.007	99.5	0.045	0.008	0.079	0.14	267
R-826	0.152	0.006	99.6	0.042	0.010	0.058	0.12	267
R-838	0.112	0.006	99.6	0.045	0.007	0.081	0.18	267
R-967	0.102	0.019	99.6	0.029	0.013	0.079	0.14	267
R-968	0.124	0.015	99.6	0.018	0.007	0.068	0.17	267
R-969	0.119	0.009	99.7	0.018	N	0.043	0.14	267
R-970	0.295	0.012	99.2	0.024	0.010	0.150	0.30	267
R-971	0.205	0.014	99.4	0.022	0.008	0.098	0.22	267
R-972	0.357	0.011	99.4	0.008	0.006	0.060	0.12	267
R-973	0.174	0.013	99.5	0.016	0.012	0.105	0.19	267
R-974	0.084	0.018	99.5	0.052	0.015	0.120	0.24	267
R-803	0.139	0.007	99.7	0.040	0.008	0.062	0.15	269
R-807	0.280	0.009	99.5	0.043	0.013	0.130	0.21	269
R-823	0.366	0.005	99.4	0.062	0.005	0.041	0.10	269
R-835	0.081	0.006	99.7	0.050	0.006	0.040	0.11	269
R-847	0.081	0.008	99.8	0.042	0.003	0.038	0.03	269
R-856	0.114	0.008	99.8	0.042	0.002	0.035	0.02	269
R-862	0.139	0.009	99.6	0.063	0.011	0.064	0.13	269
R-868	0.146	0.006	99.7	0.058	0.007	0.040	0.13	269
R-951	0.078	0.015	99.6	0.035	0.013	0.079	0.23	269
R-952	0.337	0.010	99.3	0.035	0.006	0.099	0.17	269
R-953	0.101	0.006	99.8	0.038	0.006	0.039	0.05	269
R-954	0.084	0.009	99.6	0.027	0.007	0.054	0.18	269

Table VII. Continued

Coin No.	wt % Fe	wt % Ni	wt % Cu	wt % Ag	wt % Sn	wt % Sb	wt % Pb	BMC No.
R-955	0.271	0.014	99.3	0.026	0.014	0.139	0.26	269
R-956	0.190	0.014	99.5	0.003	0.008	0.110	0.16	269
R-957	0.066	0.013	99.6	0.033	0.006	0.081	0.15	269
R-958	0.177	0.011	99.3	0.026	0.008	0.118	0.32	269
R-959	0.064	0.010	99.5	0.017	0.015	0.138	0.29	269
R-960	0.076	0.010	99.7	0.028	0.005	0.074	0.13	269
R-961	0.080	0.012	99.7	0.049	0.011	0.067	0.13	269
R-962	0.126	0.004	99.7	0.037	0.004	0.052	0.13	269
R-963	0.108	0.008	99.6	0.025	0.008	0.059	0.14	269
R-964	0.027	0.014	99.7	0.031	0.009	0.069	0.14	269
R-965	0.295	0.010	99.4	0.035	0.006	0.052	0.16	269
R-966	0.313	0.009	99.5	0.043	0.004	0.045	0.10	269
R-815	0.368	0.006	99.4	0.055	0.007	0.059	0.13	270
R-829	0.205	0.010	99.3	0.036	0.020	0.224	0.32	270
R-841	0.141	0.024R ^{b}	99.2	0.036	0.286	0.066	0.20	270
R-850	0.211	0.015	99.4	0.045	0.044	0.101	0.16	270
R-859	0.078	0.011	99.5	0.025	0.012	0.094	0.16	270
R-960	0.076	0.010	99.7	0.028	0.005	0.075	0.13	269
R-871	0.374	0.006	99.2	0.060	0.009	0.082	0.24	270
R-874	0.172	0.011	99.3	0.050	0.036	0.168	0.22	270
R-877	0.176	0.002	99.7	0.038	0.008	0.043	0.11	270
R-880	0.508	0.011	98.8	0.039	0.025	0.237	0.36	270
R-883	0.208	0.017	99.3	0.040	0.008	0.130	0.20	270
R-886	0.210	0.010	99.6	0.060	0.002	0.036	0.08	270
R-888	0.260	0.005	99.4	0.033	0.009	0.088	0.18	270
R-890	0.291	0.013	99.5	0.027	0.008	0.100	0.17	270
R-891	0.225	0.005	99.5	0.035	0.008	0.061	0.15	270
R-896	0.020	0.005	99.6	0.043	0.010	0.108	0.22	270
R-900	0.170	0.010	99.6	0.035	0.012	0.107	0.10	270
R-904	0.085	0.006	99.6	0.030	0.008	0.083	0.15	270
R-908	0.982	0.007	99.5	0.039	0.013	0.103	0.32	270
R-912	0.242	0.008	99.6	0.044	0.010	0.059	0.08	270
R-916	0.194	0.009	99.5	0.041	0.013	0.093	0.19	270
R-920	0.274	0.007	99.4	0.047	0.010	0.088	0.22	270
R-924	0.216	0.008	99.3	0.030	0.014	0.129	0.29	270
R-975	0.081	0.007	99.6	0.026	0.009	0.081	0.21	270
R-976	0.104	0.018	99.6	0.026	0.024	0.065	0.18	270
R-977	0.400	0.010	99.3	0.037	0.007	0.081	0.17	270
R-978	0.207	0.015	99.0	0.025	0.291	0.116	0.30	270
R-979	0.269	0.013	99.3	0.025	0.018	0.170	0.25	270
R-980	0.413	0.025	99.0	0.039	0.023	0.173	0.35	270
R-981	0.284	0.012	99.4	0.031	0.008	0.122	0.10	270

^{a} BBCC: Bassus, Blandus, Capella, and Catullus, moneyers.
^{b} R: reverse side of coin only.

hand-held die for the reverse (18). The dies are believed to have produced roughly 25,000 coins on the average (19, 20). Apparently a pool
of dies was used for each issue, and each day all dies were returned to
the pool (18). Consequently, a given obverse die, which lasted very
approximately about seven to 10 days on the average, would have roughly
five to 12 different reverse dies associated with it. Coins having identical
obverse and reverse dies were most likely, but not necessarily, struck on
the same day. Coins having common obverse or reverse dies were most
likely, but not necessarily, struck on different days.

Three other factors are important in considering the chemical compositions of die-linked coins: the number of batches of copper produced
each day, the homogeneity of each batch of copper, and the homogeneity
of the copper coins. A reasonable size for a batch of copper, according to
specialists in ancient metallurgy, is about 30 kg. Most likely two batches
of copper were sufficient for one day's production of quadrantes (the
weight of copper may be estimated from the probable number of dies
calculated from statistics of die links, the estimated average of 25,000
coins per die, and the average weight of a quadrante). Therefore, the
probability is somewhat less than 0.5 that coins struck on the same day
would have the same trace metal contents. Not all the copper blanks
made on a given day would necessarily be used in one day; they may
have been placed in a storage container for later use.

The metal used to produce quadrantes is fairly pure copper, about
99.5%. The individual batch of copper was very likely homogeneous
since it was common practice to stir liquid copper during refining with
green wooden poles to reduce the oxide content of the melt by the gases
coming from the green wood (21). When the blanks were made, the
copper was again agitated from ladling. It therefore seems reasonable
that a given batch of copper must have been fairly homogeneous. The
coin blanks were probably cast on ceramic slabs having small indenta-
tions (18). Thus the copper solidified relatively quickly, and the low
level of impurities and fast solidification would prevent serious segregation. At present, studies on the homogeneity of Augustan quadrantes are
underway, and the coins appear reasonably homogeneous.

The present assumptions are that the coins are reasonably homogeneous, that the melt was reasonably homogeneous, that probably two batches
of copper were made each day, and that dies were stored over night in
a container of some sort (a strong-box). Table VIII contains the analyses
for pairs of coins having double die links. Both obverse and reverse dies
are the same; in some cases this cannot be determined with certainty
since some coins are worn or corroded, and on rare occasion two different
dies are nearly the same. In all cases the die orientations of the doubly-
die linked pair are the same within experimental error, ±30 min. The

Table VIII.   Compositions of Double Die-Linked Coins[a]

Coin No.		wt % Fe	wt % Ni	wt % Cu	wt % Ag	wt % Sn	wt % Sb	wt % Pb	BMC No.
R-846	b, c	0.082	0.010	99.6	0.029	0.016	0.125	0.16	262
R-855		0.141	0.009	99.4	0.031	0.015	0.141	0.21	262
R-847	c	0.081	0.008	99.8	0.042	0.003	0.038	0.03	269
R-856		0.114	0.008	99.8	0.042	0.002	0.035	0.02	269
R-861	b, c	0.068	0.008	99.6	0.039	0.009	0.082	0.20	256
R-1012		0.091	0.009	99.6	0.025	0.010	0.090	0.22	256
R-864	b	0.041	0.014	99.7	0.048	0.012	0.056	0.10	258
R-1017		0.104	0.007	99.6	0.083	0.005	0.046	0.12	258
R-876	b	0.150	0.007	99.7	0.053	0.003	0.030	0.07	253
R-1016		0.15	0.012	99.6	0.046	0.002	0.050	0.16	253
R-962	c	0.126	0.004	99.7	0.037	0.004	0.052	0.13	269
R-966		0.313	0.009	99.5	0.043	0.004	0.045	0.10	269
R-967		0.102	0.019	99.6	0.029	0.013	0.079	0.14	267
R-972		0.357	0.011	99.4	0.008	0.006	0.060	0.12	267
R-1003		0.132	0.013	99.6	0.041	0.002	0.049	0.18	252
R-1008		0.197	0.015	99.6	0.088	0.008	0.039	0.09	252

[a] For the sake of comparison, the following standard deviations were used: Ni, 0.003% ; Ag, 0.006% ; Sn, 0.0014% ; Sb, 0.003% ; Pb, 0.05%.
[b] Probable double die link.
[c] This pair of coins has the same composition within experimental error and therefore probably came from the same batch of copper.

results show that for five pairs of doubly-die linked coins and three additional probable pairs, four pairs have compositions essentitally identical within experimental error. The only exception is the element, iron, which is highly variable, and which is assumed to vary in blanks made from the same batch of copper. Therefore, iron is not included in the comparisons. If it were, then none of the pairs would have the same composition within experimental error. The proportion of four out of eight pairs agrees with the proportion calculated from the above assumptions. Incidentally, it is highly improbable that two coins taken at random have the same compositions within experimental error. Four matching pairs out of eight is therefore most significant.

Coins having obverse die links only are presented in Table IX. There are 10 pairs, one group of three, and one group of four. Only one pair, R-862 and R-961, have the same composition within experimental error, although the group of four coins are fairly similar and could conceivably have a pair of coins from the same batch of copper.

The reverse die matches are also shown in Table IX, and of eight pairs and two groups of three die-linked coins, three groups appear to

Table IX. Compositions of Single Die-Linked Coins

Coin No.	wt % Fe	wt % Ni	wt % Cu	wt % Ag	wt % Sn	wt % Sb	wt % Pb	BMC No.
			Obverse Die Links					
R-801 }	—	0.005	99.8	0.069	0.005	0.025	0.09	260
R-1026 }	0.152	0.011	99.6	0.049	0.002	0.039	0.19	?
R-805 }	—	0.003	99.6	0.069	0.006	0.074	0.14	252
R-849 }	0.152	0.007	99.4	0.060	0.011	0.149	0.21	252
R-1006 }	0.161	0.021	99.4	0.023	0.012	0.110	0.25	252
R-814 }	0.166	0.009	99.6	0.057	0.005	0.043	0.10	253
R-1007 }	0.107	0.012	99.6	0.038	0.006	0.059	0.15	252
R-1008 }	0.197	0.015	99.6	0.088	0.008	0.039	0.09	252
R-1003 }	0.132	0.013	99.6	0.041	0.002	0.049	0.18	252
R-836 }	0.066	0.006	99.7	0.036	0.008	0.071	0.15	202
R-911 }	0.135	N	99.6	0.049	0.017	0.099	0.07	202
R-837 }	0.032	0.002	99.7	0.070	0.007	0.046	0.11	264
R-1031 }	0.142	0.007	99.6	—	0.008	0.060	0.15	264
R-843 }	0.106	0.009	99.7	0.047	0.008	0.038	0.07	260
R-870 }	0.099	0.013	99.7	0.035	0.015	0.052	0.11	var. 261
R-852 }	0.101	0.014	99.7	0.054	0.003	0.029	0.09	260
R-80 }	0.086	0.009	99.7	0.055	0.007	0.049	0.10	260
R-862 } a	0.139	0.009	99.6	0.063	0.011	0.064	0.13	269
R-961 }	0.080	0.012	99.7	0.049	0.011	0.067	0.13	269
R-917 }	0.588	9.009	99.1	0.030	0.008	0.055	0.16	200
R-928 }	0.074	N	99.6	0.053	0.009	0.040	0.26	200
R-918 }	0.170	N	99.4	0.074	0.009	0.077	0.23	201
R-933 }	0.148	0.003	99.4	0.082	0.007	0.210	0.12	201
R-969 }	0.119	0.009	99.7	0.018	N	0.043	0.14	267
R-974 }	0.084	0.018	99.5	0.052	0.015	0.120	0.24	267
R-1013 }	0.077	0.004	99.7	0.029	0.005	0.043	0.11	253
R-1014 }	0.133	0.006	99.5	0.034	0.010	0.111	0.20	253
			Reverse Die Links					
R-820 }	0.150	0.014	99.6	0.045	0.011	0.057	0.19	201
R-929 }	0.065	0.006	99.8	0.034	0.009	0.073	0.07	201
R-831 }	0.484	0.009	99.1	0.049	0.013	0.115	0.21	259
R-1021 }	0.081	0.009	99.6	0.035	0.006	0.061	0.22	259
R-846 }	0.082	0.010	99.6	0.029	0.016	0.125	0.16	262
R-1027 }	0.356	0.008	99.3	0.038	0.014	0.102	0.14	262
R-1028 }	0.131	0.010	99.7	—	0.003	0.047	0.13	262

Table IX. Continued

Coin No.			wt % Fe	wt % Ni	wt % Cu	wt % Ag	wt % Sn	wt % Sb	wt % Pb	BMC No.
R-871	}	a	0.374	0.006	99.2	0.060	0.009	0.082	0.24	270
R-920	}		0.274	0.007	99.4	0.047	0.010	0.088	0.22	270
R-894	}		0.072	N	99.6	0.064	0.013	0.061	0.23	201
R-909	}		0.063	0.003	99.7	0.064	0.011	0.057	0.06	200
R-895	)		0.164	0.006	99.7	0.050	0.006	0.109	N	202
R-899	}		0.116	0.013	99.7	0.052	0.007	0.047	0.02	202
R-911	)		0.135	N	99.6	0.044	0.017	0.099	0.07	202
R-903	}		0.159	0.002	99.6	0.058	0.014	0.130	0.03	202
R-872	}		0.177	0.002	99.5	0.054	0.008	0.122	0.12	202
R-939	}	a	0.124	N	99.5	0.064	0.009	0.048	0.23	201
R-928	}		0.074	N	99.6	0.053	0.009	0.040	0.26	200
R-995	}		0.046	0.010	99.6	0.062	0.006	0.110	0.16	207
R-997	}		0.089	0.011	99.4	0.052	0.016	0.179	0.21	207
R-1017	}	a	0.104	0.007	99.6	0.083	0.005	0.046	0.12	258
R-1018	}		0.080	0.009	99.7	0.046	0.004	0.049	0.10	258

[a] This pair of coins has essentially the same composition within experimental error and therefore probably came from the same batch of copper.

have the same composition within experimental error, and one other group is moderately close.

If a group of three die-linked coins is counted as three pairs of die-linked coins, there are 33 pairs of obverse or reverse die-linked coins, and only four pairs have the same composition within experimental error. Although most pairs of single die-linked coins should have been made on different days and therefore probably from different batches of copper, some pairs would have been made on the same day (either the obverse or reverse die would crack, requiring the use of a different die for the balance of the day), and therefore some pairs of singly die-linked coins should have essentially the same composition.

Although several assumptions have been made, they are consistent with known facts and seem reasonable. The chemical analyses of the 245 Augustan quadrantes give results that are entirely reasonable based on the above model.

Summary and Conclusions

Chemical analyses and physical measurements of a large number of Roman coins struck in four years have produced the following observations and conclusions:

(1) Weight, size, and shape of the coins change throughout the period from lighter, larger, thinner coins at the beginning to heavier, smaller, thicker coins at the end;

(2) The Pulcher, Taurus, Regulus quadrantes are probably the first quadrantes produced, rather than the Lamia, Silius, Annius coins as generally believed up to now;

(3) The mean and the range of trace metal contents change from year to year and frequently increase or decrease throughout the four-year period;

(4) The mean and the range of trace metal contents vary from issue to issue within a given year indicating that not all issues were made during exactly the same period;

(5) Half of the eight pairs of doubly die-linked coins have the same chemical compositions, evidently meaning that four of the double die-linked pairs of coins were made from the same four batches of copper;

(6) About one in eight pairs of single die-linked coins have the same composition, in agreement with hypothesized mint operations.

Acknowledgment

This reasearch was made possible by several persons and institutions: C. E. King for her original work on the quadrantes and her help in my eventual success in obtaining the coins for analysis, S. Bendall for his interest and help, A. H. Baldwin and Sons Ltd. for supplying the quadrantes for analysis, Kelsey-Hayes Research and Development Center and the Lewis Research Center of NASA for the use of their equipment. I owe a great debt to the above individuals and institutions for enabling me to do this work; to be able to analyze such an extensive group of homogeneous coins was truly a chance of a lifetime.

Literature Cited

1. Cope, L. H., "The Metallurgical Analysis of Roman Imperial Silver and Aes Coinage," in "Methods of Chemical and Metallurgical Investigation of Ancient Coinage," Royal Numismatic Society, London, 1972.
2. Riederer, J., "Metallanalysen Römischer Sesterzen," *Jahrb. Numismatik Geldgeschichte* (1974) **24**, 73–98.
3. Carter, G. F., "Compositions of Some Copper-Based Coins of Augustus and Tiberius," *Sci. Archaeol., Symp. Archaeol. Chem., 4th* (1971).
4. Carter, G. F., "Chemical Compositions of Copper-Based Roman Coins III. Tiberius to Nero, A.D. 34 to 66," *Numismatic Chronicle* (1979) in press.
5. McDowall, D., "The Quality of Nero's Orichalcum," *Schweiz. Münzblätter* (1966) **16**, 101–106.
6. Caley, E. R., "Orichalcum and Related Ancient Alloys," American Numismatic Society, New York, 1964.
7. Carter, G. F., Buttrey, T. V., "Chemical Compositions of Copper-Based Roman Coins II. Augustus and Tiberius," American Numismatic Society Museum Notes (1977) **22**, 49–65.
8. Gordus, A. A., "Neutron Activation Analysis of Coins and Coin-Streaks," in "Methods of Chemical and Metallurgical Investigation of Ancient Coinage," Royal Numismatic Society, London, 1972.

9. Carter, G. F., "Precision in the X-Ray Fluorescence Analysis of Sixty-One Augustan Quadrantes," *J. Arch. Sci.* (1978) in press.
10. Mattingly, H., "Coins of the Roman Empire in the British Museum," Vol. I, British Museum, London, reprinted 1965.
11. King, C. E., "Quadrantes from the River Tiber," *Numis. Chron.* (1975) Series 7, **XV**, 56–90.
12. Metcalf, W. E., "Roman Coins from the River Liri. II," *Numis. Chron.* (1974) Series 7, **XIV**, 42.
13. Bertin, E. P., "Principles and Practice of X-Ray Spectrometric Analysis," Plenum, New York, 1970.
14. Müller, R. O., "Spectrochemical Analysis by X-Ray Fluorescence," Plenum, New York, 1972.
15. Jenkins, R., DeVries, J. L., "Practical X-Ray Spectrometry," Springer–Verlag, New York, 1967.
16. Carter, G. F., "Reproducibility of X-Ray Fluorescence Analyses of Septimius Severus Denarii," *Archaeometry* (1977) **19**, 67–74.
17. Carter, G. F., "Preparation of Ancient Coins for Accurate X-Ray Fluorescence Analysis," *Archaeometry* (1965) **7**, 106.
18. Laing, L. R., "Coins and Archaeology," Weidenfeld and Nicolson, London, 1969.
19. Mate, M., "Coin Dies Under Edward I and II," *Numis. Chron.* (1969) Series 7, **IX**, 207.
20. Stewart, B. H. I. H., "Second Thoughts on Medieval Die Output," *Numis. Chron.* (1964) Series 7, **IV**, 293.
21. Forbes, R. J., "Metallurgy in Antiquity," E. J. Brill, Leiden, 1950.

RECEIVED September 19, 1977.

INDEX

The text of this book is set in 10 point Caledonia with two points of leading. The chapter numerals are set in 30 point Garamond; the chapter titles are set in 18 point Garamond Bold.

The book is printed offset on Mead Offset Enamel 60-pound. The cover is Joanna Book Binding blue linen.

Jacket design by Alan Kahan. Editing and production by Virginia deHaven Orr.

The text was composed by Service Composition Co., Baltimore, MD. The book was printed by Creative Printing, Inc., Hyattsville, MD; and bound by The Maple Press Co., York, PA.